PHYSIOLOGICAL BASIS OF CROP GROWTH AND DEVELOPMENT

FOUNDATIONS FOR MODERN CROP SCIENCE SERIES

PHYSIOLOGICAL BASIS OF CROP GROWTH AND DEVELOPMENT

M. B. Tesar

Editor

American Society of Agronomy
Crop Science Society of America
Madison, Wisconsin
1984

Domenic Fuccillo, *Managing Editor*
Betty Les, *Manuscript Editor*
Kristine E. Gates, final production
Cover graphics prepared by Julia L. McDermott

Library of Congress Cataloging in Publication Data

Library of Congress Catalog Card Number: 84-71222
ISBN 0-89118-037-0

The American Society of Agronomy, Inc., and
the Crop Science Society of America, Inc.
677 S. Segoe Road, Madison, Wisconsin, USA 53711

Printed in the United States of America

CONTENTS

FOREWORD

The accelerated pace of research, augmented by sophisticated instrumentation and techniques, and new opinions, imparts to crop science a rapidly changing character as new discoveries replace and/or add to former concepts. New findings force us to reevaluate and often reconstruct the foundations on which crop science rests.

The Teaching Improvement Committee of the Crop Science Society of America identified the urgent need for developing contemporary reading materials aimed at upper level undergraduate college students. A current presentation of the dynamic state of modern crop science is a formidable challenge worthy of the best talents of eminent research and teaching personnel in the field. This task necessitates assembling the most capable representatives of the various disciplines within crop science and bringing them together in teams of writers to prepare a series of publications based on contemporary research. The Crop Science Society of America and the American Society of Agronomy have undertaken this large assignment by selecting more than 100 specialists who will contribute to making the Foundations of Modern Crop Science books a reality.

The authors and editors of this series believe that the new approach taken in organizing subject matter and relating it to current discoveries and new principles will stimulate the interest of students. A single book cannot fulfill the different and changing requirements that must be met in various programs and curricula within our junior and senior colleges. Conversely, the needs of the students and the prerogatives of teachers can be satisfied by well-written, well-illustrated, and relatively inexpensive books planned to encompass those areas that are vital and central to understanding the content, state, and direction of modern crop science. The Foundations for Modern Crop Science books represent the translation of this central theme into volumes that form an integrated series but can be used alone or in any combination desired in support of specific courses.

The most important thing about any book is its authorship. Each book and/or chapter in this series on Foundations for Modern Crop Science is written by a recognized specialist in his discipline. The Crop Science Society of America and the American Society of Agronomy join the Foundations for Modern Crop Science Book Writing Project Committee in extending special acknowledgement and gratitude to the many writers of these books. The series is a tribute to the devotion of many important contributors who, recognizing the need, approach this major project with enthusiasm.

A. W. Burger, chairman
D. R. Buxton
A. A. Hanson
C. O. Qualset
L. H. Smith

PREFACE

This book is intended as a text or reference book for undergraduate students at the junior-senior level in the plant sciences, primarily in colleges of agriculture. The material in several of the chapters is comprehensive enough to meet the requirements of instruction at the graduate level.

The authors were selected from various disciplines within agronomy and horticulture from eight different universities. Each author is an authority in his field of research and teaching. Most of the authors are now actively engaged in teaching or have taught undergraduate students in various universities in the USA. Because of the wide diversity of authors, the reader will detect different styles of writing with a varying emphasis on citations or references in the written text. Some chapters have a group of references only at the end of the chapter; others may have citations referred to directly in the text. In either case, the material presented is very readable, understandable, and, I hope, stimulating enough to generate a further interest in a career in the plant sciences.

The information in the chapters should be self-explanatory but students may occasionally find the material complex—but interesting enough to warrant reading of the cited references.

All of the chapters, except Chapter 11, "Genetics and Use of Physiological Variability in Crop Breeding," are primarily physiological in nature. Chapter 11 shows the importance of physiology in plant breeding and should provide a meaningful link between the physiology and genetics of plants. This chapter, prepared initially by two authors for this book, has been modified slightly and used as a chapter in Book IV, *Crop Breeding,* in this series.

I have enjoyed editing this book and hope the authors have enjoyed the important task of writing a chapter for a reference designed primarily for undergraduate students. Each author had to put teaching or research duties aside to complete the task and to each, I am most grateful. May your efforts be reflected in better-informed students who will help solve the urgent food problems of our world in the 21st century.

M. B. Tesar
East Lansing, Michigan

CONTRIBUTORS

Allan L. Barta, Professor, Department of Agronomy, The Ohio State University, Wooster, OH 44691.

R. H. Brown, Professor, Department of Agronomy, University of Georgia, Athens, GA 30602.

A. W. Burger, Professor, Department of Agronomy, University of Illinois, Urbana, IL 61801.

Vernon B. Cardwell, Professor, Department of Agronomy and Plant Genetics, University of Minnesota, St. Paul, MN 55108.

Frank G. Dennis, Jr., Professor, Department of Horticulture, Michigan State University, East Lansing, MI 48824.

Jerry D. Eastin, Professor of Crop Physiology, Department of Agronomy, University of Nebraska, Lincoln, NE 68583-0817.

Burle G. Gengenbach, Professor, Department of Agronomy and Plant Genetics, University of Minnesota, St. Paul, MN 55108.

Kenneth L. Larson, formerly Professor of Agronomy, College of Agriculture, University of Missouri, Columbia, MO 65211; now Associate Dean for Academic Programs, Iowa State University, Ames, IA 50011.

Dale N. Moss, Professor, Department of Crop Science, Oregon State University, Corvallis, OR 97331.

C. J. Nelson, Professor, Department of Agronomy, University of Missouri, Columbia, MO 65211.

Donald C. Rasmusson, Professor, Department of Agronomy and Plant Genetics, University of Minnesota, St. Paul, MN 55108.

John G. Streeter, Professor, Ohio Agricultural Research and Development Center, The Ohio State University, Wooster, OH 44691.

Charles Y. Sullivan, Research Plant Physiologist, Agricultural Research Service, U.S. Department of Agriculture, University of Nebraska, Lincoln, NE 68583-0817.

1 Crop Classification

A. W. BURGER
Department of Agronomy
University of Illinois
Urbana, Illinois

As the crop scientist looks into nature to appraise and evaluate the many thousands of different kinds of plants growing on this planet, he or she may feel frustrated by their large number and great diversity. However, crop plants can be systematically classified in various ways. Such classification makes for orderly and relevant approaches to the discussion of the many species and their uses.

There are many alternatives in classifying crop plants. Some of the more obvious routes of classification include 1) agronomic use, 2) special purpose, 3) growth habit, 4) leaf retention, 5) structure and form, 6) climatic adaptation, 7) usefulness, 8) photorespiration type, 9) photoperiod requirement, 10) temperature type, and 11) botanical. Each of these categories of classification is discussed in this chapter. Scientific names of crop plants discussed here, unless noted otherwise, are provided in Table 1.1.

AGRONOMIC USE

One of the more obvious ways to classify plants is to list them according to agronomic use (Table 1.1). What more meaningful way is there than to designate a crop for its uses as food, fuel, fiber, oil, or drugs? Therefore, we refer to grasses grown for their edible seeds as cereals. These include, especially, barley, grain sorghum, Job's-tears (*Coix lacryma-jobi* L.), corn (also called maize), oats, pearl and proso millets, rice, rye, teff [*Eragrostis*

Published in *Physiological Basis of Crop Growth and Development,* © American Society of Agronomy—Crop Science Society of America, 677 South Segoe Road, Madison, WI 53711, USA.

Table 1.1. Some important crop plants and their characteristics.†

Common name	Scientific name	Life cycle‡	Chromosome number§	Photoperiodic reaction¶	Photorespiration type#	Temperature type††	Use (primary)
Alfalfa	Medicago sativa L.	P	16	L	C$_3$	C	Forage
Barley	Hordeum vulgare L.	A;WA	7	L;N	C$_3$	C	Food, feed, beer
Bean, field	Phaseolus vulgaris L.	A	11	S;N	C$_3$	W	Food
Bean, lima	Phaseolus lunatus Macf.	P;A	11	S	C$_3$	W	Food
Bean, mung	Phaseolus aureus Roxb.	A	11	S	C$_3$	W	Food
Beet, sugar	Beta vulgaris saccharifera L.	B	9	L	C$_3$	C	Sugar
Bentgrass, creeping	Agrostis palustris Huds.	P	14	L	C$_3$	C	Turf
Bermudagrass	Cynodon dactylon (L.) Pers.	P	15;18	—	C$_4$	W	Forage, turf
Birdsfoot trefoil	Lotus corniculatus L.	P	12	L(?)	C$_3$	C	Forage
Bluegrass, Kentucky	Poa pratensis L.	P	14;28;35	N	C$_3$	C	Forage, turf
Bromegrass, smooth	Bromus inermis Leyss.	P	21;28;35	L	C$_3$	C	Forage
Broomcorn	Sorghum vulgare technicum Pers.	A	10	S	C$_4$	W	Brooms
Buckwheat, common	Fayopyrum esculentum Gaertn.	A	8	L;N	C$_3$	W	Food, feed
Cassave	Manihot utilissima esculenta Pohl.	P	18;36	N	—	W	Food, forage
Castorbean	Ricinus communis L.	P;A	10	L	C$_3$	W	Drug
Clover, alsike	Trifolium hybridum L.	P	8	L	C$_3$	C	Forage
crimson	Trifolium incarnatum L.	WA	7;8	L	C$_3$	C	Forage
ladino	Trifolium repens L.	P	8;12;14;16	L	C$_3$	C	Forage
red	Trifolium pratense L.	P	7;14	L	C$_3$	C	Forage
sweet, white	Melilotus alba Med.	B	8	L	C$_3$	C	Forage
sweet, yellow	Melilotus officinalis Lam.	B	8	L	C$_3$	C	Forage
white	Trifolium repens L.	P	8;12;14;16	L	C$_3$	C	Forage
Corn or maize, dent	Zea mays L.	A	10	S	C$_4$	W	Feed, food
Corn or maize, sweet	Zea mays saccharum L.	A	10	S	C$_4$	W	Food

(continued on next page)

Table 1.1. Continued.

Common name	Scientific name	Life cycle‡	Chromosome number§	Photoperiodic reaction¶	Photorespiration type#	Temperature type††	Use (primary)
Cotton, upland	*Gossypium hirsutum* L.	P,A	26	N	C₃	W	Fiber, feed, food
Cotton, Egyptian	*Gossypium barbadense* L.	P;A	26	N	C₃	W	Fiber, feed, food
Cowpea	*Vigna sinensis* Endl.	A	12	S	C₃	W	Forage, food
Crotalaria, showy	*Crotalaria spectabilis* Roth.	A	8	S	C₃	W	Cover
Crownvetch	*Coronilla varia*	P	—	S	C₃	—	Erosion control
Fescue, red	*Festuca rubra* L.	P	7;21;28;35	—	C₃	C	Turf
Fescue, tall	*Festuca arundinacea* Schreb.	P	21	—	C₃	C	Forage, turf
Field pea (see pea)							
Flax	*Linum usitatissimum* L.	A	15	L	C₃	C	Fiber, food, paint
Guar	*Cyamopsis psoralides* DC.	A	7	—	—	W	Food, feed
Guayule	*Parthenium argentatum* Gray	P	—	W	—	—	Latex rubber
Hemp	*Cannabis sativa* L.	A	10	S	C₃	C	Fiber
Hop	*Humulus lupulus* L.	P	10	L	C₃	—	Beer
Kenaf	*Hibiscus cannabinua* L.	A	18	S	C₃	W	Fiber
Kudzu	*Pueraria thunbergiana* Benth.	P	22 or 24	—	—	W	Forage
Lentil	*Lentilla lens*	A	7	L	C₃	W	Food
Lespedeza, common	*Lespedeza striata* Hook & Arn	A	10	S	C₃	W	Forage
Lespedeza, Korean	*Lespedeza stipulacea* Maxim	A	10	S	C₃	W	Forage
Lespedeza, sericea	*Lespedeza cuneata* (Dum. de Cours) G. Don.	P	10	S	C₃	W	Forage
Lupine, blue	*Lupinus angustifolius* L.	A	20;24	N;S	C₃	W	Forage
Lupine, white	*Lupinus albus* L.	A	ca. 20	N	C₃	—	Forage
Lupine, yellow	*Lupinus luteus* L.	A	ca. 23	N;L	C₃	—	Forage
Maize (see corn)							
Millet, foxtail	*Setaria italica* (L.) Beauv.	A	9	S	C₄	—	Food, feed
Millet, pearl	*Pennisetum glaucum* L.	A	7	S	C₄	—	Forage, food
Millet, proso	*Panicum miliaceum* L.	A	18;21;36	S	C₄	—	Food

(continued on next page)

Table 1.1. Continued.

Common name	Scientific name	Life cycle‡	Chromosome number§	Photoperiodic reaction¶	Photorespiration type#	Temperature type††	Use (primary)
Mint (peppermint)	*Mentha piperita* L.	P	18	—	C_3	—	Flavoring, drug
Mint (spearmint)	*Mentha spicata* L.	P	18	—	C_3	—	Flavoring, drug
Oats, common	*Avena sativa* L.	A;WA	21	L;N	C_3	—	Feed, food
Oats, red	*Avena byzantina* C. Koch	A;WA	21	L;N	C_3	C	Feed, food
Orchardgrass	*Dactylis glomerata* L.	P	14	N;L	C_3	C	Forage
Pea, field	*Pisum sativum* L.	A	7	L	C_3	C	Forage
Peanut	*Arachis hypogaea* L.	A	20	S	C_3	W	Feed, food
Potato	*Solanum tuberosum* L.	P;A	12;24	L;N	C_3	C	Food
Ramie	*Boehmeria nivea* Gaud.	P	14	—	C_3		Fiber
Rape, oilseed	*Brassica napus annua* Koch	A	19	L	C_3	C	Lubrication
Rape, winter	*Brassica napus biennis* (Schubl. & Mart.)	B	19	—	C_3	C	Forage
Redtop	*Agrostis alba* L.	P	14;21	—	C_3	C	Forage, turf
Reed canarygrass	*Phalaris arundinacea* L.	P	14	L	C_3	C	Forage
Rice	*Oryza sativa* L.	A	12	S	C_3	W	Food
Rye	*Secale cereale* L.	A;WA	7	L;N	C_3	C	Food, whiskey
Ryegrass, Italian	*Lolium multiflorum* Lam.	WA	7	L	C_3	C	Forage, turf
Ryegrass, perennial	*Lolium perenne* L.	P	7	L;N	C_3	C	Forage, turf
Safflower	*Carthamus tinctorius* L.	A	12	L;N	C_3	W	Food, soap, paint
Sesame	*Sesamum indicum* L.	A	13;26	S	C_3	W	Food
Sorghum (feterita)	*Sorghum bicolor* (L.) Moench	A	10	S	C_4	W	Feed
Sorghum (hegari)	*Sorghum bicolor* (L.) Moench	A	10	S	C_4	W	Feed
Sorghum (kafir)	*Sorghum bicolor* (L.) Moench	A	10	S	C_4	W	Feed
Sorghum (milo)	*Sorghum bicolor* (L.) Moench	A	10	S	C_4	W	Feed
Sorghum (sorgo)	*Sorghum bicolor* (L.) Moench	A	10	S	C_4	W	Feed
Sorghum × sudangrass		A	10	S	C_4	W	Forage

(continued on next page)

Table 1.1. Continued.

Common name	Scientific name	Life cycle‡	Chromosome number§	Photoperiodic reaction¶	Photorespiration type#	Temperature type††	Use (primary)
Soybean	*Glycine max* Merrill	A	20	S;N	C_3	W	Food, feed
Sudangrass	*Sorghum bicolor drummondii*	A	10	S	C_4	W	Forage
Sugar beet (see beet, sugar)							
Sugarcane	*Saccharum officinarum* L.	P	40	S	C_4	W	Sugar, syrup
Sunflower	*Helianthus annuus* L.	A	17;34	N	C_3	—	Feed, food
Sweet clover (see clover, sweet)							
Teosinte	*Euchlaena mexicana* Schrad.	A	10	—	C_4	W	Feed, food
Timothy	*Phleum pratense* L.	P	7;21	L	C_3	C	Forage
Tobacco	*Nicotiana tabacum* L.	A	24	N	C_3	W	Drug (smoking)
Trefoil (see birdsfoot trefoil)							
Vetch, common	*Vicia sativa* L.	A;WA	6;7	L	C_3	C	Forage
Vetch, hairy	*Vicia villosa* Roth	WA;B	7	L	C_3	C	Forage
Wheat, common	*Triticum aestivum* L.	A;WA	21	L;N	C_3	C	Food, feed
Wheat, durum	*Triticum turgidum* L.	A	14	L;N	C_3	C	Food, feed
Wheatgrass, crested	*Agropyron cristatum* (L.) Gaertn.	P	7	—	C_3	C	Forage
Wheatgrass, slender	*Agropyron trachycaulum* (Link) Malte	P	14	—	C_3	C	Forage
Wheatgrass, western	*Agropyron smithii* Rybd.	P	21;28	L	C_3	C	Forage

† Dashes and question marks are used in the original tables. No explanation was given, but it is assumed that dashes (—) mean that the information is unknown and question marks (?) symbolize a conjectural classification. Adapted from Martin, John H. and Warren H. Leonard. 1967. Principles of Field Crop Production. The MacMillan Company, London, and Martin, John H., Warren H. Leonard and David L. Stamp. 1976. (same publication).

‡ Life cycle: A = Annual; WA = Winter annual; B = Biennial; P = Perennial.

§ Chromosome number: reduced (gametic) number (N).

¶ Photoperiodic reaction: L = Long day; S = Short day; N = Day neutral or indeterminate.

Photorespiration types: C_3 = Low net assimilation rate (high compensation point); C_4 = High net assimilation rate (low compensation point).

†† Temperature type: C = Cool-weather growth; W = Warm-weather growth.

tef (Zuccagni) Trotter], and wheat. Drug crops include mint (peppermint, spearmint), pyrethrum (*Chrysanthemum cinerariifolium* L. and *C. coccineum* L.), tobacco, and wormseed (*Chenopodium ambrosioides* L.). Many crops are grown for their fiber content. Fiber crops include abaca (*Musa textilis* Nee), broomcorn brush, cotton, flax, hemp, henequen (*Agave fourcrydes* Lem.), jute (*Corchorus capsularis* L.), kapok (*Ceiba pentandra* L.), kenaf, ramie, sansevieria (*Sansevieria* spp.), and sisal (*Agave sisalana* Perr.). Many animal feeds are derived from the vegetable matter (forage), either fresh or preserved, of various plants. Such crops are called forage crops and these include primarily grasses, legumes, and crucifers. Forage crops are used as pasture, silage, soilage, haylage, fodder, and hay. Certain legumes such as broadbean (*Vicia faba* L.), chickpea (*Cicer arietinum* L.), cowpea, field bean and mung bean, field pea, lentil, peanut, pigeonpea [*Cajanus cajan* (L.) Huth], and soybean are grown for their edible seed. Root crops are those with swollen underground roots. These include cassava, carrot (*Daucus carota* L.), mangel and sugar beet, rutabaga [*Brassica napus* (L.) Napobrassica group], sweet potato [*Ipomoea batatas* (L.) Lam.], and turnip (*Brassica rapa* L.). In contrast, tuber crops such as potato and Jerusalem artichoke (*Helianthus tuberosus* L.) are grown for their enlarged underground stems. Sugar crops include primarily sugar beet and sugarcane. Sucrose is extracted and crystallized from these two species. You might also include as sugar crops sorghum and sugarcane, as sources of syrup, and corn, a source of dextrose. Many crops are grown for their oil, either edible or nonedible. These species, known as oil crops, include castorbean, corn, cotton, crambe (*Crambe abyssinica* Hochst. ex R. E. Fries), flax, peanut, perilla [*Perilla frutescens* (L.) Britt.], rape, safflower, sesame, soybean, and sunflower. Both guayule and kok-saghyz or Russian dandelion (*Taraxacum kok-saghyz Roden*) are grown for their latex and are referenced as rubber crops.

SPECIAL PURPOSE

Many agriculturists refer to different groups of crops by the specific purpose used. Thus, a catch crop is a substitute crop planted too late for a regular crop or after the regular crop has failed. Short season crops such as millet or sorghum × sudangrass crosses are often used as catch crops. A companion crop, often referred to as a nurse crop, is one grown in association with another, sometimes for mutual benefit, and usually in order to get a return from the land in the first year of a new seeding. Thus, an oats companion to alfalfa ensures a grain or silage crop from the oats while the alfalfa crop is becoming established. The oats crop is a companion to the alfalfa and "nurses" the latter; i.e., the oats provide a soil-holding capacity while the slow-establishing alfalfa takes root. A cover crop is seeded to hold the soil temporarily or during the winter months. Cover crops such as alfalfa, clovers, vetches, soybean, cowpea, rye, and buckwheat are usually plowed under for their fertility and/or tilth values to the succeeding crop. When such cover crops are turned under, they are referred to as green manure crops. Crops harvested and preserved in a succulent condition by partial fermentation and stored in an airtight receptacle are referred to as silage crops. Corn, sorghum, sorghum × sudangrass crosses, forage grasses and forage legumes are some of the more important silage crops. Soiling crops are cut and fed green. Sorghum × sudangrass crosses, field pea, and corn are good examples of soiling crops. Trap crops are seeded to attract certain insect parasites. Such crops are usually plowed under as a green manure crop after serving their purpose.

GROWTH HABIT OR LIFE CYCLE

It is often convenient to refer to plants as summer annuals, winter annuals, biennials, and perennials, depending on their life cycles (Table 1.1). The principal food crop plants of the world are annuals, including wheat, rye, oats, barley, rice, millet, bean, soybean, pea (*Pisum* spp.), and corn. Some of these are classified as summer annuals, others as winter annuals. Thus, spring wheat, which is grown in the Dakotas, Montana, and Canada, is sown in the spring and completes its life cycle by fall. This is a summer annual. In contrast, the winter wheats, grown primarily in Kansas, Nebraska, Oklahoma, Illinois, Indiana, Ohio, Michigan, New York, Oregon, and Washington, are sown in the late summer and early fall and complete their life cycle during the following spring and summer. These wheats are winter annuals. In any case, an annual plant completes its life cycle in 1 year and perpetuates itself by seed. Biennial crop plants require 2 years to complete their growth cycle. Generally, biennials accumulate food reserves in underground storage organs during the first year and produce reproductive flowers and seed during the second year. Good examples of biennials include both white and yellow biennial sweet clover, beet, carrot, parsnip (*Pastinaca sativa* L.), onion (*Allium* spp.), cabbage (*Brassica olereacea* L.), and hollyhock (*Althea rosea* Cav.). Perennials complete their life cycle in more than 2 years and may grow indefinitely. Most forage grasses and legumes die back to the ground each year but recover during the next growing season from the crown and/or storage organs, such as rhizomes or stolons, of the previous year's growth. In this category are well known legumes such as alfalfa, red clover, birdsfoot trefoil, sericean lespedeza, and such well-known grasses as smooth bromegrass, tall fescue, orchardgrass, reed canarygrass, timothy, and bermudagrass. Some plants that are annuals in a temperate climate may perform as perennials in a tropical climate. Cotton and sorghums are good examples.

LEAF RETENTION

While not common in field crop plants, leaf drop is routine during the winter in most hardwood trees, which are said to be deciduous in contrast to evergreen, which refers to needle trees such as spruce (*Picea* spp.) or pine (*Pinus* spp.). You should remember that even evergreens have leaf drop and renewal, but they are, nevertheless, green the year around.

STRUCTURE AND FORM

Most of our field crop plants are herbaceous—soft and succulent with little or no secondary tissue. Woody plants are those that develop secondary stem tissue and considerable xylem. We think of forest trees and fruit trees as woody plants. Vines do appear in the field crop kingdom, but rather rarely. Herbaceous field crop vines include field pea, vetch, kudzu, and cowpea.

CLIMATIC ADAPTATION

Most of the crop plants in the USA are adapted to a temperate climate, which means they grow in places with a marked winter season. However, in the southern latitudes we encounter a warm climate where freezing rarely, if ever, occurs and tropical plants dominate. Tropical plants, in general, shed their leaves once a year in response to a changing season. However, they are essentially green the year around. Crop plants growing in the area between temperate and tropical climate are often termed subtropical. We often speak of hardy or tender plants, depending on their capacity to tolerate climatic extremes. For example, winter hardy alfalfas grow in the northern USA climates while the winter tender types will not survive in this northern temperate climate.

USEFULNESS

I have already discussed the agronomic use category of field crop plants. However, plants may also be classified as useful, useless, and harmful. For example, a weed is a plant out of place or a plant growing where it is not wanted. It is harmful to field crop plants because it competes for all factors of the environment, including light, nutrients, and water, and thus weakens the production potential of the crop plant. Weeds may be harmful, e.g., Canada thistle [*Cirsium arvense* (L.) Scop.] and cocklebur (*Xanthium pensylvanicum* Wallr.), because of the physical harm they may inflict, or they may be harmful when ingested because of the deadly poisonous chemical they contain, e.g., bracken fern [*Pteris aquilina* (L.)]. Perhaps poison ivy (*Rhus toxicodendron* L.) would best be classified as useless because of the irritation caused when it contacts human skin. Field crops are quite useful in many ways. They provide us feed, food, fiber, oil, or drugs.

PHOTORESPIRATION TYPE

Two types of photorespiration—respiration during daylight hours—have been identified based upon the different pathways of carbon dioxide fixation. Plants whose first carbon compound in photosynthesis consists of a four-carbon atom chain are called C_4 plants while C_3 plants are those whose first carbon compound in photosynthesis is composed of a three-carbon chain. The C_3 plants, such as soybean, cereals, and many forage crops, have a low net assimilation rate (NAR) due to a high photorespiration rate. The C_4 plants, such as corn, sorghum, sugarcane, Italian millet, crabgrass, and bermudagrass, have a low photorespiration rate and a relatively high NAR. These C_4 plants have a superior ability to utilize efficiently the sun's energy, especially at high light intensities. Thus crop plants may be classified according to their efficiency of light and CO_2 utilization. Whereas C_3 plants are quite wasteful because they have high energy losses in photoresiration, C_4 plants produce photosynthate during the day and carry on most of their respiration at night. The net assimilation rate of C_4 plants is much higher than for C_3 plants. Table 1.1 identifies some important plants by photorespiration type.

PHOTOPERIODIC REACTION

Plants can be classified according to how they are influenced by relative lengths of daylight or night (Table 1.1; see also Chapters 9 and 10). Thus, certain cultivars of crops such as wheat, oats, barley, and rye will not reach reproductive development until the day length is 14 or more hours. These plants are called long-day plants. If the day length is shorter than 14 hours, these grains remain vegetative. On the other hand, certain cultivars of crops such as corn, soybean, rice, millet, and sorghum are short-day plants, requiring day lengths of less than 14 hours to initiate the reproductive cycle. To be sure, short-day plants remain vegetative when grown under conditions of long days.

In addition to the short- and long-day classification, a third category, day neutral, is used to designate crops that will reach reproductive development under long- or short-day conditions. Thus, corn might be regarded as a short-day plant, but it is adapted to areas with long days because of breeder selection against photoperiodic sensitivity.

TEMPERATURE TYPES

Crop plants are known to prefer either cool or warm temperatures for most of their growth cycle and therefore are referred to as either cool- or warm-season species (Table 1.1). Our great crop areas have 4 to 12 months with mean monthly temperature between 10° and 20° C, and each crop species has its own optimum temperature range for growth. We refer to the cardinal temperatures for plant growth as the minimum, optimum, and maximum. At a certain minimum temperature plant growth begins and at a certain maximum temperature growth ceases. Somewhere between, at an optimum range of temperatures, plant growth is most active and efficient.

The classification of plants according to whether they prefer cool or warm seasons is somewhat arbitrary; however, many plant scientists make this distinction. Thus, the small-grain crops such as wheat, oats, rye and barley, and many forage grasses and legumes, such as smooth bromegrass and alfalfa, are referred to as cool-season plants because they germinate and thrive in the cool part of the growing season. On the other hand, such crops as corn, soybean, and sorghum, which need warm temperatures for optimum growth and production, are called warm-season species.

BOTANICAL CLASSIFICATION

Since there is a large number of field crop plants, a logical order in arranging various species is essential. While general plant morphology is necessary to describe a species, the reproductive structures (flowers) give a more valid and much more constant reflection of the broad relationships among plants. Since the basic flower structure of a plant species is relatively constant and much less subject to environmental variation, it forms the basis for the general classification of flowering plants, which include our field crop plants. While taxonomists, paleobotanists, and geneticists continue to interact to produce a more precise classification, you should recognize that the Linnaeus concept of classifying plants has remained relatively intact since the 18th century. It is safe to predict that the Swedish physician, Carl von Linne (1708–1778), better known as Linnaeus, will continue to be credited for a workable system of plant classification based on structural differences and similarities in the morphology of reproductive parts, the organs least likely to be changed by environment.

Linnaeus and his associates are also credited with the development of the binomial system of nomenclature, which designates various organisms by genus and species. In systematic taxonomy, closely allied species are placed together in a taxonomic unit called the genus. For example, the genus name for all wheats, whether hard red winter, soft red winter, durum, hard red spring or white, is *Triticum*. However, the species names *aestivum* and *turgidum* distinguish between species of the genus *Triticum*. The species

Table 1.2. The various taxa of corn, soybean, and cotton.

Taxa	Corn (dent)	Soybean	Cotton (upland)
1. Kingdom	*Plantae*	*Plantae*	*Plantae*
2. Division	*Spermatophyta*	*Spermatophyta*	*Spermatophyta*
3. Subkingdom	*Embryophyta*	*Embryophyta*	*Embryophyta*
4. Phylum	*Tracheophyta*	*Tracheophyta*	*Tracheophyta*
5. Subphylum	*Pteropsida*	*Pteropsida*	*Pteropsida*
6. Class	*Angiospermae*	*Angiospermae*	*Angiospermae*
7. Subclass	*Monocotyledonae*	*Dicotyledonae*	*Dicotyledonae*
8. Order	*Graminales*	*Rosales*	*Malvales*
9. Family	*Gramineae*	*Leguminosae*	*Malvaceae*
10. Tribe	*Maydeae*	*Phaseoleae*	*Hibisceae*
11. Genus	*Zea*	*Glycine*	*Gossypium*
12. Species	*mays*	*max*	*hirsutum*
13. Subspecies	*indentata*	—	—

is a unit comprised of similar organisms capable of reproducing themselves. Three or four characteristics are usually the minimum number acceptable for distinguishing a species. Such distinctions may be behavioral, morphological, physiological, or ecological. Durum wheats, which differ in chromosome number (Table 1.1) from the common (*aestivum*) wheats, are used for manufacture of macroni and spaghetti. Common wheats are used for cereal, bread, and pastry.

As early as the Middle Ages scientists recognized that an international, universal, scholarly language was needed so that they could speak of various organisms in the same manner and still be sure that they were talking about the same being. During this era the most important treatises were written in Latin. Today, Latin remains as the accepted language for the scientific naming of plants and animals. It is a dead language, and therefore unchanging, which makes it suitable for international use in the scientific nomenclature of plants. Though seemingly abstract to the new student of taxonomy, scientific names have meaning. For example, *Trifolium pratense* L. refers to red clover, and the word *Trifolium* means "three leaflets" while the term *pratense* means "of meadows". The letter "L." cites the author, Linnaeus, who named red clover, *Trifolium pratense*.

Several taxa besides the genus and species are quite necessary in pigeon-holing various plants. When studying Table 1.2, note the major taxa (kingdom, division, class, order, family) and minor taxa (genus, species, and sometimes subspecies, race, and subrace) or three common crop plants, namely, corn, soybean, and cotton.

The major taxa from kingdom through class are the same for all crop plants. However, beginning with the subclass, major changes occur as we classify various crop plants. For example, corn belongs to the *Monocotyledonae* while both soybean and cotton belong to the *Dicotyledonae*. For the crop scientist, it is relatively easy to remember the classification taxa of field crops since most of these plants are grasses and legumes. If you know the taxa for these two families, other members of these families obviously fall into the same classification scheme at least up to the plant family level.

SUGGESTED READING

Bailey, L. H. 1949. Manual of cultivated plants (rev. ed.). MacMillan Co., New York.

----. Hortorium. 1976. Hortus third, A concise dictionary of plants cultivated in the United States and Canada. The MacMillan Co., New York.

Core, E. L. 1955. Plant taxonomy. Prentice Hall, Englewood Cliffs, New Jersey.

Fernald, M. L. 1950. Gray's manual of botany. American Book Co., New York.

Hitchcock, A. S. 1951. Manual of the grasses of the United States. USDA Misc. Pub. 200.

Martin, J. H., and W. H. Leonard. 1967. Principles of field crop production. The MacMillan Co., New York.

----, ----, and D. L. Stamp. 1976. Principles of field crop production. 3rd ed. The MacMillan Co., New York.

Porter, C. L. 1967. The taxonomy of flowering plants. W. H. Freeman and Co., San Francisco.

2 Seed Development, Metabolism, and Composition

L. H. SMITH
Department of Agronomy and Plant Genetics
University of Minnesota
St. Paul, Minnesota

Agriculture as we know it started with the discovery that a seed planted in the soil and given water, nutrients, light, and some protection from pests would not only reproduce plants and seeds identical to that planted, but could greatly increase the number of seeds produced, which could be used for food or feed. It is apparent that seeds are the cornerstone of agriculture, and a knowledge of their development and composition is essential to increasing agricultural productivity.

A seed is a complex structure that develops from an ovule and at maturity consists of the embryo, variable amounts of endosperm, and protective layers of tissue on the surface of the seed, all enclosed by the seed coat or testa derived from the integument(s). Various external markings on the mature seed are remnants of structural components of the ovule. The point where the integuments meet at the nucellar apex is the micropyle. It may be completely obliterated on some seeds, or it may remain in the form of an occluded pore in others. The hilum, a scar on the ovule, is highly permeable to water and occurs where the seed abscises from the funiculus. The color of the hilum is widely used to identify soybean cultivars.[1]

---

[1] Scientific names of important crop plants are given in Table 1.1, Chapter 1.

Published in *Physiological Basis of Crop Growth and Development,* © American Society of Agronomy—Crop Science Society of America, 677 South Segoe Road, Madison, WI 53711, USA.

DEVELOPMENT OF SEED

Megasporogenesis

The formation of a seed begins with the development of an ovule, which first appears as a slight dome-shaped mass of undifferentiated cells upon the surface of the placenta in the flower. This group of cells differentiates into the nucellus, which consists initially of a mass of meristematic tissue. The integuments, which later form the seed coat, arise from the sides of the nucellus and grow upward enclosing the apex of the nucellus, except for the micropyle, a narrow opening at the end through which the pollen tube enters to fertilize the egg. In the young ovule all the cells that compose the nucellus are identical. However, one of the nucellar cells, usually just below the epidermis near the top, differentiates from the surrounding cells, eventually forming the embryo sac (megaspore) mother cell. Genetically this is 2N or diploid tissue. The nucleus of this mother cell undergoes two successive divisions (meiosis) forming a row of four cells (megaspores, genetically N or haploid) embedded in the nucellus (Fig. 2.1). Generally three of the four cells nearest the micropylar end disintegrate, and the remaining cell develops into the mature embryo sac. The nucleus of this remaining megaspore divides, forming a two-nucleate embryo sax; each of these two nuclei also divides, forming a four-nucleate embryo sac, and then each of the four nuclei divides to form the eight-nucleated mature embryo sac. Thus, a mature embryo sac usually contains eight nuclei (each N) consisting of one egg cell, two synergid cells, three antipodal cells, and one pri-

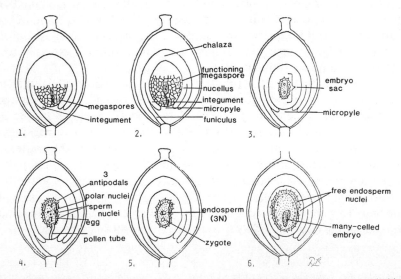

Fig. 2.1. Development of the embryo sac (1 to 3) and endosperm and embryo (4 to 6) in higher plants.

mary endosperm cell with two nuclei (Fig. 2.1). An important exception to this occurs in grasses in which the antipodal nuclei continue to divide forming a mass of 100 or more nuclei in the embryo sac. An embryo sac is considered mature when it is ready for fertilization.

Microsporogensis

Pollen is produced in the anther. In early stages of development the anther consists of a small group of meristematic cells. As it matures, four groups of microspore mother cells develop within it. Each microspore mother cell divides meiotically to produce four microspores, each of which contains the haploid, or N number of chromosomes. Each of the four microspores undergoes a mitotic division forming a structure having two nuclei, which we call the pollen grain. The two nuclei of the pollen grain are the tube nucleus and the generative nucleus. The generative nucleus subsequently divides to produce two male gametes, or sperm cells, before the pollen is shed.

Fertilization

Pollination of cross-pollinated species is accomplished by the transfer of pollen by insects or wind (Fig. 2.2), by dropping of the pollen, or by other means of direct transfer. In self-pollinated crops, pollination is accomplished by direct transfer of the pollen by dropping pollen onto the stigma of the flower containing the ovule. When the pollen grain is shed from the anther, the protoplasmic contents consist of a small amount of cytoplasm and the two nuclei. The pollen grain germinates on the surface of the stigma. The protoplasm of the pollen grain absorbs water and swells, breaking the outer membrane. The inner membrane extends through the break in the outer wall and forms the limiting membrane of a protoplasm-lined tube, the pollen tube. The pollen tube penetrates the tissue of the stigma, grows down the style, and enters the ovary, usually through the micropyle. The tip of the pollen tube then ruptures, and the two sperm nuclei are discharged into the embryo sac. One of the sperm nuclei moves toward and fuses with the egg, effecting fertilization and forming the zygote (2N). The second sperm nucleus unites with the two polar nuclei in a double fertilization forming the primary endosperm (3N). After fertilization of the ovule, the embryo and endosperm continue to grow and differentiate, ultimately forming the seed. During the development of the embryo and endosperm, part of the nucellar tissue is digested and supplies nutrients for their growth. The nucellus in the mature seed, if present, usually consists of a thin layer of cells called the perisperm, except in sugar beet where it is the primary nutrient storage tissue for the seed.

Seed Maturation

Following fertilization of the egg and endosperm, the embryo and endosperm begin a period of rapid growth and differentiation, the maturation phase of seed development, culminating in the mature seed. The maturation phase of seed development is a period of complex and active biochemical and physiological activity involving both the embryo and food storage tissues and organs, whether they be in the embryo (e.g., cotyledons) or in separate structures such as the endosperm or perisperm, which are integral parts of most seeds. It involves not only the production of cells, but their differentiation into specialized cells, tissue and organs. It also involves the synthesis and deposition of storage nutrients in the seed as well.

In the development of a seed, reserve food is deposited in the endosperm (primarily of monocotyledons), the cotyledons (primarily in dicotyledons) and occasionally in the perisperm (nucellar tissue). At physiological maturity of the cell, the endosperm and perisperm consist of dead cells that serve as "warehouses" for the storage of nutrients in the seed.

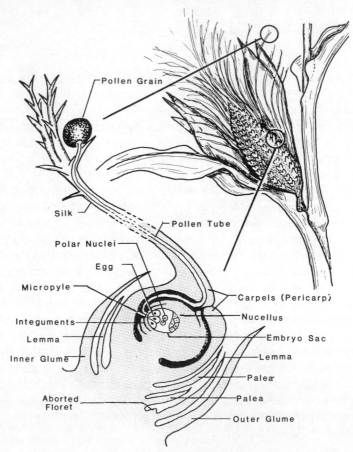

Fig. 2.2. Pollination of corn.

Corn Seed Development

As an example, let us follow the maturation of corn, a typical cereal seed, as it develops from the time of fertilization until physiological maturity (maximum dry weight accumulation) of the seed. Cell division of the endosperm is essentially complete within 28 days following pollination. This is indicated by the DNA content of the endosperm (Fig. 2.3). When cell division is complete, the DNA content will have reached a maximum, and a plot of DNA concentration vs. time will plateau. The RNA content of these cells also reaches a maximum value at this time, and the initial rapid synthesis of metabolic proteins levels off (Fig. 2.3). This initial phase of cell production and differentiation of the endosperm is also characterized by a rapid accumulation of soluble nutrients and water. The water content of the endosperm at this early stage of development may reach 90% or higher. Sugar content of the endosperm cells increases rapidly, reaching a maximum content within 20 days from pollination (Table 2.1), whereas water soluble nitrogen and amino acid contents peak around 28 days (Fig. 2.3).

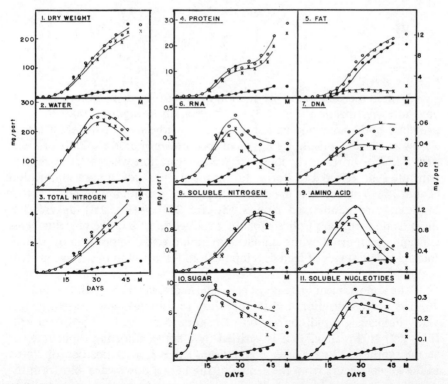

Fig. 2.3. Changes in various components of the embryo and endosperm of corn over a 46-day developmental period after pollination. M: comparable data from the analysis of mature grains; O --- O whole grain; X --- X endosperm, ● --- ● embryo. (From Ingle et al., 1965. Used by permission.)

Table 2.1. Content of starch, amylose, sucrose, and reducing sugars
during corn endosperm development.

Time after pollination	Starch		Amylose	Sucrose		Reducing sugars	
days	mg/endo-sperm	% dry wt	% starch	mg/endo-sperm	% dry wt	mg/endo-sperm	% dry wt
8	0.01	1.2	9.0	0.016	2.02	0.294	38.18
10	0.02	1.5	9.0	0.083	8.59	0.448	35.00
12	0.36	9.2	8.5	0.750	19.18	1.050	26.85
14	7.63	46.0	18.0	2.160	13.04	1.504	9.10
16	21.51	62.6	20.8	3.300	9.62	1.815	5.29
22	60.55	76.9	27.5	2.970	3.77	1.140	1.49
28	92.01	77.0	26.5	1.850	1.55	0.650	0.55

From Tsai et al. (1970).

Table 2.2. Content and concentration of metabolically active (soluble) and total protein
in the developing corn endosperm.

Time after pollination	Soluble proteins		Total proteins	
days	mg/endosperm	% dry wt	mg/endosperm	% dry wt
8	0.077	9.94	0.352	45.86
10	0.130	10.19	0.547	42.77
12	0.483	12.35	1.316	33.66
14	1.078	6.51	3.405	20.56
16	1.608	4.68	5.027	15.23
22	2.170	2.75	9.502	12.07
28	1.495	1.25	15.063	12.62

From Tsai et al. (1970).

A rapid increase in the production of storage proteins occurs about 40 days after fertilization (Fig. 2.3). It is generally recognized that different proteins, both metabolic and storage, are synthesized at different stages during seed development. Synthesis of metabolically active soluble proteins predominates in the early stages of seed maturation whereas storage proteins are synthesized and stored during the later stages of seed maturation (Table 2.2). The decrease in water soluble nitrogen and amino acid content of the endosperm observed in one study after 28 days (Fig. 2.3) suggests that these components are utilized for the production of storage proteins. This concept is supported by the similarity in amino acid composition of protein formed in late stages of seed development and that found in storage protein bodies.

The almost linear increase of endosperm dry weight over the period 30–46 days postfertilization and the decrease in soluble sugar content in this study indicate a rapid conversion of sugars to starch in endosperm cells (Fig. 2.3) at this time. This is verified by a study reporting rapid rates of starch synthesis 12 days after pollination (Table 2.1). Deposition of starch in the kernel occurs from the crown of the kernel downward and from the outside of the kernel inward. The last portion of the endosperm to solidify occurs at the base of the embryo. At this time a "milk line" is clearly visible delineating the liquid from solid endosperm and is visible until the endosperm solidifies near maturation of the kernel.

Although the DNA content of the endosperm remains relatively constant after 25–28 days postfertilization, endospermal RNA decreases very rapidly during this period to about one-third the maximum value (Fig. 2.3). This decrease in RNA is associated with a rapid increase of endospermal ribonuclease activity and suggests a significant reduction in the rate of protein synthesis during this period. Apparently the synthesis of metabolic proteins is complete at this time, and there is a shift to the synthesis of storage proteins. This shift in the type of proteins synthesized is clearly visible in Fig. 2.3 from about 35 days postfertilization to maturity of the seed.

In this study dry matter deposition in the seed proceeds in an essentially linear manner over the period 15–45 days after fertilization, and the maximum content of starch, protein, and lipid has accumulated at approximately 50–60 days postfertilization (Fig. 2.3). Increases in RNA, sugar, and soluble nucleotides observed during the final maturation period of the embryo could result from the movement of these constituents from the endosperm into the embryo.

At seed maturity an abscission layer (black layer) is formed severing the phloem tissue and cutting off further dry matter transport of photosynthate into the seed. The seed is now physiologically mature, and maximum dry weight has been achieved.

Soybean Seed Development

Let us now follow the development of the soybean seed as somewhat representative of dicotyledonous seed. Several studies have shown that rapid cell division occurs within the first 2 weeks after pollination and then gradually ceases as the maximum number of cells in the embryo is reached. The subsequent increases in seed size are a result of increases in cell size and deposition of starch, protein, and lipid in these cells. In bean the maximum number of cells of the embryo (1.2×10^6) is reached when the dry weight of the embryo is only 1/6 of its final weight. Maximum cell number in this species is reached after approximately 21 days postfertilization whereas seed maturation is not complete until about 44 days postfertilization. A similar situation also occurs in pea. Apparently, most of the cell production in the embryo is completed in legumes and many other dicot seeds prior to the rapid synthesis and deposition of the storage nutrients and starch, protein, and lipid in the cotyledon (Fig. 2.4).

Development of a soybean seed following pollination indicates that the cytoplasm of cells in the seed 15 days after pollination is devoid of definite subcellular structures except the nucleus and particles about the same size (1500 to 2500 nm), possibly ribosomes. By the 18th day postfertilization, the central portion of many cells appears as a vacuole. The cytoplasm of these cells gradually becomes organized into a peripheral area of lipid globules, immature chloroplasts with some associated starch, and other particles, with some mitochondria scattered through the cytoplasm. By 26 days postfertilization, more lipid globules, a few large protein bodies, and larger starch grains are present in the cells, but there are fewer mitochondria. Cells of 36-day-old tissue contain many distinct starch grains and protein bodies. Although most of the cells at this stage of development contain both protein bodies and lipid globules, light micrographs show that a few cells lack protein bodies. Such cells are designated lipid cells, although it should be emphasized that not only lipid cells but also parenchymatous cells may contain lipid, as indicated by a positive Sudan IV test for lipid in these cells.

The growth of the embryo initially lags beyind the endosperm but soon surpasses it. The embryo utilizes most of the endosperm for its growth, so that at maturity only a few crushed and remnant cells of the endosperm remain.

Starch is present in soybean seeds less than a week before maturity, but is generally absent from cells of mature seeds due primarily to the utilization of the endosplasm for the growth of the embryo. Except for the disappearance of the starch grains, the general appearance of the cells does not change appreciably from approximately 35 days after pollination until physiological maturity of the seed.

The sequence of events in seed development apparently remains relatively constant regardless of variety or environmental conditions, but the absolute time between the various stages of development may vary considerably depending upon the variety of the crop or environmental conditions under which it is grown.

The development of the soybean cotyledon from 1 mg fresh weight to maturity, according to one study, can be divided into five periods based upon the major changes that take place in the developing cotyledon.

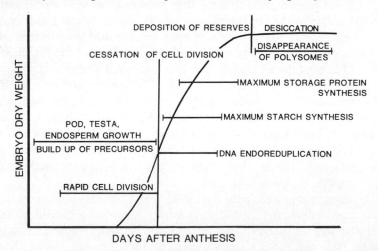

Fig. 2.4. Legume embryogenesis. (From Dure, 1975).

Period 1

During the first period (15 to 18 days after flowering) the cytoplasm is composed of what appear to be RNA granules, proplastids, and some mitochondrial preparations during this period, while low on a fresh-weight basis or when compared to those of roots and cotyledons of germinating seeds, are very high on a dry-weight basis. Lipid and protein accumulation is moderate compared to that during a similar period a week later.

Period 2

The second period of development (17 to 26 days after flowering) is a time of high metabolic activity. Fresh and dry weights of the cotyledons are increasing rapidly and the cytoplasm is undergoing many internal changes. Greater numbers of mitochondria appear in the cotyledons but oxygen consumption by mitochondria and whole seeds continues at about the same rates as during the initial period. A few storage protein bodies appear, and the number of lipid globules increases slightly in the cells of the cotyledons. Starch grains may be seen in the cotyledon at this time and persist almost until maturity, indicating that carbohydrates are entering the cotyledons faster than they can be used in metabolic activities by the embryo; consequently, they are stored as starch.

Period 3

During the period from 18 to 36 days after fertilization, the volume of an average cell increases about 10-fold, whereas the dry weight increases approximately 25-fold. A great increase in size and number of storage protein bodies is evident in the cotyledons during the third period (26 to 36 days after flowering). These protein bodies are assumed to contain mostly storage protein, as distinguished from enzymes and other metabolic protein present in mitochondria, ribosomes, etc. This is also the period of most rapid synthesis of oil and of the greatest effect of temperature on oil content of the seed.

Period 4

The fourth period (36 to 52 days) is characterized by a steady increase in the number and size of the storage protein globules. The protein content of the mitochondria fraction from the cotyledons also increases, but the metabolic activity of this fraction decreases. Present evidence indicates that a scarcity of functional mitochondria in the cotyledons during this period may limit cotyledon respiration. The relative amounts of lipid and starch remain fairly constant.

Period 5

Dehydration and maturation of the seed are completed during the fifth and final period (52+ days). Starch grains are generally converted to other components and few, if any, starch grains are present at maturity. Both the moisture content and respiration rates of the mature seed are quite low. It is postulated that the starch present during development of the seed serves as a temporary storage carbohydrate supporting lipid, protein, and other biosynthetic activity in the cotyledon. After the supply of carbohydrates from leaves is reduced or terminated by leaf senescence, synthetic activities consume the remaining starch, so that normally none is present in mature seeds.

Marked changes in cell organization and structure also occur as the seed matures. Cell nuclei, mitochondria, plastids, endoplasmic reticulum, and dictyosomes change from a relatively undeveloped state in early embryogensis to a highly active state during the accumulation of food reserves and finally become indistinct and nonfunctional in the mature seed.

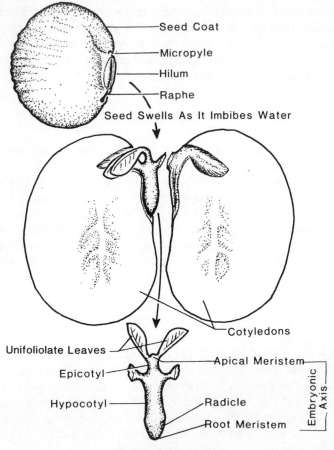

Fig. 2.5. Soybean seed.

Seed Types

Two common types of seeds are illustrated in Fig. 2.5 and Fig. 2.6. The soybean is a dicotyledonous plant in which the endosperm is absent and corn, a monocotyledonous plant in which the endosperm is present at maturity. The seed of soybean, as in most other leguminous plants, is essentially devoid of endosperm at maturity and consists of a seed coat surrounding the embryo. The seed coat is marked with a hilum or seed scar with the micropyle at one end. At the other end of the hilum is the raphe, a small groove extending to the chalaza, the point at which the integuments were attached to the ovule. The seed consists of two large, fleshy cotyledons serving as food storage organs, two well-developed unifoliolate leaves, and a hypocotyl-radicle axis resting in a shallow depression formed by the cotyledons. The tip of the radicle is surrounded by an envelope of tissues formed by the seed coat. The hypocotyl is located below the cotyledons, and a short axis epicotyl, terminating in the shoot apex, is above the cotyledons (Fig. 2.5).

Corn seed is technically a single-seeded fruit in which the pericarp (ovary wall) is tightly fused with the seed. This type of seed is called a caryopsis and is typical of grass or cereal seeds. The true seed coat or testa is often missing in corn. The seed consists of the embryo and endosperm. The endosperm is composed of an outer layer of cells, the aleurone layer, and the

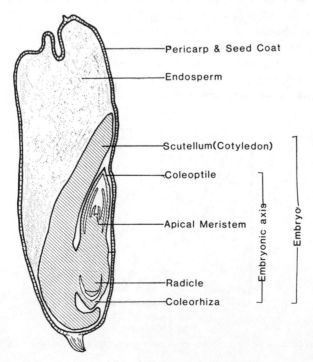

Fig. 2.6. Corn seed (caryopsis).

interior starchy endosperm. The cells of the aleurone layer contain proteins and fats, but little or no starch, whereas the cells of the starchy endosperm are filled with starch grains and protein bodies embedded in a proteinaceous matrix. The embryo lies adjacent to the endosperm and consists of a scutellum or cotyledon and embryonic axis. The shoot apex of the embryonic axis and several rudimentary leaves are surrounded by a modified leaf, the coleoptile. The rudimentary root or radicle is also surrounded by a sheath of cells, the coleorhiza. At the juncture of the shoot and root there is a very short stem structure. The cotyledon, or scutellum, comprises a relatively large part of the embryo and lies in direct contact with the endosperm (Fig. 2.6). Thus, at maturity a seed consists of the embryo, variable amounts of endosperm (sometimes completely absent), and the seed coat, or testa, derived from the integuments of the ovule.

METABOLISM OF SEEDS

Respiration in Developing Seeds

Many seeds show an increase in respiration, as measured by oxygen uptake, as the dry matter of the seed increases in early development. Then, respiratory level is gradually reduced as the seed matures, ultimately reaching a very low level at seed maturity, which remains relatively constant during the quiescent (dormant) period (Fig. 2.7, 2.8).

Marked changes also occur in the fine structure of the subcellular particles of cells as the seed matures. For example, in seeds of *Phaseolus lanatus* above a moisture content of 60%, endosperm cells are vacuolated, and contain chloroplasts with well-developed grana structure, starch grains, and polysomes associated with the endoplasmic reticulum. As the moisture

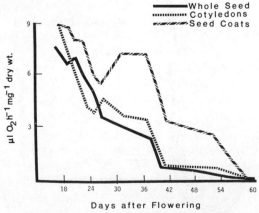

Fig. 2.7. Oxygen uptake by soybean seeds at several stages of maturation. (From Bils and Howell, 1963.)

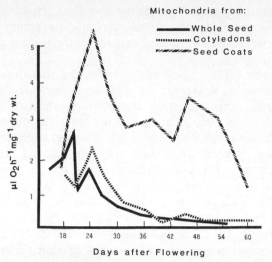

Fig. 2.8. Oxygen uptake by mitochondria from soybean seeds at several stages of maturation. (From Bils and Howell, 1963.)

content decreases below 60%, protein bodies fill the vacuoles, the endoplasmic reticulum becomes less prominent, and polysomes, which are no longer associated with it, gradually disappear. Mitochondrial structure remains more or less intact until the end of the maturation period, after which the mitochondria rapidly lose their elongated shape and become rounded during dehydration. Chloroplasts undergo the greatest changes as the seed approaches maturity. They become globular or bell-shaped with frequent invaginations; the internal membrane structure is lost and the grana disappear.

These changes in fine structure of the subcellular particles appear to result from the cessation of intensive protein synthesis in the seed as it matures. Apparently seeds acquire resistance to desiccation only after the structures associated with protein synthesis undergo degradation and become inactive. Cessation of physiological activity renders cells resistant to the effect of desiccation, rather than the loss of water being the cause of seed inactivity. Major biochemical and organizational changes within cells are required in order for seeds to withstand the extremes of desiccation, down to as low as 2% moisture, and still retain viability.

The observed gradual fall in the respiratory quotient and oxygen consumption of a maturing seed occurs while the fresh weight of the seed remains constant. Reduction in the rate of metabolic activity is paralleled by a significant reduction in water content of the seed. It is possible that the reduction in metabolic activity is related to dehydration of the seed during this period. For example, changes in mitochondrial activity have been shown to be related to the degree of swelling of the mitochondria. The activity of a number of enzymes in the seed declines with advancing maturation of the seed. These include enzymes of the glycolytic pathway, pentose phosphate pathway, Krebs cycle and the cytochrome system. In addition, there is a pronounced decrease in the activity of both the amino-acid-activating enzymes and the soluble enzymes of protein synthesis during the final stage of maturation.

Considerable evidence is available indicating that the decrease in enzyme activities during seed maturation is due to decreased synthesis of the storage nutrients—protein, starch and lipid—of the cell. This decreased synthesis is probably due to changes in mRNA metabolism and, in some seeds, to a reduction of ribosome function. Changes in endoplasmic reticulum and a loss of membrane-bound polysomes during dehydration may account for some of the qualitative alterations in protein synthesis occurring in late maturation, such as the shift from glutelin to prolamine synthesis in maturing corn seeds. In addition, increased RNAse activity, which results in decreased enzyme activity, may be a consequence of increased moisture stress as the maturing seed dries. An alternative view of decreasing enzymatic activity in seeds nearing physiological maturity is that during, or at the time of seed maturation, there is a conversion of many enzymes to an inactive form.

Proteins synthesized by maturing seeds are very different from those synthesized by germinating seeds. They consist principally of "storage" proteins (prolamines, glutelins, and globulins) rather than metabolically active proteins, such as enzymes synthesized during the early stages of seed growth and development. The mRNA's of the maturing seed are different from those of the germinating seed. During seed development there must be destruction or inactivation of mRNA involved in the synthesis of storage protein and a synthesis (and possibly storage) of new mRNA, which is essential for synthesis of metabolically active proteins (enzymes) during seed germination.

The change from synthesis of metabolically active to storage proteins during seed maturation requires at various stages during seed development and maturation a "switching-off" and a "switching-on" of the synthesis of, or translation of, some kinds of RNA and/or ribosomes and the accompanying complement of enzymes that produce these proteins. Furthermore, since storage protein synthesis coincides with membrane ribosome production, it is reasonable to assume that mRNA for storage protein synthesis is associted with membrane ribosomes. Studies on wheat indicate that during the period of rapid synthesis of storage protein, RNA increases. Research suggests that storage proteins are formed on the ribosomes of the endoplasmic reticulum in the endosperm and afterwards secreted internally, sometimes in vacuoles, to form protein storage bodies. Membrane-bound polyribosomes are particularly well-adapted for the synthesis of protein that is to be stored elsewhere in the cytoplasm. They are probably moved via the endoplasmic reticulum. The role of membranes in the regulation of seed development is poorly understood, but the marked changes in membranes that occur in seed cells during development suggest a fundamental role of these cell components in seed physiology.

These considerations suggest that RNA synthesis must occur during the development of the endosperm. This RNA could then be degraded during the maturation of the endosperm, since it has been shown that the mature corn endosperm contains a high level of ribonuclease activity and a low concentration of RNA. In rice (*Orzya sativa* L.) and wheat (*Triticum aestivum* L.), RNA synthesis occurs very early in the seed's development sequence and may be degraded as the seed matures.

Respiration In Mature Seeds

In studies of the state of water in seeds, three fractions of intracellular water were found in "dry" seeds. One fraction was tightly bound to "hydrated" starch, a more mobile fraction was hydrating protein, and a very strongly bound fraction was associated with the seed protein.

For an enzymatic process to occur at a significant rate, substrate and product molecules must be free to diffuse to and from the active sites of the enzyme. This occurs in the seed via the inter- and intracellular water. Although there appears to be no severe limit imposed on metabolic activity in dry seeds through a lack of substrates or functional enzymes, the very low water content of the seed seems to preclude any appreciable rate of metabolism.

Ultrastructural studies of dry and maturing seeds have established that the major cell organelles are present in the desiccated seed but are often in a diffuse, ill-defined state. The absence of organized endoplasmic reticulum and membrane-bound ribosomes is a striking feature of the cells of dry seeds. Electron microscopy studies on dry seeds point out their biochemical inactivity. Metabolic changes do occur, however, in dry seeds. For example, the process of afterripening, which breaks seed dormancy, occurs in such seeds and in fact often seems to be hastened by desiccation.

When a seed becomes mature, its measurable overall biological activity, as indicated by respiration, is reduced to a minimum, and probably only the basal respiration remains. Mitochondria of dry seeds are considerably less active than those of germinating seeds, and it is likely the tricarboxylic acid (TCA) cycle is not fully operative in the initial stages of germination because of this. Moreover, mitochondria from dry seeds are, at least initially, apparently incapable of carrying out oxidative phosphorylation. The ATP is absent from dry bean but rapidly appears following imbibition and increases with the initiation of germination. Aerobic respiration does not proceed actively in freshly harvested seeds but increases as germination progresses. Cytochrome oxidase has been shown to be present in several types of seeds but is inactive in dry seeds. Although complete mitochondria activity might not be measurable in a dry, quiescent seed, mitochondria can be observed in the seed by staining and can be readily seen in electron micrographs.

The aleurone cells in barley seeds are poor in α-amylase, but the interface between the aleurone layer and starch-containing cells of the inner endosperm has a high amylase content. Esterase and protease activity in wheat, rye, and barley is concentrated in the aleurone cells of the endosperm of mature seeds. These enzymes are activated by reducing agents. Proteases have been obtained from quiescent seeds of peanut, pea, and soy flour. Numerous studies indicate potential for metabolic activity in the quiescent seeds.

Quiescent seeds contain all the enzymes for amino acid incorporation into protein. When a synthetic polynucleotide is made available as a template and sources of energy are furnished, we can observe net protein synthesis in vitro. Lipases have been extracted from quiescent soybean, broadbean, wheat, barley, and several other seeds.

SEED COMPOSITION

Starch Biosynthesis

Active photosynthesis by green plants results in the formation of glucose, which is rapidly converted to sucrose and translocated to the developing seed for metabolic use or storage. Variable amounts of glucose are converted to starch in the chloroplasts. Such starch is called assimilation starch. Assimilation starch is, however, a rather transient form, being degraded and either utilized in the metabolic processes of the plant or synthesized into sucrose and translocated to storage tissues in the seed.

Starch can generally be found in all organs of most higher plants. It is easily metabolized and is, therefore, an excellent source of energy. It is produced and stored in plastids called amyloplasts, nonpigmented, specialized plastids for storing starch. Plastids multiply by division, and generally the two daughter plastids are about equal in size. However, plastids can form small "buds" that may or may not separate. In barley and wheat each plastid initially forms one starch granule that will become large and lenticular. As the plastids increase in size, they may form small buds in which starch granules of the small type are formed. These buds may separate from the mother plastid by constriction, forming amyloplasts of much smaller size than the initial amyloplasts.

Since starch is synthesized and stored in plastids, these structures must contain the enzymes and substrates necessary for its formation. In addition, starch is relatively insoluble in water at temperatures at which plants grow and therefore could not be translocated to the amyloplast for storage. Recent research indictes that synthesis of amylose and amylopectin-type starches is two separate processes regulated by different enzyme systems, as evidenced by the findings that waxy corn (storing amylopectin-type starch) lacks starch synthetase. Starch synthetase uses a sugar nucleotide, ADP-glucose, as a D-glucose donor, and is strongly absorbed to starch. It appears that the major portion, if not all, of the starch synthesized in the normal starch grain endosperm is via the starch synthetase pathway (Fig. 2.9). Since starch synthetase is capable of catalyzing synthesis of a polyglucan in the absence of a glucose primer, it is possible that initiation of starch synthesis also occurs via the ADP glucose pathway.

Sucrose is the initial substrate for starch synthesis in amyloplasts as indicated by the ready transfer of ^{14}C-glucose from $^{14}CO_2$ to the starch molecule via the reversal of the system of sucrose synthetase. This is indicated in Fig. 2.10, which shows the proportion of radioactivity in the total kernel of corn in glucose, fructose, sucrose, and starch after treatment of the entire plant with $^{14}CO_2$. This figure also demonstrates the relatively low concentrations of glucose, fructose, and sucrose in the developing seed due to their utilization in starch formation. It was found that, compared with UDP, ADP is the more effective dinucleotide involved, suggesting that ADP-glucose is the preferred substrate in starch synthesis. By attaching itself firmly to amylose molecules, starch synthetase could very well protect them from becoming branched. This would provide an effective mechanism for the separation of enzymes catalyzing the synthesis of amylose and amylo-

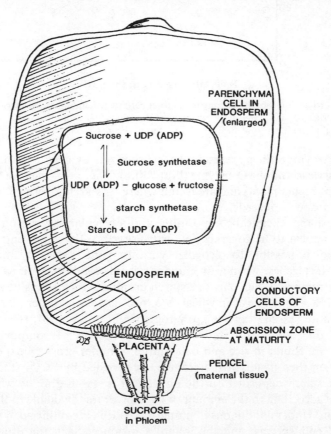

Fig. 2.9. Biosynthesis of starch in endosperm of corn seeds.

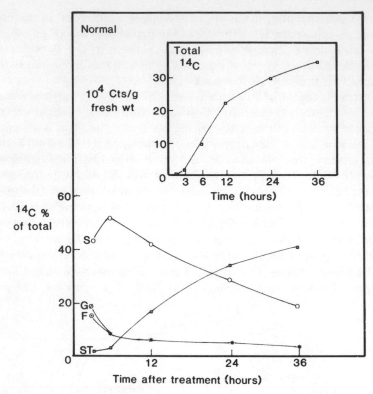

Fig. 2.10. Proportions of radioactivity in the total corn kernel and in the saccharides glucose, frustose, sucrose, and starch after treatment of the whole plant with $^{14}CO_2$. Key to the symbols: G, glucose, F, frustose, S, sucrose; ST, starch. (From Creech, 1968.)

pectin. Amylopectin appears to be synthesized by two enzymes, starch phosphorylase and the Q enzyme (branching). These enzymes are located in plastids of higher plants and are found wherever starch is formed, especially in the endosperm of seeds.

The shape, chemical composition, and molecular association of starch granules appear to be governed by the genotype of the plant, and generally this makes it possible to recognize various plant species by their starch grains. All starches, when grown under natural conditions, show a layered or shell structure. When only one starch granule is formed inside an amyloplast, it is a simple one as in wheat, rye, pea, bean, and other plants. When two or more starch granules grow together in one plastid, they form the parts of one compound starch granule. Examples of compound starch granules are found in rice and oat. Starch granules grow by deposition of material on the outside of the granule, as revealed by $^{14}CO_2$ studies. New layers of starch deposited on the outside of a starch granule will vary in thickness according to the amount of carbohydrates available at the time of formation. Under continuous illumination, starch is continuously deposited in cereal endosperms, and the granules produced lack the usual layered structure found on plants grown under field conditions.

Genes Modifying Starch Synthesis

The ability of plants to synthesize and store varying quantities of starch has long been recognized as being under genetic regulation. In fact, both the quantity and types of starch synthesized are controlled genetically. For example, a range of 9 to 74% in starch content of corn endosperm tissues has been observed in different genotypes.

The proportions of amylose and amylopectin in starch are under direct genetic control and vary from no amylose in waxy corns to slightly above 80% amylose for the highest amylose strains. Normal dent corn contains approximately 27% amylose and 73% amylopectin. Waxy starch (amylopectin) germ plasm was brought to the U.S. from China during the early years of the 20th century. Corn hybrids with waxy endosperm types were developed during World War II as a substitute for tapioca.

A number of genes are known to affect the carbohydrate composition of corn endosperm (Table 2.3). Among those investigated for their possible

Table 2.3. Concentrations of total sugar, water soluble polysaccharides (WSP), and starch in kernels of several corn genotypes at four stages of kernel development.

Genotype	Kernel age	Total sugar	WSP	Starch
	days		%	
Normal	16	17.6	2.7	39.2
	20	5.9	2.8	66.2
	24	4.8	2.8	69.2
	28	3.0	2.2	73.4
Amylose extender (ae)	16	30.6	5.7	20.8
	20	18.7	4.2	37.6
	24	11.4	3.7	48.9
	28	9.4	4.4	49.3
Dull or opaque (du) or (O$_2$)	16	24.2	4.1	25.1
	20	15.3	2.7	44.6
	24	9.0	2.4	56.5
	28	8.0	1.9	59.9
Shrunken-2 (sh$_2$)	16	28.3	5.6	22.3
	20	34.8	4.4	18.4
	24	29.4	2.4	19.6
	28	25.7	5.1	21.9
Sugary-1 (su$_1$)	16	25.7	14.3	23.3
	20	15.6	22.8	28.0
	24	13.1	28.5	29.2
	28	8.3	24.2	35.4
Sugary-2 (su$_2$)	16	16.7	3.6	39.3
	20	12.7	3.1	50.7
	24	4.5	2.5	63.9
	28	3.3	1.9	64.6
Waxy (wx)	16	19.7	3.5	34.1
	20	8.7	2.3	53.3
	24	7.0	2.8	61.9
	28	3.3	2.2	69.0

From Creech (1968).

use in changing amylose-amylopectin ratios are waxy (wx) and its allele Argentine waxy (wx²), sugary (su₁) and its allele sugary-amylaceous (suᵃᵐ), sugary-2 (suᵥ), dull (du) or opaque (O₂), and amylose extender (ae). Other genes that affect the carbohydrate composition of the endosperm are shrunken-2 (sh₂) and brittle (bt). The two genes su₁ and du interact with one another to increase the amylose content of endosperm starch to about 65%, compared to the amylose content of standard dent corn of about 25%. These two genes also increase the water-soluble polysaccharide fraction and decrease the total quantity of starch synthesized. The su₂ gene increases amylose content to 35% whereas the genes du, su₁, and su₂ interact to increase the amylose content of starch to 77%. However, no net increase in absolute yield of amylose is obtained because of a net decrease in total endosperm starch synthesized in the seed. The ae gene increases amylose content significantly (60%) without drastically decreasing the total starch content. The gene shrunken-1 (sh₁) is an endosperm mutant that reduces the total starch to a very significant degree, causing the formation of a collapsed or shrunken kernel. The genes sh₂ and su₁ increase the sugar content of the endosperm as compared to a normal dent corn. Most of the sugar produced consists of sucrose, which constitutes approximately 16% of the dry weight of the seed. Two other endosperm mutants, brittle-1 (bt₁) and brittle-2 (bt₂) reduce endosperm starch substantially and increase the content of reducing sugars and sucrose, but they also significantly reduce grain yields.

Starch exists in a colloidal or water-insoluble form in the endosperm, permitting storage of large amounts in plant cells without influencing the osmotic pressure of the cell. Starch is a high-molecular-weight polymer of D-glucose and is the principal carbohydrate stored in plant seeds. Most starches consist of a mixture of two types of polymers: amylose and amylopectin. The proportion of amylose and amylopectin varies in different starches but is generally in the range of one part amylose and three parts amylopectin.

Amylose, a linear polymer of glucose, consists of several hundred glucose units joined by α-D-1,4 linkages. Two hundred to 1000 glucosyl units are generally linked to form the molecule of amylose. The most significant properties of amylose are its ability to form strong flexible films, its value as a coating agent, its applicability to food products, and its ease of derivability to potentially useful products. High amylose corn hybrids have been developed that produce starch containing 80 to 85% amylose.

Amylopectin is a branched polymer of glucose. The molecular size of amylopectin ranges from several hundred thousand to several million representing many thousands of D-glucose units per molecule. Glucose linkages in amylopectin consist of α-D-1,4 and α-D-1,6. The latter linkages give rise to the so-called branch points in the molecule. Corn hybrids (waxy) have been developed that synthesize a starch composed primarily of amylopectin.

The general chemical and physical characteristics of the two types of starch are:

Amylose

Comprises approximately 10 to 30% of most starches
Molecular weight from 10 000 to 100 000
Phosphate free
Forms complexes with iodine and other polar agents; turns dark blue in presence of iodine
Completely hydrolyzed to glucose with α-amylase
Forms strong piable films
Crystalline X ray pattern
Insoluble in hot water (70 to 80° C)
Precipitable with n-butanol, with which it forms a crystalline complex

Amylopectin

Comprises from 70 to 90% of most starches, 100% in waxy types
Contains from 0.06 to 0.09% phosphorus as glucose phosphate of terminal glucose units
Turns red in the presence of iodine
60% hydrolyzed to glucose with β-amylase
Forms brittle films
Amorphous X ray pattern
Soluble in hot water (70 to 80° C)
Not precipitated with n-butanol

Protein Biosynthesis

Every plant is able to synthesize the 18 amino acids and two amides that form the constituents common to nearly all protein molecules. In addition, about 100 other amides and amino acids are known to have a limited distribution in higher plants, existing either as amino acids or as low-molecular-weight peptides. Except in rare instances, these compounds are not incorporated into proteins and so are conveniently termed nonprotein amino acids.

The constituents of protein (except proline) are all α-amino acids having the general structure $R \cdot CH(NH_2) \cdot CO_2H$. The group R (organic radicle) may show a neutral, basic, or acidic character and include hydroxyl, aromatic, or sulfur-containing substitutes.

Tracer techniques, including isotopic competition methods using ^{14}C-labeled metabolites, have established schemes to describe the interconversion of carbon chains involved in amino acid biosynthesis. These studies initiated the idea of amino acid families, i.e., groups of amino acids derived from similar pathways. Thus, glutamic acid, proline, hydroxyproline, and arginine are commonly termed the glutamate family. Similarly, the concept of asparate, pyruvate, and serine families has been developed.

Each major synthetic route branches from an intermediate of either the glycolytic or tricarboxylic acid cycle pathways that constitute the respiratory processes, and as expected, labeling of alanine and serine precedes that of glutamate and aspartate when ^{14}C-glucose is degraded by plants. Photosynthetic fixation of ^{14}CO$_2$ leads to a similar labeling sequence among the amino acids. Activity always appears first in alanine, serine, or glycine and only more slowly in acids like glutamate, aspartate, proline, and arginine that are metabolically more distant from the primary reactions of photosynthetic CO$_2$ fixation. In contrast, dark fixation of ^{14}CO$_2$ by leaves frequently yields aspartic acid as the earliest labeled amino acid. This presumably arises from β-carboxy-oxalacetate formed after carboxylation of pyruvate. Alanine and glutamic acid become radioactive only after label is introduced into the α-carbosyl by equilibration of malate with fumarate.

Ammonia forms the major, possibly the only, inorganic nitrogen compound utilized directly for amino acid biosynthesis. This is shown schematically in Fig. 2.11. Recent investigations using ^{15}N-labeled forms of ammonia, nitrate, and elementary nitrogen have confirmed that nitrogen rapidly enters glutamic acid and glutamine molecules; aspartic acid, alanine, arginine, and other amino acids are more slowly labeled irrespective of the type of inorganic nutrient supplied. These conclusions appear to be equally valid for higher plants, blue-green algae, nitrogen-fixing bacteria, and yeasts.

Synthesis and Deposition of Storage Protein in Seeds

All 20 of the amino acids incorporated into plant protein are first activated enzymatically in reactions involving ATP. The activated amino acids then react with transfer RNA's (tRNA's) to yield aminoacyl-tRNA's. The transfer RNA's carry the amino acids to the site of protein synthesis, the ribosome. There, successive transfer RNA's are believed to interact with a messenger RNA to determine the sequence of amino acids for formation of a specific protein. A specific messenger RNA codes for the amino acid sequence of each protein. It has been shown that storage proteins are synthesized on ribosomes attached to the rough endoplasmic reticulum, concentrated into protein droplets in dictyosomes, and transported to the vicinity of a protein storage vacuole in membrane-bound vesicles. These vesicles then empty into the vacuole, apparently by a process of membrane fusion with the vacuolar membrane, producing characteristic protein bodies of storage proteins.

At an early stage in seed development, the larger vacuoles of the cytoplasm appear to be replaced by small vacuoles in which storage proteins accumulate. Research on corn suggests that these small vacuoles arise by dilation of the endoplasmic reticulum. It is unclear just how storage proteins enter the vacuoles since no connections with the endoplasmic reticulum have been observed.

In the developing cotyledon of legume seeds, there are two growth phases: an initial one of intensive cell division, followed by a longer period of growth by cell expansion. During the period of cell expansion growth, 95% of the storage proteins are synthesized. The development of soybean seed provides an excellent example of the synthesis and storage of protein.

Classification of Storage proteins

Storage proteins in seeds normally occur in the form of definite bodies, protein bodies, protein granules or aleurone grains, or matrix proteins. Current literature has utilized the term *protein bodies* for storage protein located in cotyledons and endosperm and *aleurone grains* for protein stored in the aleurone layer. These subcellular particles contain most of the storage proteins of seeds. Protein bodies occur widely in the cotyledons and endosperm of both starch-bearing and oil-bearing seeds. Protein bodies

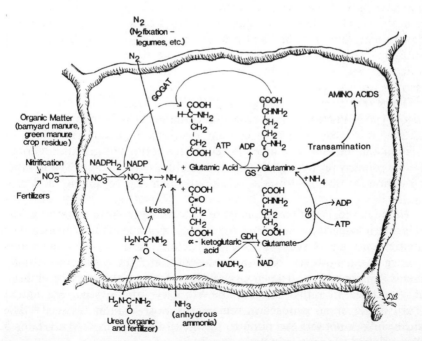

Fig. 2.11. Pathway of nitrogen incorporation into amino acids from protein synthesis (parenchyma cell of endosperm or cotyledon).

appear to have the following characteristics: 1) some contain crystalline inclusions of phytic acid; 2) they are probably surrounded by a single membrane, remaining from the vacuole in which they are stored; 3) they contain most of the cellular protein, but none of the oil; 4) their synthesis commences after the majority of the cells of the seed have been formed; and 5) they swell, coalesce, and disappear early in germination.

Most storage proteins of seeds contain large amounts of amide nitrogen. The amides are formed by incorporation of asparagine and glutamine into the protein during synthesis and not as a result of amidation of free carboxylic groups of aspartic and glutamic acid bound in proteins.

There are four general types of proteins stored in seeds of crop plants:

Protein type	Solubility characteristics
Albumin	Soluble in water and dilute salt solutions.
Globulin	Sparingly soluble in water but soluble in dilute salt solutions.
Prolamin	Soluble in 70 to 80% ethanol but insoluble in water and in absolute ethanol.
Glutelin	Insoluble in all of the above solvents but soluble in acid or alkali.

Storage proteins are often distinguished from metabolically active proteins by their disappearance during germination. Storage proteins have little or no overt biological activity and are classified primarily by their solubility in various solvents as indicated above. Enzymes are presumed to constitute the greater part of the albumin fraction, which is largely metabolically active proteins. The cell also contains other types of metabolic proteins, including membrane proteins, and proteins of cell organelles, nucleus, and cytoplasm. The sum of all these proteins—metabolic and storage—constitutes the total protein content of the seed. However, the storage proteins constitute the greater bulk of proteins stored in seeds.

These storage proteins are generally found in all seeds, but the quantity of each type of protein stored varies widely among crop species. As a generalization, the seeds of cereals contain relatively high concentrations of prolamin and gluteins with much lower concentrations of globulin. Globulin is the primary protein stored in pulses (edible seeds of legumes) and other dicotyledonous plants with relatively small quantities of glutelin and prolamine.

In the monocotyledonous plants (e.g., cereals) the primary storage organ is the endosperm, which is in close contact with the embryo proper. The scutellum may also serve as a storage structure. In dicotyledonous plants (e.g., legumes), however, both the endosperm and the cotyledons usually serve as storage organs. The endosperm is utilized by the growing embryo and "disappears" rather early in the life cycle of the seed, so that at maturity little or no endosperm remains in most leguminous seeds. The castorbean is a notable exception to this generalization in that it retains a well-developed endosperm in the mature seed.

Characteristics of Storage Proteins of Cereals

The relative concentration of storage proteins in cereals is indicated in Table 2.4 and their amino acid composition in Table 2.5. As a generalization, the quality of protein stored in cereal grains is rather poor for human and monogastric animals. This performance is due to the storage of rather large amounts of prolamin, which is deficient in the amino acids lysine and tryptophan. Glutelin-type proteins are also stored in significant quantities but are a higher quality protein for humans and monogastric animals because of a more favorable balance of essential amino acids for these organisms. Thus, rice contains a higher quality protein for humans than corn due to the type of protein synthesized for storage by these two species. Table 2.6 lists the essential amino acids for humans. The deviation of these amino acids in wheat from essential amino acid requirements of humans is listed in Table 2.7.

Table 2.4. Protein concentrations of some cereal grains. †

Crop	Protein	Protein type			
		Albumin	Globulin	Prolamine	Glutelin
	%	— % total N —			
Barley	10–16	3–4	10–20	35–45	35–45
Corn					
Normal	7–13	14	5–6	50–55	31
Opaque	7–13	25	tr.‡	25	39
Millet	7–16	tr.	10–11	57	30
Oat	12–18	10	55	10	25
Rice	8–10	tr.	about 2–8	about 1–5	85–90
Rye	9–14	5–10	5–10	30–50	30–50
Sorghum	9–13	tr.	5–6	>60	30–40
Wheat	10–15	3–5	6–10	40–50	30–40

† Compiled from several sources.
‡ tr. = traces.

Table 2.5. Percent amino acid in proteins in cereal grains. †

Amino acid‡	Wheat	Rye	Corn	Barley	Oat	Rice	Sorghum
Arginine	0.80	0.53	0.51	0.60	0.80	0.51	0.40
Cystine	0.20	0.18	0.10	0.20	0.20	0.10	0.20
Histidine	0.30	0.27	0.20	0.30	0.20	0.10	0.30
Isoleucine	0.60	0.53	0.51	0.60	0.60	0.40	0.60
Leucine	1.00	0.71	1.11	0.90	1.00	0.60	1.60
Lysine	0.50	0.51	0.20	0.60	0.40	0.30	0.30
Methionine	0.20	0.20	0.10	0.20	0.20	0.20	0.10
Phenylalanine	0.70	0.70	0.51	0.70	0.70	0.40	0.51
Threonine	0.40	0.40	0.40	0.40	0.40	0.30	0.30
Tryptophan	0.20	0.10	0.10	0.20	0.20	0.10	0.10
Tyrosine	0.51	0.30	0.50	0.40	0.60	0.70	0.40
Valine	0.60	0.70	0.40	0.70	0.70	0.51	0.60

† Source: Publication 1232, National Academy of Sciences—National Research Council, 1964.
‡ Values reported on moisture-free basis.

Table 2.6. Essential amino acids for humans.

Arginine	Phenylalanine
Isoleucine	Histidine
Leucine	Threonine
Lysine	Tryptophan
Methionine	Valine

Table 2.7. Deviation of essential amino acids in wheat proteins
from requirements of humans; FAO, 1957.

Amino acid	Deviation from requirement
	%
Lysine	− 55.0
Methionine†	− 12.5
Isoleucine	− 16.7
Leucine	+ 30.4
Tyrosine	+ 9.7
Phenylalanine	+ 40.4
Threonine	− 3.4
Valine	+ 4.5

† Adjusted provisional amino acid pattern (World Health Organization, 1965).

A major storage protein of corn is zein (a prolamin), a poor quality protein in terms of the balance of amino acids essential for monogastric animals. Zein contains a relatively large amount of alanine and leucine with little lysine and almost no tryptophan. Consequently, corn as the sole dietary source of protein is a poor quality protein for humans and other monogastrics. The protein granules of the endosperm are the main sites of zein storage.

The identification of the opaque-2 (O_2) or dull (*du*) gene in corn resulted in a significant improvement in the nutritive value of the storage protein of the seed. The opaque-2 gene is responsible for the increased synthesis of glutelin in the endosperm with a concurrent reduction in the synthesis of prolamin (zein) protein. A comparison of the synthesis and storage of proteins in corn endosperm with and without the O_2 gene is indicated in Fig. 2.12. Glutelin proteins contain high concentrations of lysine; consequently, the grain is much higher in lysine and a better quality of protein for monogastric animals and people.

A gene for altering the amino acid composition of storage proteins of barley seeds has also been found. The Hily (high lysine) gene of barley results in seeds having a higher lysine concentration than the standard seeds. Similar genes modifying synthesis of storage proteins to improve protein quality have also been reported for sorghum.

The major storage protein of wheat is gluten. Gluten is comprised of many proteins with molecular weights ranging from 25 000 to millions. The elastic nature of this protein accounts for the ability of wheat flour to "raise" and form a leavened loaf of bread. Rye also contains a relatively high concentration of gluten and is also widely used for bread. Gliadin, the other major protein class in wheat, does not have the elastic properties of gluten. Thus, it is less valuable in wheat for bread-making purposes.

Characteristics of Storage Proteins of Dicotyledonous Seeds

As a broad generalization, dicotyledonous plants synthesize primarily globulin-type proteins as the major protein stored in the seed. They are stored in the cotyledons, primarily in the form of protein bodies. There is little prolamin- and glutelin-type protein stored in the seeds of dicotyledonous plants; consequently, the quality of the protein for humans and other monogastric animals is relatively high. The relative nutritive value of soybean meal for monogastric animals is approximately 75 compared to a value of 100 for eggs, the nutrition standard for protein quality for these organisms. Wheat flour rated 50 on this nutritive scale, indicating that the quality of protein stored in soybean seeds and other dicotyledonous plants, especially pulses, is much higher than that of cereals. It is relatively easy to see why a diet consisting of cereals for energy and beans for protein is a relatively well-balanced diet nutritionally for humans and monogastric animals. Seeds of legumes (pulses) are low in methionine relative to Food and Agricultural Organization (FAO) standards.

The major types of proteins stored in the seed of same dicots is indicated in Table 2.8 and their amino acid composition in Table 2.9. In general, the total protein concentration in seeds of dicotyledonous plants is high, with seeds of soybean and other beans ranging between 25 to 50% of the total dry weight of the seed. Table 2.9 indicates that dicot seeds are relatively high in lysine with significant amounts of tryptophan and also provide an excellent source of high quality protein to balance diets high in cereal grains.

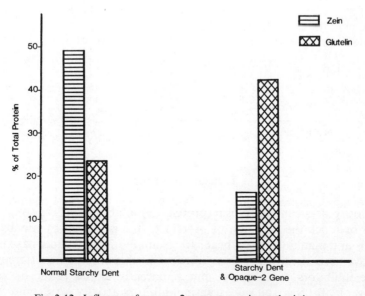

Fig. 2.12. Influence of opaque-2 gene on protein synthesis in corn.

Table 2.8. Protein concentration of some dicotyledonous crops.

Crop	% protein (dry wt)	Protein type			
		Albumin	Globulin	Prolamine	Glutelin
		% total nitrogen			
Castorbean	20	10	90	†	†
Cotton	20	†	90	†	10
Flax	30	†	max.	†	much
Lupine	40	1	78	†	16
Soybean	30–50	†	85–95	†	†
Sunflower	30–35	†	max.	†	†
Dry beans	25	†	90–95	†	†
Pea	—	21	66	12	†
Peanut	—	15	70	10	†

† Little or none detectable.

Table 2.9. Amino acid composition (expressed as percentages of total protein) in the seed proteins of some dicotyledonous plants, in parentheses.

Protein	Globulin type protein†			
	Legumin (Pea)	Vicilin (Pea)	Phaseolin (Bean)	Glycinin (Soybean)
Total N	17.9	17.5	—	—
Glycine	—	—	0.6	4.1
Alanine	—	—	1.8	3.9
Serine	—	—	0.4	5.7
Threonine	—	—	—	3.0
Valine	—	—	1.0	4.6
Leucine	—	—	—	5.8
Isoleucine	—	—	—	5.8
Methionine	—	—	—	2.6
Cystine	—	—	—	1.3
Proline	—	—	2.8	5.4
Phenylalanine	—	—	3.3	5.8
Tyrosine	4.2	4.2	2.8	3.7
Tryptophan	1.3	0.4	present	3.8
Lysine	3.5	4.6	4.6	6.9
Arginine	13.1	11.5	4.9	7.9
Hystidine	3.0	2.4	2.6	2.0
Aspartic acid	16.3	15.7	5.2	—
Glutamic acid	20.1	22.7	14.5	—
Ammonia	2.3	2.2	2.1	—

† (—) = no data available.

Trypsin Inhibitors

Most pulses contain trypsin (proteinase) inhibitors. These compounds inhibit or block the digestion of protein of the raw bean by monogastric animals and mankind. In soybean, for example, 7 to 10 proteinase inhibitors have been identified. The significance of this number of different proteinase inhibitors is unknown, but apparently is the result of genetic heterogenity for this trait since trypsin inhibitory activity varies quite widely among soybean genotypes.

In plants, most postulated functions for proteinase inhibitors fit into three categories: 1) maintain seed dormancy by preventing autolysis, 2) regulate protein synthesis and metabolism, and 3) prevent attack by predatory insects and bacteria. Because of the presence of trypsin-like enzymes in insects and bacteria that invade seeds, some specific metabolic defense mechanisms against these invaders have been attributed to plant proteinase inhibitors.

In mature soybean, the proteinase inhibitors do not inhibit the proteolytic enzyme systems. No inhibitory activity was detected in leaves, stems, and empty pods. However, trypsin inhibitory activity amounting to 50% of that in mature soybeans was already present in young (3-week-old) seeds. No loss in trypsin inhibitory activity was detected in soybeans germinated up to 1 week.

Most plant proteinase inhibitors are inactivated by heat, an effect generally accompanied by an enhancement of the nutritive value of the protein. At 100°C, only 15 min of steaming is required to inactivate the trypsin inhibitors in soybean stalks and 20 min for whole soybeans, provided the beans are tempered to approximately 25% moisture before steaming.

Efficiency of Protein Synthesis and Storage

Figure 2.13 indicates the relative efficiency of various crop species in the storage of protein per unit of photosynthate produced. It is clear from this information that soybean is an efficient species for the synthesis and

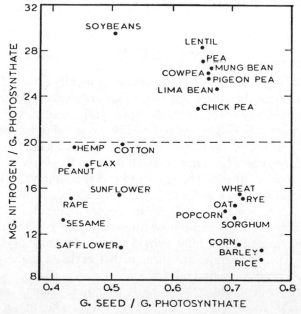

Fig. 2.13. Efficiency of seed production and protein synthesis per unit of photosynthate. The dashed line represents the nitrogen supply when the nitrogen supply rate is 5 m ha⁻¹ day⁻¹ and the available photosynthate rate is 250 kg ha⁻¹ day⁻¹.

storage of protein. However, many other species, although they store a somewhat lower level of protein, are more efficient metabolically than soybean in terms of grams of seed provided per gram of photosynthate produced. Most notable amoung these species are pea, chick pea (*Cicer arietinum* L., pigeon pea (*Cajanus indicus* Spreng.), and cowpea. Many species of dicotyledonous plants, such as sunflower, peanut, and cotton store lower quantities of protein, and the efficiency of protein synthesis and storage in the seed is much lower than the above species. The majority of cereal grains are relatively efficient in terms of grams of seed produced per gram of photosynthate produced but are relatively inefficient in terms of the amount of protein produced and stored per unit of photosynthate. It appears, therefore, that there is a wide range both in the efficiency of seed production (g seed/g photosynthate) and the efficiency of protein production (mg nitrogen/g photosynthate) among crop species. One explanation of the differences between pulses, such as soybean, and cereals, such as corn, in the efficiency of seed production per unit of photosynthate is indicated in Table 2.10. The efficiency of corn production expressed as Joules stored in seed/Joules produced via photosynthesis is approximately 87%, whereas soybean is synthesized at an efficiency of approximately 50%. The difference in efficiency of seed production between these two species is due to the amount of energy required for the synthesis of oil and protein as compared to starch, which is most economical in terms of energy required for its synthesis and storage. The synthesis of oil requires a very high utilization of energy as compared to starch, approximately 2.25 to 2.5 times as much energy per unit weight.

Lipid Biosynthesis

Lipids, being insoluble in water, cannot generally diffuse from cell to cell. It seems reasonable, therefore, that lipids are synthesized and stored in the cells in which they occur. In seeds, lipid synthesis occurs in cells of the embryo, primarily the cotyledons (especially the scutellum of cereals), and endosperm. The site of synthesis of triglycerides is located in the microsomes. The formation of α-glycerophosphate from free glycerol and ATP is essential for lipid synthesis. The glycerokinase that catalyzes this reaction occurs as a soluble protein in the cell. The thiokinases and the acylating enzyme appear to be associated with the microsomal particles. Glycerophosphate, synthesized by the phosphorylation of glycerol is acylated by acyl CoA to yield phosphatidic acid, which, in turn, is dephosphorylated to give diglycerides. The diglycerides are then further acylated in the presence of acyl to CoA to produce triglycerides.

The triglyceride stored in the majority of crop seeds may range from 10 to 50% of the tissue dry weight (Table 2.11). The major fatty acid components of triglycerides stored in seeds of higher plants are indicated in

Table 2.10. Metabolic energy required for producing soybean and corn seed.

Energy source	Com-position	Dry wt	Type of energy			Equivalent yield
			Stored	Metabolic	Total	
	% dry wt	kg		kcal/energy source		kg ha^{-1}
			Soybean			
Carbohydrates	33	7.89	32 000	0		
Oil	22	5.27	50 000	40 000		
Protein	40	9.58	54 000	29 000		
Ash	5	1.18	0	0		
Nitrogen fixation	—	—	0	33 000		
Total	100	23.92	136 000	102 000	238 000	3 029
			Corn			
Carbohydrates	84	18.39	75 000	0		
Oil	4	0.86	8 000	6 200		
Protein	10	2.18	12 000	6 400		
Ash	2	0.45	0	0		
Total	100	21.88	95 000	12 600	107 600	6 282

From Howell (1962).

Table 2.11. Oil concentration, on a dry wt basis, in oil seed crops.

Crop	Oil concentration
	%
Flax	35–40
Soybean	18–20
Sunflower	45–50
Safflower	25–32
Peanut	47–50
Cotton seed	15–25
Mustard	31–33
Rapeseed	35–45
Crambe	30–50

Table 2.12. Phospholipid and glycolipids normally represent less than 2% of the total seed lipid. In wheat, for example, lipids comprise 1 to 2% of the endosperm, 8 to 15% of the germ, and about 6% of the bran, with an average of 2 to 4% for the whole kernel. Wheat germ oil consists primarily of triglyceride, whereas lipid extracts from endosperm and bran contain much higher proportions of phospholipids and glycolipids.

The composition of oil obtained from seeds of many crops is strongly influenced by temperatures under which the crop is grown, especially during the seed-filling period. At temperatures lower than optimum for growth, a higher proportion of fatty acids present is unsaturated, as measured by a higher iodine value. These variations in fatty acid composition are generally confined to oleic, linoleic, and linolenic acids. However, temperature has little effect on the proportion of saturated acids, which is determined by the genotype of the crop being grown. Temperatures above the optimum for growth appear to result in an accumulation of oleate at the expense of linoleate (and linolenate) while sterate is unaffected.

Table 2.12. Major fatty acids stored in seeds of higher plants.

Common name	Symbol	Structure
Lauric	12:0	$CH_3(CH_2)_{10}COOH$
Myristic	14:0	$CH_3(CH_2)_{12}COOH$
Palmitic	16:0	$CH_3(CH_2)_{14}COOH$
Stearic	18:0	$CH_3(CH_2)_{16}COOH$
Arachidic	20:0	$CH_3(CH_2)_{18}COOH$
Palmitoleic	16:1(9)	$CH_3(CH_2)_5CH = CH(CH_2)_7COOH$
Oleic	18:1(9)	$CH_3(CH_2)_7CH = CH(CH_2)_7COOH$
Linoleic	18:2(9,12)	$CH_3(CH_2)_4CH = CHCH_2CH = CH(CH_2)_7COOH$
Linolenic	18:3(9,12,15)	$CH_3CH_2CH = CHCH_2CH = CHCH_2CH = CH(CH_2)_7COOH$

From Bonner and Varner (1976). Used by permission.

Crop seeds normally mature over a period of weeks, and maturation is associated with considerable increases in the size and weight of the various tissues of the seed. In oilseed crops much of this weight increase is due to the accumulation of lipid in the endosperm or embryo of the seed. The principal site of lipid storage in cereals, such as corn, is the scutellum with lesser quantities in the embryo axis. In dicotyledons, such as soybean, the principal site of storage is in the cotyledons. In soybean experiments, major changes in fatty acid composition of the triglycerides occurred during the first 52 days after flowering. During this period linolenic acid decreased from 34 to 12% of the total lipid content of the seed, whereas the percentages of linoleic and oleic acids increased, stearic acid remained fairly constant, and palmitic acid decreased slightly.

The rate at which lipid is synthesized and stored in the developing seed is seldom constant throughout the period of seed maturation. It is not uncommon for a large proportion of the total lipid content of a mature seed to accumulate during a comparatively short period near the end of seed maturation. Moreover, the component fatty acids that constitute the oil of the mature seed are not usually synthesized at constant rates throughout this period. Thus, at any given time, the oil may have a composition quite different from that of the mature seed. These variations in rate of lipid synthesis and type of fatty acid synthesized are illustrated for maturing soybean seeds in Table 2.13.

The following study with corn illustrates the changes in types of lipids and their concentrations with advancing seed maturation. Corn seeds ('Illinois high oil') harvested at immature stages in one study contained 3% total lipid. The composition of this lipid was 70% polar lipids and 10% triglyceridean neutral lipids. At maturity the lipid concentration of the seed was nearly 14%, with triglycerides comprising 92% of the lipid (Table 2.14). The rate of lipid synthesis was greatest between 15 and 45 days after pollination. Polar lipids comprised a steadily decreasing and triglycerides an increasing percent of the total lipid stored in the seed during maturation. At no period during the maturation phase were free fatty acids present in significant amounts. Large changes occurred in the fatty acid composition of

Table 2.13. Changes in field-grown Lincoln soybeans during maturation.

Days after blossoming	Moisture	Oil	Avg. weight of fatty acids			
			Saturated acid	Oleic acid	Linoleic acid	Linolenic acid
	%		mg/bean			
23	84	0.15	0.02	0.05	0.05	0.03
27	79	0.5	0.1	0.1	0.2	0.01
31	75	1.7	1.3	0.3	0.8	0.2
34	72	3.9	0.4	1.0	1.9	0.4
39	71	11.5	1.5	3.3	5.3	0.9
46	65	19.8	3.9	3.7	10.0	1.3
52	58	29.5	4.8	5.8	15.5	2.0
60	48	30.0	5.4	4.8	16.4	2.1
67	9	31.5	5.2	5.6	17.1	2.7

From Simmons and Quachenbush (1954).

the triglycerides between 10 and 45 days after pollination, the relative proportions of palmitic, linoleic, and linolenic acid falling and those of oleic acid increasing with maturation (Table 2.13).

Genetic Control of Lipid Synthesis

Breeding trials with rape, safflower, flax, and sunflower indicate that the composition of their oil is under genetic control. More specifically, it is apparent that rapeseed varieties contain a single gene that controls the chain elongation of oleic acid to erucic acid, whereas in safflower three genes at one chromosome locus govern the proportions of linoleic and oleic acids while two genes at a second locus govern stearic acid levels.

The differences in composition occurring within species attributable to genotype are far greater than those created by the environment, such as climate and nutrition, and offer great opportunities for the breeding of crops for oils of specific nutritional or industrial utilization.

Lipid Classification

As a generalization, lipids may be considered chemical compounds that are insoluble in water but soluble in fat solvents (organic solvents). They are either esters of fatty acids or hydrolytic products of such esters. Oils are lipids that are liquid, whereas fats are lipids that are solid at room temperature.

The major class of lipid stored in seeds of major crops are triglycerides (tricylglycerols), which are neutral lipids. These consist of the alcohol glycerol and three fatty acids attached to the glycerol molecule via ester linkages. The generalized formula is $RCOOCH_2CH(OCOR')CH_2OCOR''$.

Table 2.14. Variations in the lipid composition of 100 kernels of corn grain (Illinois high oil) during maturation.

Days after pollination	Corn grain		Oil	Hydrocarbons + sterol esters	Tri-glycerides	Free fatty acids	Sterols	Partial glycerides	Polar lipids
	Wet wt	Dry wt							
	g		% dry wt	% of total lipids by weight					
10	8.3	1.1	3.0	4.7	10.1	0.5	6.8	7.1	70.8
15	15.6	2.5	5.6	4.2	41.1	0.5	5.0	4.1	45.1
30	26.7	11.5	10.9	1.6	78.4	1.4	4.3	4.7	9.5
45	31.6	18.0	13.7	2.0	84.0	0.5	3.2	4.3	6.0
60	30.6	19.0	13.8	1.5	88.1	0.4	2.5	3.0	4.5
75	33.4	23.4	13.4	1.3	92.0	0.4	1.5	0.9	3.9
90	31.8	23.8	13.8	1.0	94.0	0.3	1.3	1.1	3.9

From Hitchcock and Nichols (1971).

Triglycerides are found in exceptionally high concentrations in oil-containing seeds such as soybean, flax, cotton, peanut, sunflower, and mustard (*Brassica nigra* L.). Triglyceride fats have as their major fatty acids the saturated fatty acids such as palmitic and stearic acids. Triglyceride oils have as their major fatty acids unsaturated fatty acids such as oleic, linoleic, and linolenic acids. Triglycerides are readily hydrolyzed under alkaline conditions to glycerol and fatty acids. Oleic and palmitic acids are the most common fatty acids in plant lipids. Table 2.12 lists the major fatty acids present in seeds of major crop species.

All of the unsaturated fatty acids combine with hydrogen, oxygen, or the halogens, such as iodine (giving rise to the iodine number of plant oils). The drying properties of linseed, sunflower, and other oils are a consequence of the capacity of the highly unsaturated fatty acid radicals of the oil to react with oxygen of the air, resulting in the formation of solid, waxy compounds that form the film of paint with which we are all familiar.

Mineral Composition of Seeds

A considerable amount of practical knowledge about crop production was built up over the time that humans have cultivated crops. However, the "food of plants" remained a rather vague term until the early experiments of John Woodward (1665–1728), credited with being one of the first to determine the importance of minerals for crop growth (see Chapter 7). From these beginnings until the present time, further experimentation has shown that 16 elements are essential for the growth of higher plants such as our field crops. These essential elements may be broken down into macro- and micronutrient elements based on the quantity of these elements absorbed by the plant—not their significance to the plant. These essential elements are:

Macronutrient elements	Micronutrient elements
Carbon	Chlorine
Hydrogen	Boron
Oxygen	Iron
Nitrogen	Manganese
Phosphorus	Zinc
Potassium	Copper
Calcium	Molybdenum
Magnesium	
Sulfur	

An average composition of the inorganic elements present in the seeds of several crop species is given in Table 2.15.

In addition to the essential elements, plants also absorb a wide variety of other inorganic elements from the soil. If the total ash content of a plant is analyzed, it is not uncommon to find more than 60 elements in that ash. However, the amount and kinds of elements translocated to the grain are

Table 2.15. Mineral composition of crop seeds.†

	Minerals‡									
	K	P	Ca	S	Mg	Na	Fe	Cu	Mn	B
				%					ppm	
Barley	0.63	0.47	0.09	0.19	0.14	0.02	0.006	8.6	18.0	13.0
Dry beans	1.89	0.63	0.17	0.23	0.20	0.09	0.012	--	20.0	17.0
Buckwheat	0.51	0.38	0.13	—	0.26	—	0.005	11.0	38.0	—
Corn	0.35	0.32	0.03	0.12	0.17	0.01	0.003	2.9	5.9	1.9
Cotton seed	1.20	0.73	0.15	0.76	0.44	0.02	0.059	54.0	31.0	13.0
Flax	1.24	0.47	0.40	0.06	0.58	0.11	0.020	26.4	39.4	17.0
Millet	0.48	0.31	0.06	0.14	0.18	0.04	0.005	24.0	32.0	—
Oat	0.42	0.39	0.11	0.23	0.19	0.07	0.008	6.6	43.0	—
Rice	0.17	0.26	0.05	0.05	0.07	0.05	0.004	3.7	20.0	9.4
Rye	0.52	0.38	0.07	0.17	0.13	0.02	0.009	8.8	75.0	2.9
Sorghum	0.38	0.35	0.05	0.18	0.19	0.05	0.005	11.0	16.0	—
Soybean	2.4	0.66	0.28	0.45	0.34	0.38	0.016	23.0	41.0	41.0
Sunflower	0.96	1.01	0.21	0.02	0.40	—	0.003	—	23.0	—
Wheat	0.58	0.41	0.06	0.19	0.18	0.10	0.006	8.2	55.0	1.11

† Compiled from a number of sources.
‡ (—) = no data available.

considerably less than those absorbed by the roots and translocated to the vegetative organs of the plant. The precise physiological basis for this difference is largely unknown, but the grain exhibits considerable selectivity in its absorption and accumulation of elements from the xylem fluid or, on occasion, from the phloem. This is best indicated by studies of the genetic control of mineral accumulation in grain of several crops. Studies with barley, wheat, soybean, corn, rice, and other crops reveal that three- to five-fold differences in accumulation of several mineral elements occur among varieties and genotypes of each of these crops. Studies attempting to determine the physiological processes resulting in differential accumulation of minerals among genotypes of a crop species indicate that a variety of mechanisms are involved. Among these are differential rates of root absorption, differential translocation within the plant, and differential accumulation of minerals by the grain.

Studies of the inheritance of differential mineral accumulation of strontium by barley and wheat indicate that mineral accumulation by the grain is under genetic regulation and is a quantitative character, governed by many genes. Crosses, in all combinations among crop genotypes using parents that are high and low in mineral accumulation in the grain, indicate that significant changes could be made in mineral composition of the seed by selecting the appropriate genotypes from segregated populations of these crosses. Similar results have been found in other crops as well.

Mineral deficiencies predominately affect the number of seeds produced but, unless the deficiency is severe, appear to have relatively minor effects on the mineral composition of the seed. Application of fertilizers, manures, or other sources of inorganic elements significantly increases

Table 2.16. Percent minerals of barley grains from Hoosfield experiment, Rothamsted, England.

Fertilizer element applied	Minerals					
	N	P	K	Ca	Mg	Na
None	1.54	0.29	0.51	0.065	0.10	0.39
P	1.44	0.36	0.45	0.055	0.10	0.27
K, Na, Mg	1.44	0.36	0.52	0.050	0.12	0.18
P, K, Na, Mg	1.40	0.36	0.54	0.050	0.11	0.16
N	1.90	0.28	0.45	0.055	0.11	0.47
N, P	1.65	0.34	0.42	0.055	0.10	0.47
N, K, Na, Mg	1.62	0.30	0.46	0.045	0.11	0.20
N, P, K, Na, Mg	1.35	0.32	0.53	0.050	0.10	0.23

From Roberts (1972). Used by permission.

mineral accumulation of the seed. This point is well illustrated by data from fertilizer studies conducted at Rothamsted, England since 1852 (Table 2.16). Thus an average mineral composition of various crop seeds may be misleading if specific information on a particular lot of seed is desired. In spite of this conclusion, average mineral compositions may be useful for evaluation or comparative purposes.

Vitamin Composition of Seeds

The name *vitamin* was first used in 1914 and was based on the fact that vitamins of the B group contained amino-derivatives essential for life. The Latin term *vita* (life) was utilized to describe this group of compounds. However, it was later learned that various vitamins belong to quite different groups of chemical compounds, and no all-encompassing chemical grouping could be made.

The vitamins known at present are classified into two groups: water soluble vitamins and those soluble in lipids, or fat soluble vitamins (Table 2.17).

Vitamins are organic biocatalysts synthesized by plants. They function principally as enzyme cofactors. Their absence causes severe diseases, avitaminoses, in both plants and animals. Vitamins are characterized by the fact that the amounts indispensable to the organism are minute, and therefore they are not classified as nutrients. Most higher plants are autotropic with respect to vitamins and store significant quantities of vitamins in various organs including the seeds.

The synthesis of vitamins in plants depends on an available supply of inorganic nutrient ions, particularly of macroelements. For example, the concentrations of riboflavin decrease with a deficiency of macronutrient elements. However, so far it has not been possible to demonstrate major effects of mineral elements on the synthesis of vitamins. Thus, practical fertilization, to increase vitamin composition of grains, is still empirical.

Table 2.17. Vitamin content of grains of several crop species.

Grains	Biotin (H)	Nicotinic acid	Thiamine (B₁)	Water soluble vitamins†					Fat soluble vitamins			
				Riboflavin (B₂)	Pyridixine (B₆)	Ascorbic acid (C)	Pantothenic acid	Folic acid	A	D	E	K
						ppm						
Barley	0.18	31.0	5.7	2.2	3.3	0	7.3	0.62	0.04		6.2	
Bean, navy	0.06	--	1.0			--	--	--	--			
Buckwheat	--	--	3.7	1.2	--	--	12.0	--	--			
Corn	0.09	22.0	4.4	1.3	5.7	--	7.1	0.22	4.0		0.4	
Cotton seed	--	44.0	--	3.1	--	--	--	--	--			
Flax	--	--	--	--	--	--	--	--	--			
Millet	--	--	7.3	1.8	--	--	8.2	--	--			
Oats	0.33	10.0	7.0	1.8	1.3	--	14.0	0.33	--		6.0	
Rice	0.11	46.0	3.1	0.66	7.0	0	7.1	0.24	0		7.2	
Rye	0.07	16.0	4.4	1.8	--	--	7.7	0.73	0			
Sorghum (grain)	29.0	23.0	4.6	1.5	5.9	Tr	13.0	0.33	1.2			
Soybean	0.02	58.6	100	3.1	--	--	18.6	--	33.0			
Sunflower	--	--	32.0	1.3	--	--	--	--	0			
Wheat	0.11	3.0	5.5	1.3	5.3	0	14.0	0.48	0		15.8	

† (--) =

In general, vitamins B and E are high in cereal grains, with lesser amounts of other vitamins also present. Vitamin C is notably missing in significant amounts from grains.

Gibberellins in Seeds

Immature seeds are an excellent source for studies of gibberellin because of the relatively high concentrations present. Gibberellins have also been detected in mature seeds, but the levels there are much lower. Germinated seeds contain gibberellins in the embryonic axis, the cotyledons, and the seed coat.

The sequence through which a seed begins active metabolism culminating in germination of the seed is not clearly defined. Following the imbibition of water in the seed of different species, the concentration of gibberellin rapidly increases. This increased gibberellin concentration may be due to the release of bound gibberellins, de novo synthesis, or both. Studies indicate that between 12 and 24 hours after the start of water imbibition, a significant amount of gibberellin-like substances is secreted by the barley embryo. This continues for about 48 to 60 hours after imbibition, after which there is both a qualitative change in the gibberellin spectrum and a quantitative change in gibberellin concentration. This suggests that the hormone released during the first 48 to 60 hours after imbibition of water may be derived from a bound form of the enzyme present in the dry seed, while that portion released subsequently may be synthesized during germination. Although the site of synthesis and/or release of gibberellin at early stages of germination has not been definitely identified, evidence from barley suggests that the gibberellin is derived first from the scutellar nodal area and then from the embryonic axis.

No effect of gibberellin has been observed on the starch-containing cells of the endosperm, but a striking and impressive series of metabolic events is set in motion when gibberellin reaches the surrounding aleurone layer. The aleurone cells are stimulated to synthesize amylase and proteinase de novo and release them into the central endosperm. In addition, ribonuclease, peroxidase, cellulase, and other hydrolytic enzymes cause the breakdown of endosperm reserves to the more mobile forms such as soluble carbohydrates and amino acids. Thus, through the controlled release and synthesis of gibberellins, the embryo exerts a regulating influence over the metabolic activity of a tissue spatially separated from itself. The role and significance of gibberellin as a hormone in the germination and establishment of cereals is clear. However, in dicotyledonous crops, evidence suggests that additional factors, as yet unknown, may be involved and that the mobilization of reserves in dicotyledonous seeds may be more complex and variable.

SUGGESTED READING

Bils, R. F., and R. W. Howell. 1963. Biochemical and cytological changes in developing soybean cotyledons. Crop Sci. 3:304–308.

Bonner, James, and J.E. Varner (ed.) 1976. Plant biochemistry. Academic Press, New York.

Creech, R. G. 1968. Carbohydrate synthesis in maize. Adv. Agron. 20:275.

Duffus, Carol, and Colin Slaughter. 1980. Seeds and their uses. John Wiley & Sons, Inc., New York.

Dure, L. S. III. 1975. Seed formation. Annu. Rev. Plant Physiol., 26:259–278.

Hitchcock, C., and B. W. Nichols. 1971. Plant lipid biochemistry. Academic Press, New York.

Hoveland, C. S. (ed.) 1980. Crop quality, storage and utilization. Am. Soc. of Agron., Madison, Wis.

Howell, R. W. 1962. What happens to the energy. Better Crops Plant Food 46:4–5.

Ingle, John, D. Beitz, and R. H. Hageman. 1965. Changes in composition during development and maturation of maize seeds. Plant Physiol. 40:835–839.

Kozlowski, T. T. 1972. Seed biology. Academic Press, Inc., New York.

Millerd, Adele. 1975. Biochemistry of legume seed proteins. Ann. Rev. Plant Physiol. 26: 53–72.

Roberts, E. H. (ed.) 1972. Seed viability. Syracuse University Press, New York.

Simmons, R. O., and F. W. Quackenbush. 1954. Comparative rates of formation of fatty acids in the soybean seed during its development. J. Am. Oil Chem. Soc. 31:601–603.

Tsai, C. Y., F. Salamini, and O. E. Nelson. 1970. Enzymes of carbohydrate metabolism in the developing endosperm of maize. Plant Physiol. 46:299–306.

3

Seed Germination and Crop Production

VERNON B. CARDWELL
Department of Agronomy and Plant Genetics
University of Minnesota
St. Paul, Minnesota

The seed stage in the life cycle of angiosperms and gametophytes starts with fusion of the egg and sperm nuclei. The form and potential of the plant formed from the seed are determined by the DNA complement received from each parent and present in the embryo.

Germination is the sequence of events transforming a quiescent embryo into a metabolically active, synthesizing structure. This follows a period of dormancy imposed by environmental (water, temperature, oxygen, light), physiological (immature embryo, growth inhibitors), or morphological factors (seed coat). Physiologically, germination is the process that starts with the addition of liquid water to a dry seed and terminates with the protrusion of the embryonic radicle through the seed coat. The Association of Official Seed Analysts (1978) defines laboratory germination as "the emergence and development from the seed embryo of those essential structures which, for the kind of seed in question, are indicative of the ability to produce a normal plant under favorable conditions."

From an agronomic view, germination begins when the seed is placed in a moist soil and ends when the seedling emerges above the soil surface and becomes autotrophic. The agronomic definition is more encompassing and will be the view presented in the following discussion.

Published in *Physiological Basis of Crop Growth and Development,* © American Society of Agronomy—Crop Science Society of America, 677 South Segoe Road, Madison, WI 53711, USA.

SEED QUALITY

Genetic and Mechanical Purity

Seed quality is the composite term used to reflect germination, genetic purity, and freedom from foreign material, including inert matter, other crops, and weeds. Uniform minimum standards of quality were established in 1946 by the International Crop Improvement Association (now Association of Official Seed Certifying Agencies). State and federal laws relating to seed quality are primarily labeling laws, i.e., defining information required on the seed tag.

Genetic quality of seed represents the inherited differences for specific plant characteristics such as yield, lodging, plant height, disease reaction, and chemical composition. Comparison of varieties of rice[1], wheat, corn, and other major grain crops developed during different eras indicate yield potential has increased 35-70% for varieties developed since 1900. The impact of variety choice on yield and the progress in yield potential made via plant breeding is shown for corn in Table 3.1. Genetic purity of varieties can be assured only through field inspections and laboratory tests.

State and federal seed labeling laws focus on reducing adulteration with foreign matter and weed seeds. If a seed lot is labeled as a specific variety, not more than 5% of another variety may be present without indicating its kind and percentage by weight. Individual states specify the primary noxious weeds that, if present, prohibit the sale of a seed lot. In addition, each state can establish limits on the number of restricted or secondary noxious weed seeds allowed per kilogram of seed and the total weight of weed seed present in a seed lot. Restricted weeds are commonly limited to 4-7 seeds per kg, and total weed seed must be less than 1% by weight.

Seed labeling laws require variety name (if known), lot number, origin for some crops, percent weed seeds by weight, kind and number of restricted noxious weeds, percentage of other agricultural crops, percentage inert matter by weight, name of seller, germination percentage, and the date the tests were performed. Seed laws may or may not establish minimum standards for germination, varietal purity, or mechanical purity.

Table 3.1. Yield of corn hybrids of different eras grown in 1971–73.[†]

Period hybrid developed	With poor conditions		With good conditions	
	kg ha⁻¹	% increase over 1930	kg ha⁻¹	% increase over 1930
1930	3709	--	6538	--
1940	4464	20	7544	15
1950	4778	29	7670	17
1960	4902	32	8550	31
1970	5972	61	8990	38

† Adapted with permission from Russell (1974).

[1] Scientific names of important crop plants are given in Table 1.1, Chapter 1.

Drill-box surveys indicate that the quality of small-grain seed planted is frequently substandard. In North Dakota, 20% of small grains and 50% of flax seed sampled had not been cleaned. Evaluation of 268 samples of oat collected from drill boxes in Pennsylvania revealed that only 13% of the seed lots met or exceeded minimum standards for certified seed, varietal identity was not known or was incorrect in 18% of the samples, and germination was below 90% in 42% of the samples. Kansas noted lower wheat yields for each year seed was removed from certification.

Seed and Seedling Vigor

The germination capacity of seeds determined under optimum conditions, as in a standard laboratory germination test, is intended to indicate the probable field emergence. However, field conditions are more unfavorable, and any lack of seed vigor results in delayed and/or reduced seedling emergence.

The concept of vigor and its definitions are as diverse as the people working on the problem. Seed vigor is highly complex. At the biochemical level it involves energy and biosynthetic metabolism, coordination of cellular activities, and transportation and utilization of reserve foods. Perry (1978) notes that the International Seed Testing Association (ISTA) Vigor Committee, after years of debate, defined seed vigor as, "the sum total of those properties of the seed which determine the potential level of activity and performance of the seed lot during germination and seedling emergence."

The following tests are frequently used to assess vigor:

1) *Accelerated* aging of seeds at high relative humidity (near 100%) and high temperature (40–45°C) for a few weeks. High vigor seed lots should show less loss in germination than low vigor lots, indicating greater tolerance to adverse conditions.

2) *Cold testing* of seeds is used to assess tolerance to soil pathogens in a cold, wet soil. Seeds are usually planted into nonsterilized soil at 70% of water-holding capacity. The cold test for corn is conducted at 10°C for 7 days, and seeds are then transferred to 25°C and germination counts made 4 days later.

3) *Conductivity tests* are used on pea, bean, and corn to measure the loss of soluble sugars, starches, protein, etc., from 50 seeds following soaking in 250 ml deionized water for 24 hours at 20°C. The conductivity of the filtered liquid is measured in decisiemen per meter (dS m^{-1}) with a conductivity bridge. This measures membrane and seed coat integrity.

4) *Cool germination* testing is commonly used to estimate low temperature tolerance of cotton seeds. Seeds are germinated on blotters at a constant 18°C temperature rather than the standard 30°C with germination counts made 6 or 7 days after planting.

5) *Seedling growth rate* is frequently used as an index of vigor to show differences in seed lots due to genotype, seed size, production location, and freeze damage. Germination is usually on paper towels placed in a darkened germinator at 25°C for 7 days. Seedling dry weight separated from the seed or cotyledons is the most reproducible. Excess moisture is a frequent problem causing growth irregularity.

6) *Seedling vigor classification* is based on physical examination of seedlings for prompt development and freedom of defects of the root system, hypocotyl, and epicotyl. Seeds are germinated in a standard manner with precautions taken to prevent obstruction of root and epicotyl growth. Seedlings are classified abnormal or normal, with the normal further categorized as strong or weak. Weak seedlings are those with missing primary root; breaks, lesions, necrosis, twisting or curling of the hypocotyl; one cotyledon missing; or partial decay of primary root or epicotyl. Test results may be seriously affected by microorganism activity and/or careless handling of the test material.

7) *Root elongation* after 96 hours of germination has been used as an index of vigor for lettuce (*Lactuca sativa* L.). The test works best where seeds can be oriented with radicle end downward. Toxicity of blotter paper must also be avoided for most reproducible results.

8) *Tetrazolium chloride* (commonly 2,3,5-triphenyl tetrazolium chloride) tests the internal integrity of the embryo by staining those tissues exhibiting dehydrogenase enzyme activity. Since dead tissues do not stain, staining pattern and intensity of coloration indicate respiration rate. However, the test is usually used as a quick test of viability rather than a test of vigor. The tetrazolium chloride test measures the viability of dormant and nondormant seeds, is rapid and inexpensive. Its disadvantage is the special training needed by the technician to interpret seed staining patterns.

Commerical seed testing laboratories generally perform only the standard germination test, and for some classes of seed will conduct the cold, cool germination, and tetrazolium tests as measures of vigor.

Seed Maturity

Three aspects of seed quality are affected by the stage of maturity at harvest: viability, seedling vigor, and storage life. Partial viability may be achieved in as little as 4 days, and full viability is commonly achieved in approximately one-half the time required to reach physiological maturity. However, at this stage the seed lacks vigor due to inadequate accumulation of storage reserves. Germination and vigor both increase with increased size and advanced maturation of the seed. Immature seeds exhibit more rapid deterioration in storage than physiologically mature seeds.

Seed maturity at harvest is frequently a compromise between maximum yield, maximum vigor, and seed quality. This is particularly true for species with an indeterminate growth habit, such as sugar beet and many forage legume species that set seed over several weeks or more. Seeds from early flowers may begin to shatter while seeds from late flowers are still immature. Prolonged delay in harvesting after seeds reach physiological maturity may decrease seed quality as a result of increased weathering and increased mechanical damage associated with lower harvest moisture.

Environmental Stress Effects During Maturation Seed Quality and Vigor

Climatic conditions during seed maturation influence seed quality and vigor. Generally, factors favoring normal seed growth and development also produce high quality seed.

Temperature

Soybean and cotton have lower field emergence and greater disease problems when seeds mature during temperatures above 30°C. Best quality seed for future planting of cotton and soybean in warm, temperate regions is obtained from late seeding, when maturation occurs during the later, cooler part of the growing season.

High temperatures during maturation produce reduced seed size with altered seed coat characteristics, resulting in more hard seeds in legumes and some increase in seed-borne diseases. However, faster seed germination rate in Italian ryegrass, perennial ryegrass, and meadow fescue (*Festuca pratensis* Huds) was obtained at 25/20°C than at 15/20°C day-night temperatures. Similarly sugar beets germinated more rapidly when matured under 35 than 30°C.

Low temperature effects on seed quality are closely related to the seed's moisture content and the intensity and duration of the low temperature. The higher the moisture content, the greater the injury. Corn seed viability is reduced to zero after 24 hours exposure to −15 to −18°C with 20–25% moisture, −13, to −16°C with 30–35% moisture, and 0 to 3°C with 60–65% moisture. Most seeds with less than 15% moisture will survive −20°C and at 10–12% moisture will tolerate temperature of −192°C. Frost before physiological maturity reduces seed fill, and may cause shrivelling and seed coat damage. Reduction in viability is thought to be due to disruptive effects of low temperature on seed membranes.

Tolerance of numerous crops to high artificial drying temperatures increases as the seed becomes drier. Wheat, oat, and barley can tolerate a range of 43°C at 30% moisture to 67°C at 18% moisture without being injured. Normally, seed drying temperature should not exceed 35°C for many vegetable and garden seeds and 45°C for other seeds. Rapid drying rates increase the number of stress cracks and future susceptibility to mechanical damage. Seed corn is commonly dried on the cob because of the slower moisture loss, longer drying time, and lower tendency to develop stress cracks, as well as the difficulty in shelling high-moisture corn without seed damage.

Moisture

Numerous reports indicate the adverse effects of moisture stress during the reproductive period upon yield. Early stress may reduce the number of seeds but have minimal effect on seed size. Mid- and late-season stress reduces seed size, which in turn has a negative effect on seed vigor (see section on seed size). Studies on malting barley noted that water stress during anthesis increased seed dormancy, reducing germination from 98 to 68%. Seeds stored 17 weeks germinated 100% for the nonstressed and 85% for the stressed seeds. Removal of the hulls (lemma and palea) resulted in 96% germination for the water-stressed seeds, indicating that the covering layers of the seed restrict gas exchange to the embryo.

Fertilizer

The effect of the macroelements (nitrogen [N], phosphorus [P], and potassium [K]) on seed vigor may be due to their influence on maturation or to change in the chemical composition of the seed. When fertilizers hasten or delay maturity, temperature and moisture conditions under which the seeds mature may change and, as noted previously, seed quality may be affected.

Evaluation of mineral composition of seed suggests only minor direct effects of macroelements on seed and seedling vigor. Studies with orchardgrass and timothy show increased seed weight with N applications. Nitrogen applied to perennial ryegrass increased seed weight and seed N content. Seed N content correlates with seedling dry weight in wheat and perennial ryegrass. Conversely, high N application to timothy has no effect on seed quality, and it decreases germination on sugar beets, primarily due to delayed maturity and higher germination inhibitor concentration. High P levels applied to mother plants likewise give contradictory results, increasing radicle growth of Chinese cabbage (*Brassica rapa* L.) and increasing germination rate and seedling emergence of tomatoes (*Lycopersicon esculentum* Mill.) but reducing rate and amount of germination in garden beet.

Potassium deficiency effects on vigor are less pronounced, but are associated with small seeds, reduced vigor, and increased incidence of stem canker [*Diaporthe phaseolorum* (Cke. & Ell.)], a major cause of low seed quality in soybean in the eastern USA. Diseased seed decreased from 75% with no K fertilization to 27% with 490 kg/ha of potassium chloride.

Micronutrient effects on seed vigor vary. Copper, zinc, manganese, and molybdenum (Mo) have adverse effects on seeds produced under extremely deficient conditions. When soybean is grown in areas of severe Mo deficiency, seed lots produced in areas with high Mo content significantly outyield the local seed of the same variety. Treating seeds with Mo resulted in no significant difference in seed yield among seed lots produced in low or high Mo regions. Generally, environmental stress that adversely affects yield by influencing seed size or disease will have some negative influence on seed viability and quality.

Mechanical Damage Effects on Seeds and Seedling Vigor

Loss of seed viability and seedling vigor from mechanical damage is well documented, but the physiological basis is not well understood. The physical damage caused during threshing, conditioning, treating, bagging, transport, and seeding may result in microscopic breaks in the seed coat or injury to the growing points if the impacts are in the embryo region.

Decreased moisture content of the seed significantly increases seed mechanical damage. Maximum resistance to mechanical injury occurs at 14–18% moisture for large-seeded crops while small-seeded legumes and grass can be threshed at moistures as low as 8–10% with minimal injury. Damage due to bruising and crushing of seeds, however, increases at higher moisture contents. Dicots are more susceptible to injury than monocots and large seeds more susceptible than small. Flat seeds, such as flax, are more susceptible than other seed shapes, and the small, round seeds of *Brassica* spp. appear most resistant to mechanical injury. The cumulative effect of handling damage is reflected in germination. A soybean seed lot with 84% germination for uncleaned seeds was cleaned 5 times with a progressive decline in germination to 84, 82, 77, 74, and 71% after each successive cleaning.

Mechanical damage during threshing also increases with increasing combine cylinder speed and decreasing concave spacing. Soybean seeds at 13.5 and 12.5% moisture exhibited only 4 and 5% damage seed at proper cylinder speed, but 12 and 48% injury occurred when the cylinder speed was increased 60%.

Maintaining Seed Quality in Storage

Good storage practices should aim at maintaining high seed quality by eliminating damage caused by insects, mites, fungi, and protecting seeds from rodents. The rate of quality loss depends primarily on moisture and temperature conditions.

On-farm storage of seeds in uncontrolled temperature, humidity, and oxygen conditions requires initially dry seeds (approximately 10% moisture content) for long-term storage. However, moisture of stored seeds fluctuates with the relative humidity (RH) of the surrounding air. The equilibrium point varies with temperature and seed composition. For corn, peanut, soybean, and wheat at a constant temperature of 25°C, respective seed moisture contents in equilibrium with 15 and 90% RH are 6.4 vs. 19.1, 2.6 vs. 13.0, 4.3 vs. 18.8 and 6.8 vs. 19.7.

Some general rules for seed storage are: 1) for long term storage the sum of the percentage RH and the mean temperature in °F should total 100 or less; 2) each 1% reduction in seed moisture between 4–14% doubles seed life; and 3) each 5°C reduction in seed temperature between 0–50°C doubles the seed life. These guidelines give a quick index of potential seed viability loss under farm storage conditions in different climatic regions and seasons. Deterioration of stored seeds is greater during the warm, humid summer months in the North Central States than during the winter. Similarly, storage problems are less likely to occur in the arid regions of the western states than in the more humid southeastern areas of the USA. In Colorado, barn-stored wheat, oat, and barley lost only 10% germination in 10 years. After 15 years of storage, germination was approximately 72% for wheat, 86% for barley, and 86% for oats; after 20 years of storage, germination was 24, 51, and 62%, respectively. Soybean maintained 90% or better germination for 5 years with viability reaching zero after 14 years. Corn germinated 87% after 5 years, 70% after 10 years, and 36% after 20 years.

Deterioration in more humid regions is associated with activity of storage fungi, which become active at seed moistures in equilibrium with 60% or higher RH. Storage of grass seeds at 32°C and 90% relative humidity (simulated warm, wet, tropical conditions) resulted in zero germination within 8 to 10 weeks. The effect of temperature and humidity on seed viability for dry beans is illustrated in Fig. 3.1. Similar patterns of germination loss have been shown for most grain and forage crop seeds.

Insect activity results in respiration, consumption of seed parts and oxygen, and production of carbon dioxide, water, and heat. The respiratory heat produced results in a localized hot spot, causing moisture migration to cooler areas in the bin, resulting in more favorable conditions for fungal growth.

Most seed-damaging insects are of subtropical origin, and thus most have a temperature optimum of 28–36°C for growth and reproduction, showing limited tolerance to temperatures below 17–20°C. An exception are mites, which are capable of reproducing at temperatures down to 5°C in grain above 13% moisture. Heating dry seed to 60°C for 10 min is an effective method of treating small quantities of insect-infested cereals.

Seed storage insects depend on the moisture of their food supply to sustain their life process. Insect activity is very low below 8 or 9% seed moisture and increases rapidly at 12–14% moisture. At higher moisture

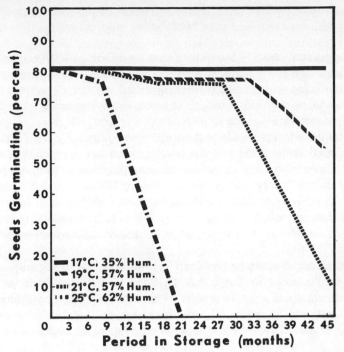

Fig. 3.1. Progressive effects of four storage conditions on % germination of bean seed stored 45 months. Each value is the mean of two seed lots of four varieties. Adapted with permission from Toole and Toole (1953).

levels, microorganisms develop, which destroy the insect populations and also destroy seed viability via respiratory heating.

Following harvest, seed viability can only decline. The rate of decline is conditioned by the many aforementioned factors. For long-term germplasm storage (10 years) seed moisture of most crops should be 4–7%, and they should be maintained at 4–10°C and about 40% RH.

Seed Size, Seed Density, and Vigor

Seed size is positively correlated with seedling vigor among the smaller-seeded species. Among cereal crops, some studies show that heavier seeds produced higher yield than lighter seeds when seeded by number of seeds per area but not when seeded by weight per area. For the cereal crops and other large-seeded crops, there is no consistent relationship between seed size and vigor over the normal range of seed sizes sold. Occasionally, data indicate superiority of the intermediate and smaller seed sizes. This failure of large seeds within a seed lot may be associated with their greater mechanical injury during harvesting and processing. High seed density has been associated with seedling vigor in wheat, cotton, and wild rice (*Zizania palustris* L.). Seed density may be an indication of maturity at harvest.

Seed Age

The gradual changes associated with aging are collectively reflected in seed vigor. Rate of aging is strongly influenced by temperature and humidity during storage and further modified by species, variety, stage of harvest, and mechanical damage. The accelerated aging test has been developed to estimate the overall integrity of the seed, which may indicate potential failure when placed in unfavorable environments in the field. Loss of vigor is sufficiently great in some species, such as soybean, that seeds are seldom stored more than 1 year for planting purposes. The aging process is thought to be a function of increased levels of DNAase, which cause breakage of the DNA molecules within the nucleus. Subsequent transcription of RNA is then imperfect, leading to reduced protein synthesis, reduced ATP synthesis, decline in respiratory activity, and visual reduction in seedling vigor.

The effect of aging on five corn hybrids with two age groups for each hybrid is illustrated in Table 3.2. Differences between varieties prevent standardizing some tests as predictors of vigor across all varieties. Valid comparisons can be made only for different seed lots within a hybrid or variety and should not be made between seed lots of different varieties.

Table 3.2. Differences in germination and vigor tests between and within cultivars of corn with different storage conditions.[†]

Cultivar	Age[‡]	Germi- nation[§]	Cold test[¶]	Root length[#]	GADA test[††]	Days of adverse storage required to kill 50% of the seeds
		%	%	mm	mm³	
1	N	99	96	195	168	34
	A	77	93	150	119	14
2	N	99	91	147	159	34
	A	98	94	166	105	20
3	N	98	96	177	137	32
	A	97	92	140	71	6
4	N	100	95	214	197	30
	A	98	97	156	139	10
5	N	99	94	195	122	24
	A	96	91	156	77	10

[†] Adapted with permission from Grabe (1966).
[‡] N = normal storage; A = artificially aged.
[§] Standard laboratory germination at 27–31°C.
[¶] 7 days at 10°C followed by laboratory germination at 27–31°C.
[#] After 5 days at 27–31°C.
[††] Glutamic Acid Decarboxylase Activity: evolution of CO_2 after 30 min from mixture of standard quantities of ground seeds and glutamic acid. Note that only the GADA test forecasts differences in storability, but cultivars need individual calibration.

REGULATION OF GERMINATION

Germination is controlled by dormancy and various environmental factors acting upon the seed. Given a viable, nondormant seed, the environmental conditions favorable for germination include adequate supply of water, suitable temperatures and oxygen, in some cases light, and the absence of external inhibitory factors such as inorganic salts or organic substances.

Dormancy

Some distinction between quiescence and dormancy should be made. Quiescence is arrested seed development imposed by unfavorable environmental conditions such as inadequate water supply or low temperature. Dormancy describes arrested development owing to structural or chemical properties of the seed that prevent germination when environmental conditions are favorable.

Dormancy is a widespread phenomenon among noncultivated species, as it is a survival mechanism spreading germination over time and space. Among cultivated seed crop species, genetic selection has occurred for uniform and rapid germination.

Types of dormancy found in agronomic crops and weeds are as follows:

1) *Impermeability of seed coats to water.* These seeds generally are referred to as hard seeds and occur among members of the Leguminosae, Malvaceae, Chenopodiaceae, Liliaceae, and Solanaceae. This characteristic is inherited and is influenced by environmental conditions during maturation. Hard seeds generally are undesirable in annual grain crops when uniform germination and emergence are important but are desirable to ensure self-reseeding of annual and winter annual forage crops, such as crimson clover, Persian clover (*Trifolium resupinatum* L.), and hop clover (*T. aureum* Poll.). Low soil moisture during seed maturation increases the frequency of hard seeds in alfalfa. Low RH and high temperature contribute to the occurrence of hard seeds in soybean, and the drier the seed, the higher the percentage of hard seeds. Dormancy owing to hard seeds can be broken by chemical and mechanical methods. The most common treatment for legumes is scarification of the seed by rubbing the seeds against an abrasive surface. Cotton typically is scarified with sulfuric acid, as it is delinted, to improve germination and ease in planting.

2) *Mechanical resistance of the embryo to growth.* Failure to germinate because of excessive strength of the seed coats has been re-

ported in several weed species: plantain (*Plantago* spp.), pigweed
(*Amaranthus* spp.), and cocklebur (*Xanthium strumarium* L.).
Cocklebur seeds have two embryos. The dormant upper embryo is
able to generate only 41 g of imbibitional and growth force while the
testa requires a force of 56 g to pierce. By contrast, the lower, nondormant embryo generates 84 g of force and the testa requires 67 g
to pierce.

3) *Low permeability of the seed coat to gases.* Germination of most
agricultural seeds is inhibited in the absence of oxygen. Notable exceptions are rice and wild rice.

4) *Dormancy due to a metabolic block.* The most notable examples
among agricultural crops are the postharvest dormancy of some
cereals, the stratification requirement of crops such as wild rice, and
the requirement of light for the germination of some forage grass
and weed species.

Water

Water is the activating agent that starts the germination process. The
absorption of water by the seed is differential; i.e., the embryo absorbs 3–10
times more moisture on a dry-weight basis than do the cotyledon and endosperm of corn. However, the relative size of principal storage tissue dominates the total moisture uptake.

Water is absorbed in three stages: an initial period of rapid uptake, a
lag period in which little water is absorbed, and a second uptake period
associated with embryo growth (Fig. 3.2). Factors limiting the rate and
amount of water imbibed include composition of the seeds, permeability of
the seed coat to water, availability of water in liquid or gaseous phase, and
temperature.

Imbibition is a physical process, which accounts for the similar absorption curves for living and dead seeds, and the rate is closely related to
the colloidal properties of the seed (Fig. 3.2). Protein represents the major
colloidal constituent of seeds, although celluloses and pectic substances also
exhibit some swelling. Starches and lipids do not contribute appreciably to
seed swelling. The extent of seed swelling and total amount of water imbibed generally reflects the seed storage constituents.

The type of lipids in the seed coat control permeability and entry of
water into the seed. In hard-seeded legumes, moisture penetration appears
to be through the micropylar area of the seed, where the seed coat is
thinnest. In pea, the chalaza end (point of attachment of ovule) of the seed
is twice as permeable as the smooth part of the testa.

Imbibition of corn, soybean, and cotton seed with and without epoxy
glue-covered micropyles at 20°C for 24 hours resulted in 88, 179, and 85 mg
of water per seed for the check, while epoxy-covered seeds imbibed 82, 156,

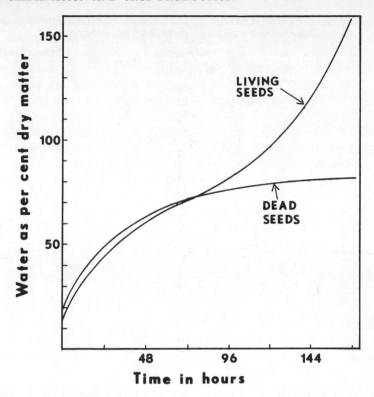

Fig. 3.2. Water uptake by living and dead wheat seeds at zero water potential. Adapted with permission from Owens (1952).

and 78 mg of water per seed, respectively. For these crops, water is clearly not absorbed preferentially through the micropyles. Further, as seeds become more hydrated, their permeability to water and gases increases.

The availability of water in the gaseous or liquid phase influences the imbibition rate. Seed moisture moves toward equilibrium with the RH of the air, and the RH of air-dried soil is usually higher than 98%. However, once the seed is placed in the soil, liquid moisture is usually the most important source of water. Soybean can obtain enough moisture from within a 1-cm radius of the seed to raise the moisture content to the minimum required for germination. Availability of water to the seed decreases as matric (suction) and osmotic (solute) tensions increase. Matric potential is the most important in nonsaline soils. In saline soils, not only is the osmotic effect important, but also the toxicity of the ions in solution.

Stresses greater than 0.38 MPa result in reduced water uptake by seeds of vetch, chickpea (*Cicer arietinum* L.), and pea. At stresses greater than 2.0 MPa most seeds are unable to imbibe sufficient water to initiate embryo growth. Species vary widely in the reported minimum moisture content required for germination. Minimum seed moisture required for germination is

Table 3.3. Minimum moisture content of seed and minimum soil moisture tension at which
radicle emergence has been reported from many sources.

Species	Seed moisture	Moisture tensions
	% dry weight	MPa
Alfalfa	118–130	−1.2 to −1.5
Barley	35	
Bean	86	
Bulbous canarygrass	35	−1.0
Chickpea	72–75	
Common vetch	75	
Corn	30	
Cotton	75	−1.0 to −1.2
Faba bean	82–86	
Lima bean	70	
Orchardgrass		−1.0
Pea	150	
Perennial ryegrass	50–60	−1.6
Rice	26	−0.8
Sorghum	25–32	−0.8 to −1.5
Soybean	50	−0.7
Subterranean clover	130	−1.0
Sugar beet	31	−0.4
Sunflower		−0.8
Wheat	46	−1.5 to −2.0
White clover	160	−1.0

apparently not related to minimum soil moisture required for a seed to germinate but is related to seed composition (Table 3.3).

Water uptake by the seed is influenced primarily by temperature and to a lesser extent by the initial seed moisture content (Fig. 3.3). The temperature quotient, Q_{10}, of imbibition is 1.5–1.8 in numerous studies, which is indicative of a physical process.

Temperature

The seeds of each species and cultivar have a minimum, optimum, and maximum temperature for germination. The minimum temperature may be near freezing and the maximum approaches 50°C, at which some plant proteins become denatured if they are hydrated.

The imbibitional phase of germination is primarily physical and shows low sensitivity to temperature. The second and subsequent phases of germination are temperature dependent due to the biochemical processes involved. Percent germination at a given temperature yields a sigmoidal or S-shaped curve characterized by a lag phase, which is longer at sub-optimal temperatures, a maximal rate of germination (tangent at the inflection point), and a final percentage germination.

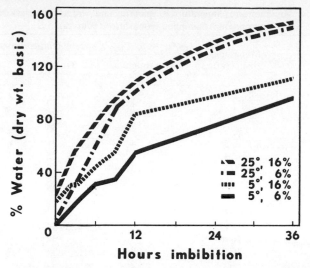

Fig. 3.3. Water uptake of Hawkeye soybean seed equilibrated to 6 or 16% initial seed moisture before imbibition at 5 or 25°C. Adapted with permission from Obendorf and Hobbs (1970).

Many biological processes respond to temperature according to the following relationship:

$$R_2 = R_1 \cdot Q_{10}{}^a$$

where $a = (T_2 - T_1)/100$, R_2 = rate at the second temperture (T_2), R_1 = rate at the first temperature (T_1), and Q_{10} is a constant reflecting the effect of a 10°C increase in temperature. The exponent, a, permits calculations of R_2 when the temperature difference is less than 10°C for $T_2 - T_1$. Q_{10} typically is 2–3 for biological systems between 0 and 50°C.

Any field condition that increases the length of the lag phase usually results in lower final emergence. A North Dakota study showed 60, 72, and 77% emergence of corn when the days to final emergence were 12, 10, and 8, respectively. Under laboratory conditions final germination may be similar over a range of lag phase durations.

The literature on temperature effects on germination rate and total germination dates back to the work of Sachs (1860) and Haberlandt (1875–77). Sachs proposed the concept of cardinal temperatures, i.e., minimum, optimum, and maximum temperature for germination. Cardinal temperatures for major crops are presented in Table 3.4. These values should be viewed with caution. The temperature range at which a given seed lot will germinate is a function of seed quality, genotype, and the duration of the germination period as Fig. 3.4 illustrates. Minimum temperatures reported for seed germination are frequently based on studies of short duration and constant temperature. In short duration studies, the range is narrowed, and

Table 3.4. Minimum soil temperature for planting and cardinal temperatures for germination (compiled from various sources).

Crop	Minimum soil temperature	Cardinal temperatures		
		Minimum	Optimum	Maximum
		°C		
Alfalfa		1	30	38
Rye		2	25	35
Timothy		3	26	30
Wheat, spring	3	4	25	32
Barley	5	4	22	36
Oat	6	3	28	34
Millet (foxtail)		5	20	30
Flax	7	4	28	40
Sugar beet	10	4	25	30
Pea		4	30	35
Soybean	10–13	9	30	41
Corn	10–13	9	33	42
Cotton	17	15	34	39
Peanut	18	13	20	38
Field bean	18	--	--	--
Sorghum	21	9	33	40
Rice	21	11	32	41

the optimum and minimum temperatures are higher. The minimum and maximum temperatures for germination under field conditions depend upon the seed's ability to survive moisture stress, pathogens, insects, and the kinetic stability of the seed membranes and enzyme systems. Further, the minimum temperature required to initiate germination may not be the same as the minimum temperature that the species can endure without injury. High quality seed will have wider temperature limits than seeds of lower quality.

Failure of imbibed seeds to germinate at below freezing temperature is presumably due to ice crystal formation and membrane disruption. Germination failure at high temperature is due to denaturation of nucleic acids, proteins, and membranes. Imbibition at low temperatures at 5–15°C and at initial seed moisture below 12–14% causes injury and reduces germination in warm-season crops, e.g., lima bean, garden bean, cotton, sorghum, soybean, and corn.

Soil temperature is determined by the balance between incoming and outgoing radiation at the soil surface and the redistribution within the soil. Two cyclic temperature patterns occur, diurnal and annual. The amplitude of diurnal and annual soil temperature variation decreases with depth, and the below-surface maximum and minimum temperatures lag behind those at the soil surface. The soil temperature at 5.7 cm from April to mid-June in North Central States is very close to the mean of the daily maximum and minimum air temperatures. Thus, minimum soil temperatures for germination (Table 3.4) are temperatures at the seeding depth and must be

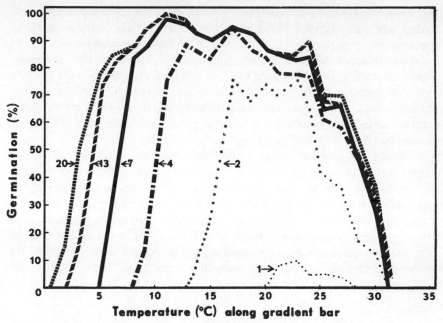

Fig. 3.4. % germination of English catchfly (*Silene gallica* L.) on successive days after seeding at different temperatures. Adapted with permission from Thompson (1972).

viewed with caution, since seedings made when soil temperatures are near the minimum must be done with the expectation of continued warming trends. Soil conditions, weather patterns, and time of year are considerations in deciding when to plant.

Soil surface temperature is seldom constant. Some seeds germinate more quickly with varying temperature than with constant temperature. Alternating temperatures favor germination of bermudagrass, tobacco, redtop, Kentucky bluegrass, and numerous weeds. In one study, redtop germinated 29, 53, 72 and 79% at constant temperatures of 12, 21, 28 and 35°C, while at alternating temperatures of 12/21, 21/28, and 21/35°C germination was 69, 95, and 95%. Kentucky bluegrass germinated 15, 12 and 63% at 20, 30 and 20/30°C.

High temperaure will break postharvest dormancy in some dry seeds, such as rice, where exposure to temperatures of 40–47°C for up to 7 days allows normal germination at 30°C or lower.

In many temperate-region species, the embryo is mature when the seed is shed or at the time of harvest but requires afterripening at low temperature (0–10°C), preferably near 5°C, for normal germination and seedling development. The low temperature pregermination treatment of seeds is called stratification. This was originally a horticultural practice of layering bulbs between layers of peat and sand and watering periodically. Under field conditions stratification insures that seeds will not germinate during the fall when temperatures and moisture may be favorable but an insufficient growing season remains for completion of the life cycle.

Seeds requiring chilling often contain growth inhibitors and promoters, and chilling alters the balance of the growth regulators. Gibberellic acids (GA) appear to be promoters and abscisic acid (ABA) a major inhibitor.

Stratification, while a common laboratory procedure for increasing germination of freshly harvested seed, is not used commonly with agricultural seeds because of the problem of handling wet seeds in planting. An exception is wild rice, which has a stratification requirement of 90 days and must be stored and handled wet because these seeds lose their viability if dried to less than 30% moisture.

Finally, temperature ranges in the field would be expected to affect the mean number of seeds germinating, the rate of germination, and the uniformity of emergence. Variation in seed lots of the same cultivar to temperature effects may be due to a number of factors, e.g., seed maturity at harvest, preharvest weather conditions, threshing damage, seed size, seed age, seed composition, and storage conditions. Within the wide range of temperatures encountered in a seedbed, only at the extreme limits do temperatures alone limit the final emergence percentage. The greatest effect is on emergence rate.

Oxygen

Viable seeds are living organisms requiring oxygen for energy release. Germination is an energy-consuming process requiring the aerobic release of energy from storage materials in the endosperm (grasses and castorbean), perisperm (sugar beet), or cotyledon (most dicotyledons). The nonimbibed seed requires very little oxygen, and less oxygen is needed at lower temperatures. Oxygen supply to the embryo is a function of external concentration, solubility of oxygen in water, length and resistance of the diffusion path (primarily seed coat), and affinity of the seed enzymes for oxygen. The resistance to oxygen diffusion increases with imbibition due to increased thickness of water film over the seed and embryo surface. The oxygen diffusion rate in air at 20°C is 10^4 times the rate in water, 10^5 times the rate through the testa and pericarp, and 10^3 times the rate of embryo tissue. As germination proceeds, the resistance values of these barriers and the oxygen requirements of the embryo change. The major initial barrier to oxygen uptake is the testa, which persists until the radicle emerges.

Oxygen levels in the seed environment have no effect on imbibitional water uptake but do affect subsequent growth. Excess water around the seed may be a barrier to oxygen diffusion. Oxygen consumption for germinating soybean was reduced from 323 to 215 mole kg^{-1} $hour^{-1}$ when submerged in water. Corn exhibited an even greater reduction from 681 to 244 mole kg^{-1} $hour^{-1}$. Oxygen requirements of the imbibed embryo increase directly with temperature, but oxygen solubility in water decreases with temperature increases. With the exception of seeds of cattail (*Typha latifolia* L.), wild rice, rice, and a few other aquatic plants, germination in submerged water is limited by low oxygen availability.

Germination depends more on moisture than oxygen or carbon dioxide. Only when soil moisture is near saturation and oxygen supplies are below 9–10% is germination restricted in corn, soybean, and wheat. However, 4 range grasses—crested wheatgrass, tall wheatgrass [*Agropyron elongatum* (Host) Beauv.], bulbous canarygrass (*Phalaris aquatica* L.), and Indian ricegrass (*Oryzopsis hymenoides* Ricker)—exhibit a significant reduction in germination at oxygen concentration below 15%.

Oxygen diffusion rates greater than 3 to 4×10^{-5} mg m^{-2} s^{-1} are required for maximum germination of corn. Higher oxygen diffusion rates are required for emergence of the shoots from soil than to initiate germination and root growth. The depressing effects of increasing carbon dioxide on germination is generally slight compared to the effects of low oxygen concentration.

Thus, the oxygen requirement for germination varies with the moisture content of the seed, temperature, seed vigor, and the energy substrate utilized for respiration, i.e., starches, proteins, and lipids.

Light

Photomorphogenesis is a light-controlled developmental process. Three broad classes of germination response are: seed promoted by white light, seed inhibited by white light, and seed that germinates in both light and dark. The light sensitivity of seed has been known since the end of the 1800s, but the mechanisms of photocontrol were largely determined since the initial work of Borthwick et al. (1952). This work demonstrated that different spectral regions of white light are responsible for the stimulation and inhibition of germination. Red light (560–700 nm) stimulates germination while ultraviolet (less than 200 nm) and far-red (greater than 700 nm) inhibits germination.

Numerous studies conclude that phytochrome, a protein-bound chromophore, is the photoreceptor responsible for the light-stimulated promotion or inhibition of seed germination. The mechanism of phytochrome-mediated germination is not thoroughly understood, but phytochrome in its active form has been shown to alter membranes, cause permeability changes, activate enzymes, and cause new enzyme synthesis, any one or all of which may trigger germination. The phytochrome pigment becomes biologically active in red light, while far-red light converts the pigment to an inactive form, preventing germination. Table 3.5 illustrates the effect of exposure to 5 min of red (R) and far-red (FR) light on the germination of dormant tall fescue. Germination was 30% greater when exposure was to red light rather than far-red light.

The response of seed to light depends on many factors. The phytochrome is commonly located in the seed coat or pericarp and must be hydrated to 17–19% moisture to be activated. The requirement for light is

Table 3.5. Germination of dormant Alta tall fescue seeds at 15 and 25 °C for 7 days
after successive 5 min treatments with red (R) and far-red (FR) light.†

	% germination	
Light treatment‡	15 °C	25 °C
R	91	41
R + FR	66	8
R + FR + R	90	45
R + FR + R + FR	67	8
R + FR + R + FR + R	91	36
R + FR + R + FR + R + FR	66	7
Dark conrols	63	3

† Adapted with permission from Danielson and Toole (1976).
‡ Seeds were irradiated immediately after removal from the dormancy induction period at
35°C.

generally greatest in freshly harvested seed and decreases with seed age, although old, buried weed seeds frequently exhibit light requirements. The
sensitivity and effect of phytochrome are greater under stress than nonstress
environments. High temperature seems to increase the light sensitivity
(Table 3.5). Similarly, low oxygen availability increases light requirements.

In the laboratory, nitrate, nitrites, nitric acid, thiourea, and ammonium salts have been used to negate light requirements in dormant seeds.
Alternating temperatures, sub- or supra-optimal temperatures frequently
overcome light requirements. More species are stimulated by light than inhibited by light, with most grain and large-seeded food crops exhibiting
minimal response to light.

The ecological benefit of light sensitivity for survival of weed species or
native plants is obvious. Buried seeds will not germinate until the soil is disturbed and the seeds are placed near the soil surface where the chance of
seedling establishment is improved. Very little data are available on the
depth of light penetration into the soil, but 1 study reports quartz sand
transmitted 10% of the 400-nm wavelength and 30% of the 900-nm wavelength incident to the sand surface at a depth of 10 mm. Light-requiring
seeds seldom germinate when placed deeper than 3 to 5 mm in the soil. Light
quality affects seed germination under a crop canopy. The ratio of red to
far-red light in direct sunlight is 1.3 and gives better germination of *Chenopodium* species than under a green canopy with ratios of 0.70–0.12. Leaves
are relatively transparent to far-red light while absorbing most of the red as
well as blue wavelengths.

Light intensity generally has little or no effect on germination of agricultural seeds, since most germination photo-responses occur at light intensities of 1/50 to 1/100th of full sunlight. Only a few species, such as
colonial bentgrass (*Agrostis tenuis* Sibth) show increased germination with
increasing light intensity.

The germination process is highly interactive with environmental
factors. Temperatures determine the limits and the rates of germination.
Water uptake is required before dry seeds resume growth, but uptake may

be prevented by hard seed coats and other dormancy mechanisms. Oxygen seldom limits germination except under waterlogged conditions. Light quality may promote, inhibit, or have no effect on germination, depending on species and cultivar. Germination is a critical stage in the life cycle, and numerous factors can interfere in the establishment of an autotrophic seedling under field conditions.

PHYSIOLOGY OF GERMINATION

The sequence of germination in its simplest form is: water imbibition, enzyme activation, hydrolysis and catabolism of storage material, initiation of embryo growth, anabolism and formation of new cell structures, rupture of seed coat, and emergence of the seedling. In reality, the process continues to be one of the great mysteries of life with many unknown steps in a process that is complex and highly interactive with the environment. The following is a general discussion of the biochemical and physiological events that seem to be common to germination, which involves both oxidative and synthesis pathways.

Water Imbibition

Reactivation of metabolic systems, conserved since seed maturation and storage at a moisture content of 10–14%, commences with the imbibition of water and the rehydration of proteins, enzymes, and cellular organelles. The water becomes the solvent in which reactions occur; the media for transport of enzymes, cofactors, and coenzymes; and a substrate in hydrolytic reactions, such as the enzymatic breakdown of the polymers of starches, proteins, and lipids into their basic units. In addition, water provides turgor pressure for increase in cell volume and accounts for the 30–40% increase in volume upon initial hydration of the seed.

Hydration of the seed to levels above 16–18% moisture results in a rapid rise in mitochondrial activity and activation of phytochrome. Respiratory increases following initial imbibition may be noted in as little as 10 min. The initial respiratory rate increase continues for 2 hours in crimson clover at 20°C, 6 hours in rye, 8 hours in corn, and 10–16 hours in dry bean at 25°C before a plateau or steady state is reached (Fig. 3.2). This respiration depends on substrates stored in the embryonic axis.

Hydration and Enzyme Activation

During the steady-state period when initial moisture uptake rates do not change, the various enzyme systems are activated, and cellular activity increases. Enzyme activity following hydration may be from enzymes formed during maturation of the seed or de novo (newly synthesized) enzymes, distinguishable by the rapidity with which their activity appears. Previously formed enzymes are of two types. First are those requiring only hydration to become active, and this activity is reversed by drying. Triosephosphate isomerase, cytochrome reductase, and adenylate cyclase are a few examples of these enzymes. Second are those enzymes requiring the action of a hormone or other enzymes to gain activity. These enzymes become activated within minutes to several hours.

De novo synthesis is also of two types. First, there are enzymes produced by preexisting mRNA utilizing a pool of preexisting amino acids. Activity of these enzymes may occur in 2–4 hours. Second are those enzymes produced from newly synthesized mRNA. These enzymes utilize amino acids derived from degradation of storage proteins, which begins as early as 6–12 hours after hydration.

The following mechanisms of enzyme activation are suggestive rather than conclusive because of the diversity found among plant species and the conflicting information in the literature.

Endogenous hormones control many facets of germination, and the precise regulating mechanism for most species is still open to speculation. GA appears to have the dominant activator role following hydration of the seed through its effects on membrane permeability, ATP synthesis, and the interaction with cytokinin and ABA.

There is evidence that endogenous cytokinin is located in nonembryonic regions of the seed. Studies show that cytokinins influence membrane permeability, overcome the effects of inhibitors such as ABA, function in the binding site of some tRNA or mRNA ribosome complexes, and interact with membrane-bound pigments such as phytochrome to mediate the germination process. Thus, cytokinin has a controlling role in the translation of genetic code into new proteins.

ABA appears to be a common inhibitor of germination by blocking DNA synthesis and translation by mRNA for key enzymes such as isocitratase and phosphorylcholine glyceride transferase. Frequently the inhibitory effects of ABA can be overcome either through degradation of ABA over time, as with wild rice, or by an increase in the level of GA and/or cytokinin. Germination of wild rice is usually less than 15% until stored for 90 days, when ABA levels show a drop of about 50%. Exogenous applications of GA and cytokinin cause the dormant seeds to germinate after 30–60 days of storage.

One of the more thoroughly studied growth regulator systems is the role of GA in germination. Over 40 different isomers of GA have been identified. The initial synthesis of GA occurs in the scutellum, but the embryonic axis also produces GA by the 3rd day of germination. GA effects on germinating seeds can be divided into two roles: germination (embryonic activity) and substrate mobilization. The GA regulation of germination is thought to occur primarily by changing membrane permeabilities and by influencing the level of initial ATP synthesis. Dormant seeds with low initial ATP levels and requiring stratification can be induced to germinate given a GA treatment, which changes membrane permeability and energy levels in the embryo.

Substrate mobilization effects of GA have been extensively studied in barley. GA increases the level of poly (A) RNA which is believed to contain the mRNA for the enzyme α-amylase synthesized by the aleurone cells. This enzyme is responsible for the hydrolysis of starch to maltose. Ribonuclease and protease synthesis in aleurone cells is also stimulated by GA. In addition, GA enhances the release of β-1, 3-glucanase, which dissolves the cell walls of the aleurone, permitting an increased supply of nutrients from the endosperm to the embryo. Figure 3.5 presents a diagrammatic representation of GA and α-amylase production and solute utilization and regulation in barley seed.

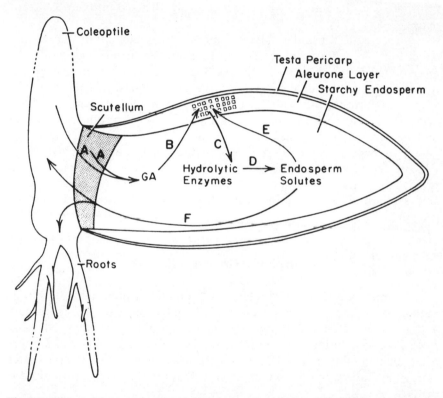

Fig. 3.5. Diagramatic representation of the relationship between gibberellic acid (GA) production, α-amylase production, and solute accumulation in germinating barley seeds. GA produced by the coleoptile and scutellum (A) migrates into the aleurone layer (B) where hyrolytic enzyme synthesis and release is induced (C). These enzymes hydrolyze the reserves in the endosperm (D), producing solutes that can inhibit further hydrolytic enzyme production (E), and function to nourish the growing embryo (F). Adapted with permission from Jones and Armstrong (1971).

Enzymatic Degradation of Storage Materials

Seed component utilization during germination begins with simple sugars, free fatty acids and amino acids, and is followed by starches, lipids, and proteins (Fig. 3.6). Starch catabolism to glucose is primarily by the hydrolytic amylase enzymes, which account for 90% of the degradated starch. The oxidative pentose phosphate pathway appears to function during the first germination phases of cereals but is quickly replaced by the hydrolytic pathway as synthesis of α- and β-amylase enzymes occurs 24–48 hours after the imbibition of water. Sugars utilized by the growing embryo are principally glucose, sucrose, raffinose, and stachyose.

Protein catabolism involves protease degradations (Fig. 3.6) of insoluble storage proteins. The soluble proteins are further hydrolyzed by peptidases to the component amino acids. The released amino acids are transported to the embryo where they are resynthesized into new cell proteins.

Oil stored as fat bodies (glyoxysomes) in the scutellum of grasses and cotyledons or endosperm of dicots are degraded from triglycerides to glycerol and fatty acids by acid lipases (Fig. 3.6). Monoglyceride degradation is carried out by alkaline lipases in the glyoxysomes. In some oil seeds, such as castorbean, this hydrolysis occurs early in germination. The fatty acids are oxidized via the β-oxidation pathway, which produces 2-carbon units of acetyl-CoA capable of entering the tricarboxylic acid (TCA) or the glyoxylate bypass, where fats are converted to sugars via reverse glycolysis. Glucose can be converted via the oxidative pentose phosphate pathway to provide the pentose (5-C) sugars needed for nucleic acid synthesis. Nucleic acid synthesis is a prerequisite for cell division associated with embryo growth.

Synthesis and Growth of Embryo

The metabolic pattern of ATP, RNA, and DNA production by germinating wheat embryo is shown in Fig. 3.7. Initial protein synthesis occurs within 30 min after imbibition begins. This protein is synthesized using RNA produced during seed maturation. New RNA synthesis begins 3 hours after initial imbibition, and DNA synthesis has been observed 15 hours after start of hydration. Studies on onion show that the first RNA synthesized is mRNA.

Emergence of the radicle in cereals usually precedes cell division; however, radicle emergence in some species is associated with the concurrent processes of cell division and enlargement. Radicle emergence usually precedes shoot emergence.

Germination represents a dynamic period in the life cycle of plants as a seed makes the transition from a metabolically quiescent to an active and growing entity. The processes involved in rehydration and activation of the

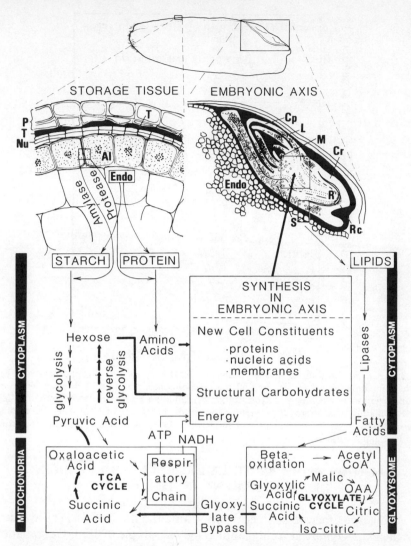

Fig. 3.6. Schematic representation of seed storage product degradation (light arrows), the participating organelles, the pathways of transformation, and synthesis (heavy arrows) in germinating seeds. (P-pericarp, T-testa, Nu-nucellar tissue, Al-aleurone, Endo-endosperm, Cp-coleoptile, L-embryo leaf, M-apical meristem, Cr-coleorhiza, R-radicle, Rc-rootcap, S-scutellum).

embryo are very complex and not fully understood, even in the most thoroughly studied systems. In addition to hydration, most seeds have a simultaneous release of growth regulators that trigger enzyme activity, new RNA, and finally new DNA synthesis associated with cell division and growth. Sustained growth of seedlings, until they become autotrophic, depends on storage reserves in the seed. Clarifying our understanding of these processes is prerequisite to understanding many aspects of seed vigor and seedling performance.

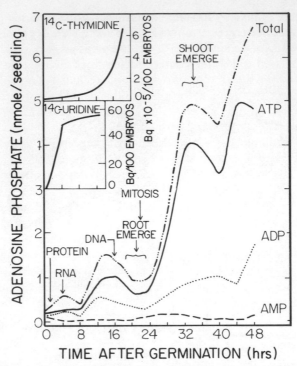

Fig. 3.7. Changes in total adenosine phosphates (TOTAL), adenosine triphosphate (ATP), adenosine diphosphate (ADP), and adenosine monophosphate (AMP) in wheat embryos during germination (Ching and Kronstrad, 1972), and time courses of DNA synthesis using ¹⁴C-thymidine as substrate (upper insert, Mory et al., 1972), and of RNA using ¹⁴C-uridine as substrate (lower insert, Rejman and Buchowicz, 1973). Adapted with permission from authors and publishers.

FACTORS REGULATING EMERGENCE AND SEEDLING ESTABLISHMENT IN THE FIELD

Numerous factors interact to determine the number of healthy auto-trophic plants obtained from planting a given sample of seed. Germination under laboratory conditions frequently is a poor simulation of the rigors encountered in the field. Factors influencing seedling establishment under field conditions include the physical, chemical, and biotic properties of the soil; method, date, depth and rate of seeding; and seed treatments. In addition, each of these factors interacts with the environmental factors of water, temperature, oxygen, and light that regulate the rate of germination.

Soil Moisture

Soil moisture conditions exert a dominant influence on stand establishment because of the modifying effects of moisture on soil properties. Soil structure, soil water potential, and seed-soil surface contact determine the rate of moisture uptake by the seed in soil. Increasing seed surface contact with liquid water decreases germination time and increases germination percentage. Any factor reducing soil hydraulic conductivity (determined by pore diam) or seed-soil-water contact (determined by aggregate size and matrix potential) reduces rate of water uptake and delays germination. Based on laboratory studies, the general guideline for preparing seedbeds in the row zone is to obtain a mean soil aggregate size of $\leq 1/5$ the seed diam to insure rapid water uptake and germination.

Percent seedling emergence under field conditions varies with soil moisture content, soil type, and plant species (Fig. 3.8). Work with many crops shows that total emergence is affected only slightly by moisture tension from 0.05–0.3 MPa. The number of emerging seedlings decreases rapidly at moisture tension greater than 0.7 MPa (Fig. 3.9), and the time to maximum emergence increases as moisture stress increases. Many crop seeds are able to imbibe sufficient water at moisture near or slightly below the permanent wilting point to initiate germination but not elongate. Insufficient aeration presumably limits germination at very high soil moisture conditions, since diffusion of oxygen in air is 10 000 times greater than in water at 20°C.

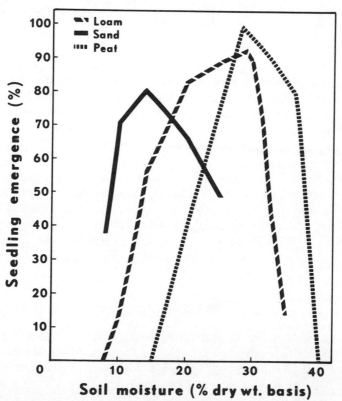

Fig. 3.8. Effects of soil water content and seedling emergence on *Beta vulgaris* L. (sugar beet). Adapted with permission from Longden (1972).

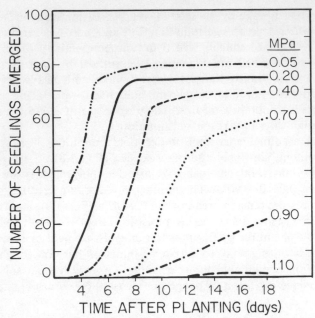

Fig. 3.9. Cotton seedling emergence per 100 seeds planted, as a function of days after planting and soil water tension. Adapted with permission from Jensen (1971).

Soil Strength

Soil strength is influenced primarily by bulk density (texture and soil particle aggregation) and moisture content, with some modification by organic matter and base saturation. As soils dry, soil strength increases, with soil strength restricting emergence before soil moisture limits seedling cell division and elongation. Seedlings growing horizontally under a crust reflect these phenomena.

Soil crusting is an increase in soil strength due to soil compaction and loss of soil structure. Crusting increases the resistance to shoot and root penetration and is most frequently a problem in stand establishment on fine-textured soil with poor soil aggregation and low organic matter. Seedling emergence through crusts is controlled by seedling size, number and spacing; soil cracking pattern; and crust strength around the emerging seedling.

Seedling Emergence Strength

Emergence strength of individual seedlings varies from a low of 0.15 newton for alfalfa to a high of 2.9 newtons for corn. Multiple seedlings in a group are able to rupture higher-strength soil crusts. For example, studies show that the maximum thrust of one, two, and three cotton seedlings is 3.8, 5.8, and 8.5 newtons, respectively. Similarly, subterranean clover (*Trifolium subterraneum* L.), when seeded 2-cm deep in clumps of 1, 2, and 5 seeds, has 28, 72, and 85% emergence.

Large seeds exhibit greater emergence thrust but also encounter more soil resistance due to the greater surface area of their emerging seedling structures. Emergence data for many species and cultivars generally favor the larger-seeded types when all other factors are equal. This is particularly true for the smaller-seeded grasses and legumes where emergence from deeper plantings is promoted by large seed size, even though emergence force per gram of seed weight is slightly greater for small seeds. Large-seeded dicots occasionally fail to establish themselves as well as smaller seeds of the same cultivar. However, this is usually due to differences in seed quality. Equal seed quality for comparison of seed size is difficult to obtain from machine-harvested seed lots because of mechanical injury problems. Large seeds frequently exhibit more evidence of injury than small seeds.

In cotton, hypocotyl elongation is more sensitive than the radicle to increasing soil impedance and decreasing soil moisture (Fig. 3.10). Increased physical strength and decreased soil water reduce the depth from which seedlings can potentially emerge from over 7 cm at 0.03 MPa moisture and 0.02 MPa soil strength to less than 2 cm at 1 MPa moisture and 0.3 MPa soil strength.

Soybean exhibits genetic variation in hypocotyl length, with temperatures of 21–28°C inhibiting elongation of sensitive varieties. Thus, poor emergence of late plantings may be due to lack of hypocotyl elongation.

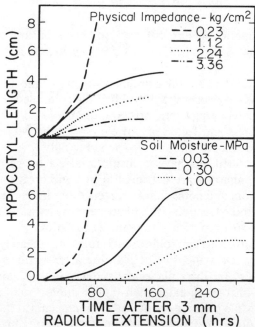

Fig. 3.10. Hypocotyl length of cotton at various times after radicle emergence at 32°C under 4 levels of physical impedence (compaction) at 0.03 MPa soil moisture (upper) and 3 levels of soil moisture at 0.23 Kg cm^{-2} physical impedance (lower). Adapted with permission from Wanjura and Buxton (1972).

Soil Salts and Fertilizer Effects on Emergence

High concentrations of fertilizer and soluble salts in soil solution restrict germination and field emergence due to osmotic and, in some cases, ion toxicity effects. Field crop establishment problems are greatest in arid and semiarid soils where soluble salts are more prevalent. A similar problem may occur in subhumid and humid regions when heavy fertilizer rates are concentrated in the seed zone.

Crops exhibit a wide salt tolerance range both among and within species. Saline soils contain more than 0.2% soluble salts and have conductivities greater than 4 dS m^{-1}. Conductivity greater than 6 dS m^{-1} reduces emergence rate and final emergence percentage of sorghum, sugar beet, field bean, pigeon pea (*Cajanus indicus* Spreng), and cowpea. Salt tolerance is usually less during germination than for established plants. Sugar beet and alfalfa generally are classed as high-salt-tolerant crops, while corn is intermediate. However, corn exhibits much higher salt tolerance than either sugar beet or alfalfa during laboratory germination and emergence.

Salt concentration in a saline soil increases as moisture content decreases. Thus, high moisture availability is required for satisfactory stand establishment. The interaction of soil moisture and osmotic (salt) potential on germinating seeds is expressed by the following equation:

$$OP = (SP/PW)\,0.36 \times EC_e \times 10^3$$

where: OP = osmotic pressure, SP = % soil water at saturation, PW = % soil water at the tension studied, and EC_e = electroconductivity of soil solution in dS m^{-1}.

Emergence is related to the combined stresses of moisture and osmotic potential. Emergence declines rapidly when the combined forces exceed 0.8 MPa. Approximately equal total stress and emergence is obtained at 0.4 MPa moisture stress and 1.4 dS m^{-1} of electroconductivity or at 0.05 MPa of moisture stress and 6 dS m^{-1} of electroconductivity.

Commercial fertilizers have an osmotic effect on germination. Urea and anhydrous ammonia are detrimental at concentrations of 0.0005 to 0.001 mole fraction, and potassium fertilizers cause injury at concentrations of 0.005 to 0.02 mole fraction. Phosphates are the least toxic, with concentrations of 0.05 to 0.1 mole fraction required for injury.

Applying anhydrous ammonia at high rates immediately before seeding may cause injury. Delaying time of planting after fertilizer application or increasing depth of fertilizer placement reduces injury. Concentrations of $(NH_3 + NH_4^+)$-nitrogen greater than 0.0006 mole fraction inhibits corn radicle growth, while plumule growth inhibition occurs at 0.0009 mole fraction. Concentrations greater than 0.001 mole fraction reduced field stands. Applying 112 kg ha^{-1} immediately prior to planting can result in 0.0014 mole fraction ppm of $(NH_3 + NH_4^+)$-nitrogen in the seed zone.

Seed and seedling damage from anhydrous ammonia, urea, and ammonium phosphates apparently is due to the mobile ammonia ions. In a fine sand at 3% moisture, ammonia moved 7.5 cm in 3 hours. In silt loam at 50% moisture saturation, ammonia from diammonium phosphate moved 2.5 cm in 24 hours and 7.5 cm in 27 hours. Ammonia moved at the rate of 2.5 cm per hour in a saturated soil.

Seeding Depth

A rule of thumb for planting depth of many crops is 4 to 5 times the average seed diam (Table 3.6). Other factors influencing seeding depth include type of emergence, i.e., hypogeal or epigeal, soil texture, date of planting, and available moisture supply.

Rate of emergence depends more on soil temperature and moisture conditions than on planting depth. Percent final emergence frequently is associated with depth in addition to numerous other factors. Corn emergence is delayed 1–2 days with each 2- to 3-cm increase in depth, particularly with early-season planting. Very shallow plantings may fail due to rapid soil drying near the surface, since the top 1.25 cm of soil dries to below the permanent wilting point within 24 hours after wetting. Rate of soil drying varies with soil depth, and the drying front advances linearly with time. Thus, slight increases in planting depth may make large differences in amount of moisture available to the seed. A seed planted 2 cm deep, as shown in Fig. 3.11, may require 10 days to emerge and could dry out before emerging, whereas planting 4 cm deep could avoid the problem. Soil compaction can improve emergence from shallow planting depths because rapid drying of the soil surface is avoided due to better capillary water movement.

Successful establishment of small-seeded grasses and legumes is often a function of planting depth. Thus, the benefit of large seed size among small-seeded species becomes important as average seed size becomes

Table 3.6. Planting depth for various crops.†

Normal depth of seeding	Usual maximum depth for emergence	Representative crops
cm	cm	
0.5–1.25	2.5–5.0	Bermudagrass, birdsfoot trefoil, bluegrass, carpetgrass, redtop, timothy, tobacco, alsike clover, white clover
1.25–2.0	5.0–7.5	Alfalfa, crimson clover, foxtail millet, lespedeza, red clover, sweet clover, turnip
2.0–4.0	7.5–10.0	Bromegrass, broomcorn, crotalaria, flax, sudangrass, sugar beet, sunflower
4.0–5.0	7.5–12.5	Barley, buckwheat, hemp, mungbean, oat, peanut, rice, rye, sorghum, soybean, vetch, wheat
5.0–7.5	10.0–20.0	Corn, cotton, pea
10–12.5		Potato, Jerusalem artichoke, sugarcane

† Adapted with permission from Martin et al. (1976).

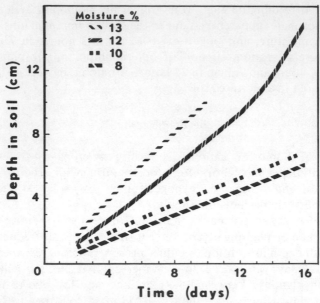

Fig. 3.11. Rate of drying to given moisture contents in columns of Hidalgo sandy clay loam
     soil. Adapted with permission from Wiegand (1962).

smaller. Increasing seed size within a species generally increases final
emergence at all depths of planting, with the greatest benefit occurring at
maximum depths.

Recommended seeding depths generally are greater on sandy than clay
soils due to lower water-holding capacity and faster drying rate at a given
depth. In regions where planting occurs during periods of low rainfall and
low soil moisture availability, seeding depth may be double or more the
depth commonly used in more moist regions. Increased soil compaction
also increases emergence in drier regions. Wheat in the Palouse region of
eastern Washington and Oregon may be planted more than 10 cm deep
while seeding depth in the North Central States seldom exceeds 5–7 cm.
Maximum depth for wheat planting is related to the coleoptiles ability to
elongate and the mechanical resistance of the soil. Semidwarf varieties
typically have shorter coleoptile lengths, averaging 6.7 vs. 9.1 cm for
normal height varieties. This difference resulted in emergences of 22 and
49% when seeds were planted 10 cm deep.

Soybean may emerge satisfactorily on sandy loam soils when planted as
deep as 7.5 cm, while depths greater than 2.5 cm may reduce final
emergence on clay soils. Also, soil temperatures near 25°C inhibit
hypocotyl elongation of some soybean varieties, limiting emergence from
depths greater than 7.5 cm.

Emergence of corn from 2.5–17.8 cm depends on moisture conditions.
During dry years, placement up to 12.7 cm deep produces better stands than
shallower placement, while no difference in final emergence is noted for
2.5–7.5 cm in moist years.

Date of Planting

Soil temperature and moisture are the major environmental factors influencing seedling establishment with various planting dates. Early plantings in temperate regions usually are associated with favorable moisture supplies but low soil temperatures. Cool, moist conditions favor pathogen development on seeds low in vigor by increasing days to emergence due to low temperature and increasing time for pathogen activity and ultimately reduce percent emergence (Table 3.7). Planting date in nonirrigated warm climates frequently is determined by the onset of rains. In addition, soil temperature above 35°C, as well as the low moisture availability, may restrict germination of cool season species.

Early spring soil temperatures in temperate regions decrease rapidly with depth. Early plantings should be shallower than later plantings, because temperatures are warmer near the soil surface and because soil moisture conditions usually are better early in the season. Rate of emergence is closely correlated with early season soil temperatures measured at the seed depth. The emergence rate increases linearly from 5–25°C for cool-season crops and from 10–35°C for warm-season crops.

Seedbed Preparation

Final seedling emergence under field conditions usually is higher with better prepared seedbeds, better soil-seed contact, and more precise planting depth. Traditional seedbed preparation techniques of plowing, field cultivation, discing, and harrowing once or twice before planting generally result in a firm seedbed giving good soil-seed contact and uniform, rapid emergence under favorable temperature and moisture conditions. Minimum tillage seedbed preparation frequently results in lower plant populations when older equipment not designed for the rough field conditions is used. New planting equipment designed for minimum tilled seedbeds produces final plant populations equivalent to well-prepared seedbeds.

Table 3.7. Days to emergence and % emergence of five bean varieties when planted at six different dates, average of 2 years.†

Date	Days to emergence	% emergence
1–8 April	42.4	51.1
14–22 April	30.1	64.9
28 April–6 May	22.8	73.1
12–20 May	15.0	75.8
27 May–3 June	12.8	65.4 (dry)
9–17 June	8.3	80.3

† Adapted with permission from Hardwick (1972).

Planting Equipment

In semiarid regions, shovel drills are used for small grains and listers or row crop planters equipped with furrow openers are used to scrape dry soil to the side of the row, permitting the seed to be placed in moist soil without burying the seed too deep. In addition, drills and planters generally are equipped with press wheels to insure good soil-seed contact and enhance capillary movement of soil water to the seed.

Soil-seed contact, uniformity of seed distribution, seed placement, seedling emergence, and seedling survival are all influenced by the method of seeding. Broadcast seeding, as with an endgate broadcaster, is the least expensive method, but it may result in poor spacial distribution, poor soil-seed contact, and large variation in depth placement, depending on the method used to incorporate and cover seeds. Poor seedling survival may occur from very shallow plantings or from very deep plantings. In Iowa studies, 108 kg ha^{-1} of oat broadcast yielded results similar to those obtained with 36–72 kg ha^{-1} drilled. Yields were about 10% lower where broadcast seeded than where drilled at comparable seeding rates.

In Montana and North Dakota studies, planting winter wheat in furrows 10 cm deep resulted in better survival in the spring and higher yields than planting with surface drilling. Presumably, this is due to increased snow cover and protection of seedlings from desiccating winds. As the environment becomes more severe, the advantage of furrow planting increases.

Small-seeded grasses and legumes require shallow seed placement for establishment. Establishment is improved by a well-prepared seedbed and the use of cultipack seeders on all but very fine-textured soils with poor soil structure, which are susceptible to crusting. Punch planting, e.g., placing seeds in holes punched in the soil, has been used with success with small-seeded vegetables on soils prone to crusting.

Biotic Factors Influencing Stand Establishment

Pathogens inhabiting the seed or soil may reduce stand establishment whenever seedling emergence is delayed by low temperatures, high soil moisture, or other conditions more favorable for pathogens than for germination and emergence of seeds.

Field invasion of seeds by fungi can occur as the seed is maturing but requires 90–95% atmospheric RH and at least 22–25% seed moisture in cereal grains. Fungi genera commonly found in field-infected seeds included *Alternaria, Aspergillus, Penicillium, Fusarium,* and *Helminthosporium. Alternaria* does not reduce germination, although it does cause discoloration. All of the other fungi may reduce germination.

The major storage fungi influencing seed viability include a large number of *Aspergillus* spp. and *Penicillium* spp. Optimum conditions for development of these species are 25–30°C and 60–90% RH. "Safe" grain moisture to prevent microbial deterioration of stored seeds is less than 12.5% for cereals and below 8.5% for some oilseed crops.

Soil-inhabiting microorganisms influencing seedling establishment include those causing damping-off (*Pythium, Rhizoctonia, Botryis,* and *Phytophthora*). The symptoms may differ depending on the stage of germination or growth when infections occur, i.e., 1) preemergence damping-off associted with seed and seedling rot, 2) postemergence damping-off associated with stem rot near the soil surface, and 3) root rot associated with plant stunting and eventual death. Damping-off due to *Pythium* or *Rhizoctonia* generally is greatest in warm-season crops planted early. The optimum temperature range for these organisms is 20–30°C. Planting seeds requiring temperatures of 25–30°C at a time when the soil temperature is near 20°C may tip the balance in favor of pathogens. Cool season crops avoid these pathogens due to their faster growth rate at temperatures below 20°C.

Moisture conditions favorable for germination may be the most favorable for pathogen development. In beets, capillary wetting of the soil provides better aeration and conditions for germination than a saturated soil, resulting in more emerged beet seedlings. The incidence of *Phoma beta*-caused seedling death also increases from saturated soil conditions to the capillary-wetted soil.

Seed Treatments

Fungicides

Fungicides for seed treatment are of three common types:
1) Disinfestants that kill pathogens located under the seed coat, such as loose smut (*Ustilago nuda* [Jens.] Rostv.) of barley
2) Disinfectants effective on seed surface-inhabiting organisms such as covered smuts
3) Protectants that protect the seed after placement in the soil from soil-inhabiting fungi causing seed rot and seedling blights, and from insects such as seed maggots and wireworms.

Seed treatments with fungicides are used most commonly with species exhibiting hypogeal emergence. The potential for invasion of seedling blight organisms is greater when the food storage structure remains below the soil surface in an environment more uniform in temperature and RH than the above-ground environment. Air-dried soils seldom have RH below 97–98%. Seed lots seeded into unfavorable environments or those of initial low quality, whether epigeal or hypogeal, show greater response to seed treatment than do vigorous, high quality seed lots planted at more favorable temperature and moisture conditions.

Scarification

Scarification is the mechanical process of abrasively etching or scratching the seed coat to reduce the number of hard seeds. Hard seed problems occur frequently in alfalfa and other small-seeded legumes where viable seeds fail to germinate immediately when placed in moist soil due to impermeable seed coats. Generally, seed specialists and agronomists agree that more than 20–30% hard seeds are undesirable with the possible exception of self-reseeding annuals, where 50–60% hard seeds are considered acceptable.

A seed lot scarified mechanically cannot be stored safely for more than 1 year due to rapid loss of viability. Techniques developed to scarify without mechanical damage include the use of radio-frequency (10 MHz) waves for a few minutes or infrared exposure for a few seconds. Short exposures to high temperatures appear to change the seed coat permeability but have very little effect on seed longevity. Sulfuric acid has been used commercially to delint and scarify cotton seed but is not popular because of problems in handling acids and treated seeds.

Preplant Hardening

Preplant hardening to hasten germination and precondition the seed to stress environments has been reported repeatedly in the Russian literature. The procedure consists of repeated inbibitions of the seed, short of radicle emergence, followed by redrying. The cycle may be repeated 3 to 5 times. The process reportedly activates latent physiological processes. Most studies indicate more rapid germination and seedling emergence using this technique, but no change in germination percentage has been reported in wheat, rye, beet, or carrot. More rapid water uptake of previously soaked and dried seeds of Lehmann lovegrass (*Eragrostis lehmanniana* Nees) than for untreated seeds has been shown. This may explain the reported improved tolerance to seeding under stress conditions. Seeds should not be soaked for more than 6–8 hours at 25°C before drying because DNA synthesis, which may begin within 6–8 hours, is adversely affected by drying the seed.

Use of AMP

Adenosine 5′-monophosphate (AMP) has stimulated germination and emergence under low temperatures associated with early planting in cotton.

Other germination-stimulating chemicals include potassium nitrate (KNO_3), hydrogen peroxide (H_2O_2), thiourea, gibberellins, auxins, cytokinin, and ethylene, but these are used primarily in seed laboratories.

Electromagnetism

Electromagnetic seed treatments have been marketed for several years for "energizing" or "exciting" seeds of low vigor. Seeds either are placed in an electromagnetic field for a specified time, or the seed is passed through an electromagnetic force field via an auger. Results have been variable and yield increases usually non-significant.

Herbicides

Herbicide protection by banding 165–330 kg ha^{-1} of activated charcoal directly over the seed row at planting and prior to preemergence herbicide application has been effective in preventing movement of germination-inhibiting herbicides. The technique improves the selectivity of the chemical and/or permits the use of chemicals to control similar species, such as grassy weeds in a grass-seed field.

Herbicide antidotes may be applied as a seed treatment prior to planting as with 1,8-naphthalic anhydride or as a herbicide spray tank mix, e.g., N, N-diallyl-2,2-dichloroacetamide. Eradicane® and Sutan +® are herbicides with the latter antidote in a 12:1 ratio of herbicide to antidote. The antidotes permit the use of higher rates of herbicide for the control of difficult weeds with minimal damage to the germinating and growing crops. However, the antidotes may also alter the selectivity of herbicides.

Inoculation

Inoculation of legume seeds with appropriate strains of *Rhizobium* bacteria is necessary for symbiotic nitrogen fixation when natural populations effective for the crop species planted do not inhabit the soil or are present in low numbers. Inoculation of seeds with *Rhizobium* bacteria has little or no effect on germination. However, problems of effective nodulation do occur when it is coupled with chemical seed treatment, because most fungicides are toxic to the rhizobia. Fortunately, epigeal-emerging legume species seldom show responses to fungicide application unless seed lots of low initial seed vigor are planted.

Preinoculation of legume seeds by procedures such as pelleting, lyophilization, and vacuum infiltration are used in an attempt to avoid the time squeeze at planting. In general, the seed surface is a hostile environment for rhizobia bacteria, and some research indicates the need for over 5000 rhizobia per seed if inoculation occurs 2–3 months before planting. Pre-inoculation is successful only where it is done during the cold periods preceding planting in areas such as the Upper Midwest of the U.S. Coating with a peat-base inoculant has been most effective. Rhizobia will not survive prolonged elevated storage temperatures regardless of inoculation method.

Seed Coatings

Pelletizing has been practiced in the vegetable industry to facilitate precision mechanical planting of small seed or irregularly shaped seeds. Pelleting consists of coating the seed with small quantities of inert matter, usually bentonite clay, to change the size and shape of seeds to some standard size for precision planting. Pelleting sugar beet seeds has become widespread with the development of monogerm seed and precision planting.

Coating seeds with lime to protect rhizobia is a common practice in aerial seeding of some legumes on acid soils in Australia. United States studies show no benefit from lime-coating alfalfa seed when evaluated by seedling counts, nodulation, or yield on nonacid soils and only slight increases in yield on acid soils of pH 4.6–5.0.

Pelletizing seed provides the opportunity to incorporate fertilizers, growth regulators, and fungicides in close proximity to the seed and in rather precise amounts, but it is not a commercial practice for most field crops.

SUMMARY

Stand establishment is a critical period in the life cycle of a plant. Stand survival for large-seeded species may exceed 90%, while only 10% survival may be obtained with many small-seeded species. The seed is a dynamic structure changing in vigor and viability from before maturation until death of the embryo. The dominant environmental factors affecting establishment, temperature and moisture, are in a constant state of flux, which places substantial stress on the germinating and emerging seedling. The interaction of the seed and its environment determines the success or failure of a planting. Selecting cultivars, seed lots, seeding methods, and seed treatments that permit more rapid establishment of uniform high-vigor plants will help avoid stand failures and insure maximum yield under existing environmental conditions.

REFERENCES

Association of Official Seed Analysts. 1978. Rules for testing seeds. J. Seed Technol. 3:29.

Borthwick, H. A., S. B. Hendricks, M. W. Parker, E. H. Toole, and V. K. Toole. 1952. A reversible photoreaction controlling seed germination. Proc. Natl. Acad. Sci. U.S. 38: 662–666.

Ching, T. M., and W. I. Kronstad. 1972. Varietal differences in growth potential, adenylate energy level, and energy charge of wheat. Crop Sci. 12:785–788.

Danielson, H. R., and V. K. Toole. 1976. Action of temperature and light on the control of seed germination in Alta tall fescue (*Festuca arundinacea* Schreb.). Crop Sci. 16:317–320.

Grabe, D. F. 1966. Significance of seedling vigor in corn. Ann. Hybrid Corn Conf. 21:39–44.

Haberlandt, F. 1875–77. Uber-die untere, Grenze der Keimungstemperatur der Samen unserer Culturpflanzen. Wiss-Prakt. Gebiete Pflanzenbaues 1:109–116.

Hardwick, R. C. 1972. The emergence and early growth of French and runner beans (*Phaseolus vulgaris* L. and *Phaseolus coccineus* L.) sown on different dates. J. Hort. Sci. 47:395–410.

Jensen, R. D. 1971. Effects of soil water tension on the emergence and growth of cotton seedlings. Agron. J. 63:766–769.

Jones, R. L., and J. E. Armstrong. 1971. Evidence of osmotic regulation of hydrolytic enzyme production in germinating barley seed. Plant Physiol. 48:137–142.

Longden, P. C. 1972. Effects of some soil conditions on sugarbeet seedling emergence. J. Agric. Sci., Camb. 70:543–545.

Martin, J. H., W. H. Leonard, and D. L. Stamp. 1976. Principles of field crop production. p. 204. MacMillan Publishing Co. Inc., New York.

Mory, Y. Y., D. Chen, and S. Sarid. 1972. Onset of deoxyribonucleic acid synthesis in germinating wheat embryos. Plant Physiol. 49:20–23.

Obendorf, R. L., and P. R. Hobbs. 1970. Effect of seed moisture on temperature sensitivity during imbibition of soybeans. Crop Sci. 10:563–566.

Owens, P. C. 1952. The relation of water absorption by wheat seeds to water potential. J. Exp. Bot. 3:276–290.

Perry, D. A. 1978. Report of the vigor test committee 1974–1977. Seed Sci. Technol. 6:159–182.

Rejman, E., and J. Buchowicz. 1973. RNA synthesis during the germination of wheat seeds. Phytochemistry 12:271–276.

Russell, W. A. 1974. Comparative performance for maize hybrids representing different eras of maize breeding. Ann. Corn Sorghum Res. Conf. 29:81–101.

Sachs, J. 1860. Physiologische Untersuchungen uber die Abhangiakert der Keimung von der temperatur. Jahrb. Wiss. Botan. 2:388–377.

Thompson, P. A. 1972. Geographical adaptation of seed. p. 31–58. *In* W. Heydecker (ed.) Seed ecology. Proc. 19th Easter School in Agric. Sci., Univ. of Nottingham. Pennsylvania State Univ. Press, University Park, Pa.

Toole, E. H., and V. K. Toole. 1953. Relation of storage conditions to germination and to abnormal seedlings in beans. Proc. Int. Seed Test. Assoc. 18:123–129.

Wanjura, D. F., and D. R. Buxton. 1972. Hypocotyl and radical elongation of cotton as affected by soil environment. Agron. J. 64:431–434.

Wiegand, C. L. 1962. Drying pattern of a sandy loam in relation to optimal depth of seeding. Agron. J. 54:473–476.

SUGGESTED READING

Copeland, L. E. 1976. Principles of seed science and technology. Burgess Publishing Co., Minneapolis, Minn.

Hartmann, H. T., and D. E. Kester. 1968. Plant propagation, principles and practices. 2nd ed. Prentice-Hall, Inc., Englewood Cliffs, N.J.

Hebblethwaite, P. D. 1980. Seed production. Butterworth Publishers Inc., Boston, Mass.

Heydecker, W. 1972. Seed ecology. Proc. 19th Easter School in Agric. Sci., Univ. of Nottingham. The Pennsylvania State Univ. Press, University Park, Pa.

Khan, A. A. 1977. The physiology and biochemistry of seed dormancy and germination. Elsevier/North-Holland Biomedical Press, Amsterdam, The Netherlands.

Kozlowski, T. T. 1972. Seed biology. Vol. I. Importance development, and germination. Academic Press, Inc., New York.

----. 1972. Seed biology. Vol. II. Germination control, metabolism and pathology. Academic Press, New York.

Mayer, A. M., and A. Poljakoff-Mayber. 1975. The germination of seeds. 2nd ed. Pergamon Press, New York.

----, and Y. Shain. 1974. Control of seed germination. Annual Rev. Plant Physiol. 25:167–193.

Roberts, E. H. 1972. Viability of seed. Syracuse University Press, Syracuse, N.Y.

4 Seedling Growth

C. J. NELSON AND K. L. LARSON
Department of Agronomy
University of Missouri
Columbia, Missouri

Seedling growth covers the period in the life cycle of green plants from emergence of the radicle through the seed coat until the appearance of enough green leaves to make the plant independent of stored energy. Thus, seedling growth is an arbitrarily delimited phase of the total growth period, encompassing that part of ontogeny between initiated growth that is entirely dependent on stored energy and the fully autotrophic phase of vegetative growth. The major activity of seedling growth is the establishment of root and shoot tissue for autotrophism. This is accomplished largely at the expense of stored food energy and is primarily influenced by the soil environment.

In terms of crop productivity, seedling growth can be very critical. The rigors of becoming self-sufficient can cause a high mortality among seedlings, and often this loss must be overcome by higher seeding rates or improved management practices during establishment and early stages of seedling development. These problems usually are more critical for small-seeded crop species such as forage grasses and legumes, rather than for large-seeded crop species such as corn, cotton, and soybean.[1] For example, recommended seeding rates are often about 15% higher than desired plant number in corn and soybean. In contrast, with forages such as alfalfa, 40–50% emergence and eventual establishment are considered excellent, and seedling survival as low as 20% is not uncommon. Thus, alfalfa is basically seeded in excess in order to assure survival of sufficient plants for

[1] Scientific names of major crop plants are given in Table 1.1 of Chapter 1.

Published in *Physiological Basis of Crop Growth and Development,* © American Society of Agronomy—Crop Science Society of America, 677 South Segoe Road, Madison, WI 53711, USA.

the desired density. Typically 10–12 kg ha^{-1} are planted, but good stands can be obtained with as little as 1.5–2 kg ha^{-1} with excellent management practices and optimum growing conditions.

Another important aspect of seedling development is seedling vigor. The abilities to become autotrophic rapidly and to develop a large seedling in a short time are ecologically advantageous. At the seedling stage, competition with weeds may be great, and seedlings need to be as vigorous as possible to help withstand early insect and disease infestation. Further, with a rapid early growth rate, plants are able to grow extensive root and shoot systems. In contrast with less vigorous seedlings, the larger and more extensive root system in contact with soil particles provides a greater absorption area for uptake of nutrients and water. The large shoot system of vigorous seedlings has more green tissue exposed to intercept solar energy, which is converted to chemical energy in the form of carbohydrates to supply growth needs of the plant. This latter concept is often likened to the compound interest law, wherein each increment of shoot growth allows interception of more radiation than the prior increment. The additional energy formed by the larger shoot can support a larger growth increment during the next period.

Oftentimes additional seedling vigor does not cause a marked change in final yield, especially under good management practices, but it may become a limiting factor if planting rates and environmental factors are not optimum. The major advantage of seedling vigor is to enhance plant survival and development of leaf area to intercept radiation.

Growth has three basic aspects: an increase in plant substance (mass and volume); production of specific cells, tissues, and organs (development); and reproduction (new individuals). Seedling growth involves the first two of these aspects, which are characteristic of vegetative growth in general. The embryonic axis of the seedling consists of the root and shoot. That axis, which usually comprises only a small percentage of the dry weight of monocot seeds (e.g., 2–3% of wheat or corn kernels), grows in size and weight at the expense of stored food energy in the endosperm. In dicot seeds such as soybean or peanut, the axis includes the cotyledons and thus comprises a high percentage of the dry weight of the seed. Energy for growth is derived from the cotyledons in most dicots. Mobilization of these energy sources was discussed in the previous chapter.

Figure 4.1 places seedling growth into perspective with processes preceding it—seed formation, dormancy, and germination—and those following it—autotrophic growth. This relationship suggests that the total energy supply available at initial stages of seedling development partially depends on environmental conditions during seed development in the parent plant and on the length of storage time from harvest to germination. Favorable environmental conditions during development and storage, in combination with a short storage period, allow a high supply of energy for germination and seedling development, whereas unfavorable conditions and a long stor-

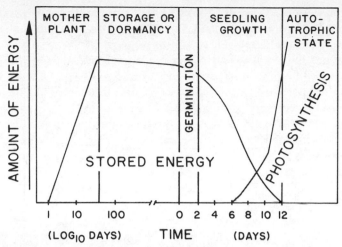

Fig. 4.1. Energy changes in seeds during seed development on the mother plant, storage previous to germination, the germination period, and seedling growth to the autotrophic state.

age or dormancy period decrease the supply. As such, this relationship is important in understanding rapidity of seedling growth and transition time from dependence on stored energy to energy provided through photosynthesis.

Heterotrophism and autotrophism are illustrated at the biochemical level in Fig. 4.2. For autotrophic growth, the environment must provide CO_2, H_2O, light, O_2, essential nutrient ions, and a suitable temperature. The first five requisites are extensive properties of the environment and are consumed in measurable quantities during growth. Temperature (not to be confused with heat) is an intensive property that is not consumed but plays a regulatory role. The biosynthesis underlying growth is accomplished by absorbing oxidized elements from the environment, reducing them, and incorporating them into specific components of the plant. A simple illustration is synthesis of the amino acid cysteine.

$$CO_2 + SO_4^{2-} + NO_3^- \xrightarrow{\text{energy}}$$

(environment)

$$\begin{array}{ccc} H & H & \\ \backslash & | & \\ HC & - C & - C \diagup\!\!\!\!\diagdown O \\ \diagup & | & \diagdown \\ SH & NH_2 & OH \end{array}$$

(plant)

Reduction of carbon (C), sulphur (S), and nitrogen (N) requires a form of reducing energy, illustrated in Fig. 4.2 by [H]. Bonding energy is depicted by the biochemist's symbol for the high energy bond, $\sim$. For most of the biosynthetic processes of growth, these energy forms are derived from respiration.

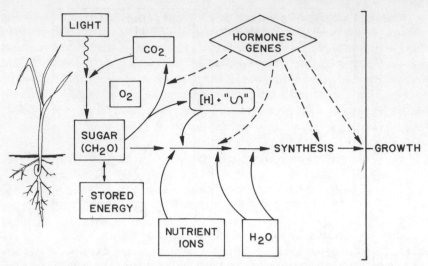

Fig. 4.2. Energy flow of autotrophic (light energy) and heterotrophic (stored energy) systems showing similarities once beyond the energy source. A suitable temperature is necessary for the processes to function.

Heterotrophic seedling growth is preceded by biosynthetic processes powered by photosynthesis within the parent plant and by other processes that accumulate seedling food reserves containing reduced C, N, S, and the remaining essential mineral elements. Inheritance of a stored supply of food and nutrients (Fig. 4.1) reduces the initial environmental requirements for growth of the embryonic axis to H_2O, O_2, and a suitable temperature. Mobilization of stored chemical energy in the form of food reserves substitutes for photosynthesis and mineral absorption during early seedling development. As the root system of the seedling extends, absorption of nutrient ions from the soil increases. As the green stem and leaf tissues develop above soil level, available energy increases through photosynthesis. Thus, a gradual transition occurs until the stored supplies are exhausted (Fig. 4.1). Figure 4.2 can be misleading in not showing that growth involves more than biosynthesis. Many hundreds of compounds must be synthesized in the proper proportions and then assembled in exact order and arrangement to produce plants.

STRUCTURE AND GROWTH OF CELLS

Regulation of seedling growth occurs at the cellular level. Indeed, every aspect of plant growth and development occurs in cells. It is essential, however, to recognize that individual cells do not behave as autonomous units, but rather that they are highly integrated into the activity and function of other cells in close proximity. Even so, before one can understand the dynamics of these cell groups one must carefully understand the basic composition and structure of these units.

Principal cellular structures are shown in Fig. 4.3. Discussion of the roles of the parts labeled in the diagram will be oriented toward growth. Many anatomical differences exist among plant cells because each matures and becomes specialized for specific functions. Even so, all cells are initiated by a meristem and through early stages of development follow the same pattern. Location and function of specific meristems for different phases of growth will be detailed later in this chapter.

Cell Walls

At first glance, the cell wall may seem to constitute a static factor, but this is not true. The chemical and physical properties of the wall are paramount in determining the rate and extent of cell growth. Cell walls are initially created as partitions in the final stages of cell division in the meristem. They are subsequently extended during cell enlargement and are variously thickened and strengthened during maturation, depending on how the cell differentiates.

The outer wall, which is the first to form during late stages of mitosis, is the primary wall. This relatively thin wall is flexible and possesses a degree of plasticity and elasticity. As the cell attains its maximum size and approaches maturity, stiffer and less elastic secondary cell wall layers are deposited on the inner side of the primary cell wall. Thus, while it is obvious that the increase in cell surface area during maturation does not simultaneously cause a decrease in cell wall thickness, some secondary thickening must also occur. Once the cell has grown to the dimension required for its

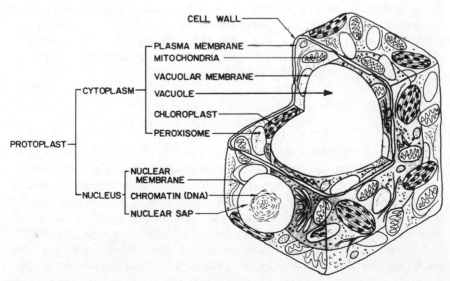

Fig. 4.3. Generalized diagram of plant cell showing major parts. Adapted with permission from Larson et al. (1975).

function, or to that allowed by its environment, secondary wall tissue is laid down rapidly, and the cell loses its elasticity and ability for continued or renewed growth.

Plasma Membrane

The plasma or cytoplasmic membrane circumscribes the inner surface of the cell wall and primarily serves as a differentially permeable barrier to solute and water movement. During cell growth it functions as a molecular sieve to maintain a high solute concentration and thus allows a positive turgor pressure. In most instances the plasma membrane tends to react to rather than initiate plant growth, but, although it plays a passive role, it is very essential for cell, tissue, and seedling growth.

The membrane is composed of layered lipid and protein components that collectively allow the membrane to be selectively permeable to both polar and nonpolar solutes. The exact mechanism for this selectivity has not been firmly documented, but it is well known that selectivity and permeability may be changed by growth regulators, pesticides, and environmental conditions such as cold or drought.

Vacuole

The vacuole of a plant cell is a large, liquid inclusion separated from the cytoplasm by the vacuolar membrane. The solution in the vacuole is usually acidic, with a pH of 4.0 or lower compared with the cytoplasm's pH of 6.5–7.0. The vacuolar membrane allows substrates within the vacuole to be compartmentalized away from active metabolism of the cytoplasm. Undoubtedly this gradient must be maintained by the expenditure of energy at the membrane interface. Young cells typically do not have a distinct vacuole. The vacuole usually enlarges rapidly as the cell grows, so that it eventually is the largest component of the cell.

Mitochondria

Mitochondria are organelles in the cytoplasm that serve as sites of aerobic respiration, allowing the cell to metabolize reduced carbon compounds such as carbohydrates. This oxidation process releases chemical energy useful in energy-requiring processes in the cell. Located on the cristae (folds) in the mitochondrial membrane are the enzymes associated with the Krebs cycle for sequentially oxidizing organic compounds to CO_2 and H_2O. This oxidation process is coupled with reduction of pyridine nucleotides such as NADP (nicotinamide adenine dinucleotide phosphate) and NAD (nicotinamide adenine dinucleotide), which can later be used in synthesis and reduction reactions in other parts of the cell. Perhaps more obvious is the production of high energy phosphate bonds ($\sim$P) in ATP that also can be transferred to other parts of the cell for use as an energy source to drive reactions.

Chloroplasts

These structures in the cytoplasm of green tissue or of cells capable of photosynthesis are highly organized and contain the grana and stroma where the light and dark reactions, respectively, of photosynthesis occur (see Fig. 4.3). Juvenile cells contain proplastids that reproduce by fission and evolve into chloroplasts under proper environmental conditions. Chloroplasts, like mitochondria, have a membrane and are considered prokaryotic cells that originally were separate entities, existing independently of the remaining cytoplasm. However, during early evolution, mitochondria and chloroplasts invaginated the protoplast of larger cells to eventually become mutually synergistic. They allowed the larger cells to utilize aerobic respiration and also become independent of outside energy supplies. The nature and structure of chloroplasts are discussed further in the chapter on photosynthesis.

Endoplasmic Reticulum

The endoplasmic reticulum is a filamentous material made from protein and lipid that connects the nucleus to the cell membrane, and also forms many strands within the cell. While communication appears to be a major function, this tissue also serves as a site for protein synthesis. Peptide bond formation along the endoplasmic reticulum terminates the long and elaborate process of protein synthesis. Nuclear DNA is transcribed to messenger RNA (mRNA), which codes the ribosomal RNA (rRNA) on the endoplasmic reticulum. The matching of the rRNA code with transfer RNA (tRNA) aligns the amino acids in the correct order for peptide bonding to complete protein synthesis.

Maintaining integrity of the endoplasmic reticulum is paramount for efficient cell growth and metabolism. The nature and function of the endoplasmic reticulum are influenced by environmental stresses such as heat, drought, cold, mineral deficiency, or atmospheric pollutants. When not functioning at near maximal capacity, protein synthesis, which is primarily enzyme synthesis, is reduced, and the plant may be deficient in certain proteins. Thus, one mechanism to expose deficiencies of N or other elements is through reduced activity of the endoplasmic reticulum.

Nucleus

The nucleus is the governing body of the cell and the coarse tuning mechanism for function and metabolic control. Enclosed within the nuclear membrane are the strands of DNA that swell to become distinguishable as chromosomes during cell division (mitosis). However, when cells are not undergoing mitosis, the strands function as relays of the genetic information for protein (enzyme) synthesis that provides a degree of metabolic control. During differentiation and maturation of tissue such as water-conducting xylem cells, the nucleus degenerates and is lost. These cells have died and do not function metabolically. They still retain an important role, however, as structural and conducting tissues.

Cytoplasm

The liquid portion of the cell that contains the organelles is the cytoplasm. Within this viscous medium, the organelles are suspended, and this is where many of the metabolic reactions take place. The cytoplasm has very specific behavior for avoiding freezing injury to the cell, maintaining turgor pressure for cell enlargement, and resisting CO_2 diffusion from the external environment to the chloroplast. The cytoplasm, along with the nucleus, comprises the protoplasm, commonly referred to as the living substance of the cell.

Peroxisomes

Peroxisomes are cell organelles that are usually intermediate in size between chloroplasts and mitochondria and believed to be involved in photorespiration (see Chapter 5). Early research showed that glycolate metabolism was active in these organelles. A closely related particle, the glyoxysome, is involved with fat degradation in storage tissue through pathways similar to those in the peroxisome. Metabolic importance of these organelles in crop plants is only beginning to be explored and evaluated.

GROWTH

Plant growth occurs through cell division and cell elongation. Growth of the whole plant is compartmentalized into areas called meristems where cell division and elongation are localized. Therefore, the growth factors that require an understanding of cell biochemistry and physiology include those that initiate and maintain cell division, those that initiate and limit cell expansion, and those that delineate differentiation into the commonly observed visual features.

As meristem cells undergo repeated division via mitosis, DNA and nuclear material replicates and cell mass increases. Adjacent to this area of meristematic activity is the region of cell elongation, which phases into the region of differentiation or maturation (Fig. 4.4). These latter regions do not have sharp spatial boundaries and cannot be easily distinguished. Thus, the processes act as a continuum. The continuum may be separated partially in time as in the growth of determinate organs such as leaves, flowers, or seeds, in which cell division may predominate in early stages, and cell elongation and differentiation may occur more simultaneously during later stages. An example is in epigeal-emerging species (cotyledons emerge above ground) such as soybean where cell division and some elongation and differentiation occur in seed cotyledons during development and maturation on the parent plant. Later, during imbibition of water during germination, and continuing until after the seedling emerges, cell elongation and differentiation resume, producing a noticeable size increase in the cotyledons.

The growth of meristems, which appears at first sight to be a simple phenomenon, is one of nature's most complex processes. Plant scientists are attempting to learn how meristems function, and the environmental, cultural, and genetic mechanisms that govern them.

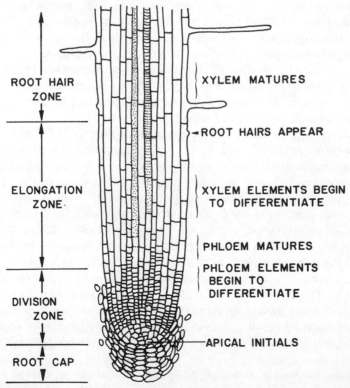

Fig. 4.4. Diagram of growing zone of a root. Number of cells is normally much greater than shown in this diagram. Adapted with permission from Ray (1972).

CELL DIVISION AND VACUOLATION

When growth and meristematic activity in apical meristems of the root and shoot are continuous or indeterminate, some general principles of cell dynamics can be described. In the area of the meristematic region (Fig. 4.4) where cell division predominates, cells are relatively small and have a cytoplasm that is continuous throughout the volume of the cell. The prominent, spherical nucleus is located near the center of the cell, and the cell wall is comprised largely of primary cell wall tissue and is relatively thin.

The two daughter cells formed in mitosis are initially about half the size of the original meristematic mother cell. Thus, growth of both daughter cells must take place, in one case to regain the original size of the cell that is to remain meristematic, and in the derivative cell to begin the diversion toward differentiation into a functional unit. Growth of both cells involves synthesis of cytoplasm and cell wall material, but vacuolation does not yet occur. The number of cells capable of dividing remains relatively constant within the meristem. The exact reason that one of the two cells formed remains meristematic while the other does not is unknown.

Once the meristem region has grown away, the derivative cell begins the process of vacuolation and enlargement. As cells enlarge they may increase in volume by 30–150 times. During enlargement the cell wall becomes plastic, and large amounts of water are absorbed. The vacuole is enlarged, and additional cytoplasm and cell wall material are synthesized.

The required force for cell enlargement is turgor pressure (ψ_p), or wall pressure in older literature, that occurs due to water uptake by the cell. However, another controlling factor is the degree of plasticity of the cell walls, which is mediated by the growth regulator auxin.

Due to the presence of solutes, the osmotic potential of the cell sap (ψ_s) is lower (more negative) than that of water located outside of the cell. The net water potential of a cell (ψ_{cell}), or the negative of the diffusion pressure deficit in older literature, is usually negative and is equal to the sum of ψ_s and ψ_p, the latter normally being a positive pressure. The ψ_{cell} needs to be lower (more negative) than that of surrounding tissue for water to be absorbed. Therefore, water uptake can be promoted either by increasing the concentration of solutes in the sap (making ψ_s more negative), by decreasing ψ_p (making it less positive), or both. In fact, during enlargement the cell wall loosens and the solute concentration decreases as it is diluted by incoming water. The cell enlarges because ψ_s increases less than ψ_p decreases, probably because solutes continue to be synthesized in the cell. These concomitant activities result in a sustained low ψ_{cell}, which allows water to be absorbed while retaining a positive ψ_p to keep pressure on the walls to cause enlargement.

As the cell wall is stretched, new cellulose and other cell wall constituents are synthesized and incorporated. These latter processes require energy, and therefore high rates of respiration per unit of tissue occur.

Moreover, growth requires aerobic conditions and an adequate supply of carbohydrate, both as an energy source and as a supply of structural material. Collectively, you can visualize the influence on cell growth of environmental factors such as drought, heat, or energy stress, which are reflected in observable features such as leaf size, plant height, and grain weight.

CELL WALL GROWTH

The main structural framework of the cell wall consists of a mass of cellulose microfibrils oriented in much the same manner as a random brush heap (Fig. 4.5). The microfibrils are long strands 10–30 nm wide and 5–10 nm thick. The cellulose polysaccharide chain is imbedded with hemicelluloses, which are noncellulose polysaccharides composed mainly of residues of pentoses, such as arabinose and xylose; and hexoses, such as glucose, galactose, and mannose. Pectins, which contain large amounts of galacturonic acid and small amounts of protein and lipid, comprise the remainder of the matrix. Cell walls in oat coleoptiles have been found to consist of 51% hemicellulose, 25% cellulose, 12% protein, 5% pectin, and 7% waxes, pigments, and other minor constituents.

Lignin is not normally found in young cell walls but occurs in heavily thickened walls of mature cells. Although lignin is associated intimately with the cellulose framework of the wall, being deposited in the microscopic spaces between the cellulose strands, the two substances are not joined to form a chemical compound. Lignin gives the cell wall rigidity, but also interferes with digestion of the cellulose by ruminant animals.

Growth of enlarging cells of higher plants occurs rather uniformly over the inner surface of the entire cell wall. This gives a growth pattern different from tip-growth, which occurs in specialized cells such as pollen tubes and root hairs where only the tip is extended. Fibrils in young cells are usually

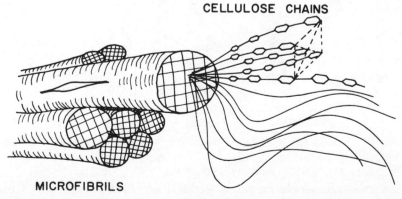

Fig. 4.5. Diagram showing how cellulose chains are associated to form microfibrils of the cell wall. Adapted with permission from Roelefsen (1965).

oriented transversely to the direction of elongation. During cell elongation the fibrils are stretched and become oriented in the direction of the cell axis. As more fibrils are synthesized to thicken the wall, the fibrils are laid down transversely, so that the final cell wall appears strongly layered with fibrils oriented in many offsetting directions (Fig. 4.6).

Originally, it was thought that the cell wall was merely stretched during cell elongation. We now recognize that other processes—including slippage between closely associated fibrils, a continual synthesis of cell wall constituents, and hydrolysis of the glucose-glucose bonds—form the basic links within the microfibril chain. However, for growth or distension to occur, some critical force (turgor pressure) must be exerted on the reversibly elastic cell wall to cause, additionally, an irreversible plastic stretching (Fig. 4.7). This critical force does not act independently, since many experiments have shown that auxin increases the plasticity of cell walls.

There are two hypotheses regarding the influence of auxin on growth of cell walls. One is that auxin influences the calcium (Ca) linkage between the Ca-pectate compounds of the cell wall, therefore allowing for structural change in the wall configuration. This may occur in response to an auxin-Ca^{2+} reaction or an auxin-mediated reaction that controls the formation of pectin from pectic acid. Calcium would not be able to form linkages with pectin to the same degree as with pectic acid, and thus the plasticity of the cell wall could be changed.

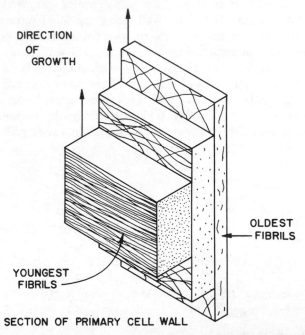

Fig. 4.6. Diagram showing the change in orientation of microfibrils as sequential layers of the cell wall develop as the cell enlarges. New microfibrils are laid down perpendicular to direction of growth, but are reoriented in the older sections as the wall is stretched during growth. Adapted with permission from Roelefsen (1965).

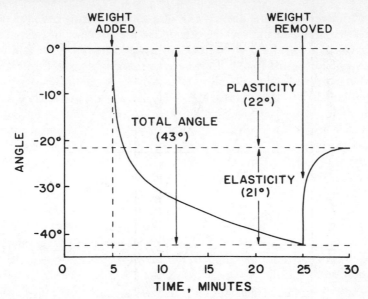

Fig. 4.7. Deformation over time of an oat coleoptile suspended horizontally following application of mechanical force to the tip. Weight was removed 20 minutes after it was applied. The difference between the original and final position of the coleoptile was taken as a measure of plasticity. Adapted with permission from Tagawa and Bonner (1957).

Another hypothesis, with perhaps greater credibility, is that cell wall materials have molecular bonds that are continuously being hydrolyzed by cellulases and then reformed. Auxin is known to influence the activity of such enzymes. During the dynamic process of cleavage and resynthesis of bonds, some slippage and rearrangement of microfibrils would likely occur, particularly if the wall was being stretched by an internal force.

In reference to these two theories, auxin is known to increase the thickness and deposition of new cell wall material. This is accompanied by a rapid increase in rate of respiration and other cellular activities. Therefore, it is difficult to define precisely the role of auxin in cell enlargement. However, the role auxin plays and its interaction with cell growth and elongation are nevertheless of paramount importance.

DIRECTIONAL CONTROL OF GROWTH

Now that mechanisms for cell growth have been described, it is necessary to consider how growth of the total plant is controlled. Tropisms are growth movements that occur in response to a unidirectional stimulus, changing the position of a plant part toward or away from the stimulus. If growth is toward the stimulus, the response is positive, and if away from the stimulus, the response is negative. External stimuli causing tropisms include gravity, light, temperature gradient, touch, and atmospheric gases such as CO_2 and O_2. Often a plant part is not reacting to a single stimulus, but

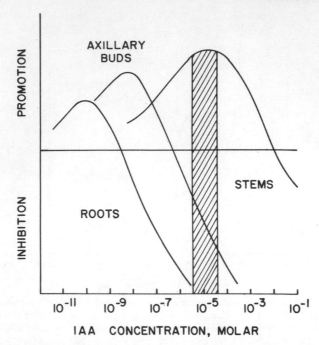

Fig. 4.8. Differential effect of auxin (IAA) concentrations (log scale) on growth of roots, axillary buds, and stems. Optimum stem growth occurs at 10^{-5} M, a concentration that inhibits axillary buds and roots.

rather several stimuli, at a given time. Thus, the final shape and form of a plant will be the result of several stimuli acting collectively.

Auxin is the growth hormone usually accountable for the tropism when the growth response occurs in the meristematic region, particularly in the zone of cell elongation. Geotropism, the response to gravity, and phototropism, the response to light, are examples of this type of movement. Alternatively, leaf rolling in response to drought and sleep movement (leaves dropping to a vertical position at night) are largely mediated by water through turgor pressure changes.

Geotropism and phototropism are probably the best understood of the auxin-mediated growth movements. Both of these responses depend on an unequal concentration of auxin in the cross section of the zone of elongation of the meristem. This differential concentration gives rise to a differential growth rate of the cells, presumably by affecting the plasticity of the cell walls. To fully understand this, you must recognize that growth response depends on auxin concentration and that roots and shoots react differently to the same auxin concentration. For example, a 10^{-5} M concentration of auxin inhibits root growth but stimulates stem growth (Fig. 4.8). However, this response is usually limited to the primary stem and root. Secondary roots and shoots are often plagiotrophic (grow at angle to stimulus), while tertiary roots and shoots often are little influenced, merely growing at right angles to the secondary tissue (Fig. 4.9).

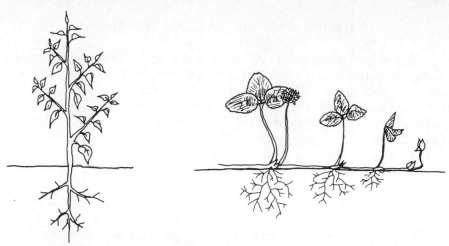

Fig. 4.9. Geotropic growth responses influence the shape of the plant. The plant on the left shows positive geotropism of the root, negative geotropism of the main stem, plagiotropism of the branches and secondary roots, and non-geotropism of the tertiary roots. The stolon of ladino clover on the right shows diageotropism.

Geotropism is mediated by gravity, which causes auxin concentration to be higher on the lower than on the upper side of the tissue. Auxin molecules themselves are too small to be influenced by gravity, and, therefore, other means of accounting for their movement must be considered. Some evidence suggests that the location of starch grains is influenced by gravity and that the grains tend to accumulate on the lower sides of the cells. In turn, these grains influence the permeability of the cell membrane at that location, which may allow auxin to preferentially penetrate through the membrane to the cell wall.

When gravitational pull affects auxin concentration in horizontal shoots, auxin levels on the lower sides of stems are increased above the 10^{-8} M level causing increased cell elongation on the lower side, and the stem grows upward. The stem will continue to grow upward because, once oriented vertically, auxin levels will be similar over the entire area of elongation. Conversely, when auxin levels in roots are increased on the lower side of horizontal tissue, cell elongation is inhibited on that side, and the root grows downward. For these reasons seed orientation is usually not of great concern in agronomic crops because, even if the axis of the embryo is not oriented vertically, geotropism during emergence allows it to reorient and emerge properly.

Phototropism also depends on the unequal concentration of auxin in the zone of elongation, but in this case the external stimulus is light, and, more specifically, light at 400 or 480 nm wavelength. The receptor molecule for light has not been clearly identified, but evidence suggests it may be a carotenoid. It is unknown whether the differential auxin concentration occurs because auxin synthesis is inactivated on the lighted side of shoot tissue, or if processes occur that cause lateral transport of auxin to the less-illuminated side.

GROWTH IN THE NATURAL STATE

Several growth stimuli are obviously acting on a seedling as it emerges and grows naturally in the field (Fig. 4.9). Further, tissues differ in their response to external stimuli. For example, leaf blades of some species, such as red clover and dandelion (*Taraxacum officinale* Weber), tend to be oriented horizontally while grasses often orient their leafblades more vertically. Growth responses of the petiole and petiolules of a compound leaf are usually the controlling factor in leaf orientation.

Growth regulation continues as the seedling develops into its mature shape. Rhizomes of many plants tend to grow at a given depth in the soil. If the rhizome is disturbed, it will grow upward or downward to reachieve the normal depth. This response is generally regarded as a diageotropic response, where growth is perpendicular to gravity, the dominant stimulus. However, little is known as to how O_2 or CO_2 levels, which influence some tissues, might also be involved in these responses. Likely, stolons are also under the control of a diaphototropic response (Fig. 4.9).

Plant tissues may differ markedly in their response to external stimuli. This is most obvious in reproductive organs where the flower bud, the open flower, and the fruit each may influence the tropic response of the pedicel. An example is the peanut, where the pedicel supporting the flower is negatively geotropic but after flowering becomes positively geotrophic and grows downward so that the developing fruit is buried in the soil.

SEEDLING EMERGENCE

Depending on the growth activity of the hypocotyl, dicotyledonous plants emerge in two different ways. In epigeal emergence, which includes more than 90% of the dicot species, the hypocotyl is active and pulls the cotyledons above (epi = on) ground during its growth (Fig. 4.10). In early stages of development, and after the primary root has emerged, the hypocotyl forms a prominent arch with a heavy-textured area on its top to assist in penetrating the soil. Growth is largely by cell expansion in the area between the hypocotyl arch and the primary root. This expansion occurs until the arch reaches the soil surface and is exposed to light. At this time, the phototropic response causes a change in auxin levels that stops hypocotyl growth. An auxin response (phototropism) causes the arch to straighten and the cotyledons to open, exposing the epicotyl.

The other emergence mechanism of dicotyledonous plants is termed hypogeal because the hypocotyl remains inactive and the cotyledons remain below (hypo = under) ground (Fig. 4.11). In this type, the epicotyl begins active growth soon after the primary root emerges. Activity of the terminal meristem as well as elongation of lower internodes forces the epicotyl arch above ground. When the epicotyl arch reaches the soil surface and is exposed to light, the lower internodes stop growth, and the arch straightens due to an auxin-mediated phototropic response.

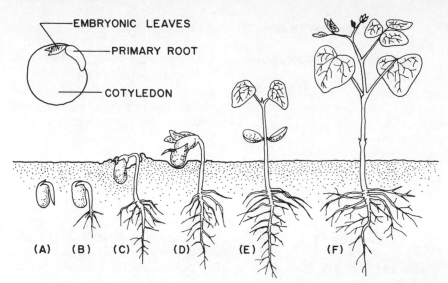

Fig. 4.10. Seedling development of an epigeal emerging dicot such as soybean. Germination proceeds with emergence of the radicle to form the primary root (A), development of secondary (branch) roots (B), and elongation of the active hypocotyl with the hypocotyl arch penetrating through the soil surface (C). Seedling becomes erect due to action of light on auxins (D), with cotyledons attached to the first node providing photosynthate in addition to stored energy for a short period of time (E), prior to drying and falling from the autotrophic seedling (F). Inset of enlarged seed with one cotyledon removed shows the primary root and the embryonic leaves that develop into the first true leaves attached to the second node.

Plants with epigeal emergence have advantages over those with hypogeal emergence in that their terminal bud is tightly enclosed and protected by the cotyledons while it is being drawn through the soil. Further, the cotyledons can be photosynthetic after emergence, and thus provide a greater quantity of energy to the developing seedling than the amount provided only by storage. Cotyledons remain photosynthetically active for a few days in crops such as soybean but can persist for several weeks in radishes (*Raphanus sativus* L.), cotton, and other crop species.

On the other hand, hypogeal-emerging dicots have an advantage in that their stored energy supply and some axillary buds remain below ground, and can provide regrowth if the tops are removed by insects, frost, or other means. Epigeal-emerging dicots have no buds or stored food below the point of cotyledon attachment (first node, Fig. 4.10) and die if the tops are removed below that point. Also of interest in comparison is that epigeal-emerging dicots have hypocotyl tissue, whereas hypogeal-emerging dicots have epicotyl tissue at the soil surface. Many seedling diseases infect plants at the soil surface and one type of tissue may be more vulnerable than another, although this has not been firmly substantiated.

Along with the advantage of being photosynthetically active, cotyledons of many epigeal-emerging dicot species increase in area during the after emergence. Some dicots such as crownvetch, cotton, and many others can increase cotyledon area by over 10-fold, thus providing an advantage in

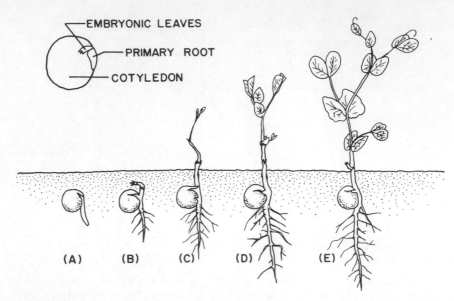

Fig. 4.11. Seedling development of a dicot with hypogeal emergence such as garden pea. The seed germinates with the radicle emerging to form the primary root (A), and with an inactive hypocotyl, the cotyledons remain stationary in the soil as the epicotyl elongates (B). Seedling emergence (C) occurs as the apical meristem continues to form internodes and nodes, the latter from which leaves develop and grow (D). As energy is depleted from the cotyledons the plant becomes fully autotrophic (E). Inset of enlarged seed with one cotyledon removed shows the primary root and embryonic leaf tissue of the epicotyl.

quickly increasing leaf area. This advantage is most apparent in seedlings with small amounts of reserve substances, where dependence on photosynthesis occurs more rapidly. Differences in seedling vigor among cultivars of birdsfoot trefoil, a species with relatively low seedling vigor, have been related directly to differences in growth rate of surface area of the cotyledons after emergence.

Contractile growth of the hypocotyl region occurs in sweet clover, alfalfa, red clover, and ladino clover during the 8th to 10th week of seedling growth. Hypocotyl cells that had grown long and narrow during seedling emergence begin to grow wide, thus shortening the vertical dimension and pulling the node where cotyledons were attached below the soil surface. This aids in winter survival because the axillary buds are protected by the insulating properties of the soil. Sweet clover, a very winter-hardy species, is capable of contracting over 3 cm, alfalfa and red clover contract 1.5–2 cm, and ladino clover about 1 cm.

All grasses have hypogeal emergence as the hypocotyl is inactive and the scutellum (cotyledon) remains below ground (Fig. 4.12). In contrast with dicots, the main function of the scutellum is not to store food, but to secrete amylase and other enzymes that digest energy-storing constituents in the endosperm. Emergence of most grasses is largely dependent on elongation of the coleoptile and the first internode. The coleoptile is actually a

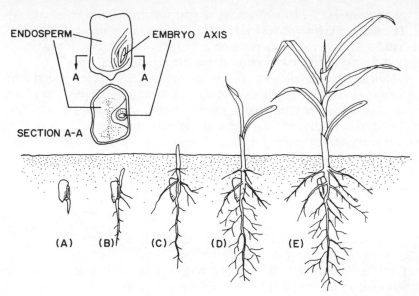

Fig. 4.12. Seedling development of a grass plant such as corn. Emergence of the radicle through the coleorhiza forms the primary root (A). The coleoptile and first internode elongate, followed by initiation of seminal roots from the base of the first internode (B). The coleoptile penetrates the soil surface (C) where action of light inhibits further growth of the coleoptile and first internode. The developing seedling initiates coronal roots at base of lower internodes as true leaves grow (D) to support an autotrophic plant (E). Inset shows transverse and cross section through the caryopsis to show location of embryo axis which contains 2–3 initiated leaves.

hollow cylinder (modified sheath with two prominent veins) attached to the second node. When the primary root (radicle) has emerged, the pointed coleoptile elongates to form a hollow tube to the soil surface. In some grasses, such as corn, the first internode (located below the coleoptile) is very active and helps push the entire coleoptile upward. When the coleoptile reaches the soil surface where it is exposed to light, elongation of it and the first internode stop due to the auxin-mediated phototropic response.

Corn grown by Indians in southwestern USA was characterized by a very long first internode. This allowed emergence from deep soil depths as seed was planted to be in contact with soil water. In some grasses such as wheat the first internode is not very active. In that case emergence depth is affected largely by coleoptile elongation.

Lack of activity of the first internode may be a disadvantage to plants emerging from deep seeding depths. This was a particular problem when semidwarf wheats were developed in the late 1960s. These plants had shortened internodes resulting in reduced plant height, and they also had shorter coleoptile growth, which prevented their emergence from as deep in the soil as the taller cultivars previously grown. For example, Sunderman (1964) compared emergence of several wheat cultivars that showed a wide range in coleoptile lengths. When seeded 7.5–12.5 cm deep, the emergence percentage in the field was correlated positively with coleoptile length.

In other cases, the lack of an active first internode may be advantageous. The shortened internodes lead to deeper crown development from axillary buds. Further, root systems emerging from the lower internodes are deeper and close to soil moisture for developing seedlings.

Hypocotyl length has been studied in soybean. Studies show that inhibition of hypocotyl elongation increased as seeding depth increased, perhaps due to soil resistance and a temperature response. Burris and Fehr (1971) ranked cultivars for final hypocotyl length and reported that at 25°C hypocotyls of 'Amsoy' averaged about 90 mm while those of 'Lindarin' and 'Hawkeye' were about 130 and 160 mm in length, respectively. Further, long and short types were indistinguishable up to 7 days after germination began, suggesting that duration of growth period and not rate of growth gave the differences.

ROOT DEVELOPMENT

Roots are vitally important in seedling development, both as a means of anchorage in the soil and for the absorption of minerals and water. A firm seedbed helps to insure that these requirements are met, especially in supplying adequate water for cell elongation and mineral uptake. As the seedlings mature into plants, the roots of perennial plants also serve as a storage site for carbohydrates and other energy constituents.

Plants are often classified as fibrous-rooted or tap-rooted. Root systems of cereals and other grasses are fibrous. Such roots are slender, fine, and fiberlike, and no one root is more prominent than the others. Plants with a root system consisting of a prominent main root, such as found in red clover, sweetclover, sugar beet, alfalfa, and flax, are known as tap-rooted. Species with a taproot generally can penetrate more deeply into the soil than species with fibrous roots.

During germination the radicle, a part of the embryonic plant, emerges as the primary root. Branch roots develop from the pericycle within the primary root, and these roots are called secondary roots. In tap-rooted plants the primary root usually remains the principal root throughout the life of the plant.

Developing roots of most higher plants tend to follow similar patterns, insofar as the structures comprising each root are the same (Fig. 4.4). The root cap of a growing root is at the tip, and the cells of this structure are constantly being sloughed off. They are replaced by activity of the apical meristem located directly behind the root cap. The root cap protects the growing region as it is forced through the soil, and its absence would be very injurious to the apical meristem. Cells of the apical meristem are actively dividing and adding new cells to the root cap and to the zone of elongation (Fig. 4.4). There the cells enlarge, resulting in increased root length. The zone of maturation is composed of cells that have already enlarged and are differentiating into specialized functions.

Root hairs are found in the zone of maturation. These extensions of the epidermal cells are the principal organs for the absorption of water and minerals. For most species, the life of a given root hair is very short, lasting for only a few days. As the root develops in the soil, new root hairs are formed, aiding greatly in bringing the root surface into contact with newly reached soil particles.

Epigeal- and hypogeal-emerging dicots have a taproot system that remains relatively intact throughout the life of the plant. In contrast, the fibrous roots of the hypogeal-emerging monocots arise adventitiously from lower internodes and are constantly being replaced. These roots are called adventitious roots. Generally, all roots arising from organs other than roots are known as adventitious. This includes the seminal, coronal, and brace roots found in corn.

In contrast with tap-rooted plants, grass seedlings have a temporary root system before the major mass of permanent roots is developed. Figure 4.12 (a) shows the radicle emerging, followed by development of three to five seminal roots that arise from the base of the first internode. These roots are generally near the soil surface, especially in species in which the first internode is active and has elongated. This characteristic makes close and deep cultivation of corn plants hazardous, as pruning of these roots greatly reduces the capacity of the young seedling to absorb minerals and water. By the time the seedlings reach stage (d), the permanent coronal roots are emerging from the bases of the second through sixth internodes. As the seedling progresses in size, the coronal roots continue to grow rapidly while the primary and seminal root systems deteriorate and die. Thus, the grass seedling develops from total dependence on a temporary root system to total dependence on the permanent root system.

Even though they are temporary, the significance of the seminal roots should not be overlooked. These roots are critical to water and mineral absorption early in the life cycle. Furthermore, most corn planters place starter fertilizer about 5 cm below and 5 cm to the side of the planted seed. The seminal roots, which tend to grow at an angle to gravity, intercept the fertilizer band quickly, allowing efficient mineral uptake early in the life cycle when seed supplies of essential elements are becoming exhausted.

Factors such as soil moisture content, available fertility, soil aeration, and soil structure can greatly influence root development. Generally speaking, best root growth occurs when soil moisture is near field capacity, soil fertility is near optimum, and oxygen availability is sufficient for normal aerobic respiration. Respiration is essential for the release of energy useful in performance of work, such as the absorption of minerals from the soil solution and the enlargement of cells. Soil structure is also critical, as poor soil structure can impose a shortage of oxygen and cause a physical barrier to the development and growth of the roots.

Roots of plants have not been researched to the same degree as above-ground tissues and systems. This is largely due to their underground position, which makes it difficult to study them in a natural condition. Nevertheless, the plant is extremely dependent on its root system for survival and productivity.

VEGETATIVE SHOOT DEVELOPMENT

The terminal meristem is always located at the tip of the stem and has primary responsibility for initiating leaves, developing axillary buds, and laying down nodes and internodes. In dicot seedlings the terminal meristem also develops cells involved in elongation of the internode. However, in most grasses the internodes do not elongate until the plant is nearing reproductive growth, at which time the terminal develops into the inflorescence, and the intercalary meristem at the base of each internode becomes active. That meristem develops and elongates cells to elongate the internode.

Grasses and dicot crop plants differ slightly in the organization and phyllotaxis, or leaf positioning, of the terminal meristem (Fig. 4.13). Terminals of grasses have a distichous phyllotaxis, which means that leaves originate on opposite sides of the apical dome with only one leaf per node. Dicotyledonous crop plants usually have spiral phyllotaxis, which means that one primordia per node is initiated in a rotating manner around the apical dome. This system is then subclassified according to the angles between successive primordia or the ratio between number of leaves and number of revolutions needed to obtain a new leaf directly above one formed previously. Thus, in grasses new primordia develop 180° from the previous one, while in the dicot subterranean clover (*Trifolium subterraneum* L.) the angle is about 165°.

A platochron index is often used to describe and identify specific stages in the development of terminal meristems or shoots. A plastochron is defined as the time interval between initiation of any two successive leaves. More broadly, it can be defined as the time interval between any two leaves at a given reference stage when it is called a phyllochron. The exact time of leaf initiation, or the point when the protuberance from the meristem is called a leaf instead of a primordium, is difficult to assess and is most often described as the time when the primordium reaches 10 mm in length. More commonly, phyllochron development of shoots of agronomic crops is described using a defined point of leaf maturity. Examples include the day of collar formation in grasses, or the day of a specific stage of leaf maturity in dicots.

Time between maturation of successive leaves is affected by temperature, with longer durations occurring between leaves when temperatures are below or above optimum. Typically, when actively growing at near optimum conditions, cool-season grasses such as perennial ryegrass and small grains produce a fully developed leaf every 6–9 days. In contrast, corn and other warm-season grasses produce fully developed leaves every 4–6 days,

Trifolium subterraneum

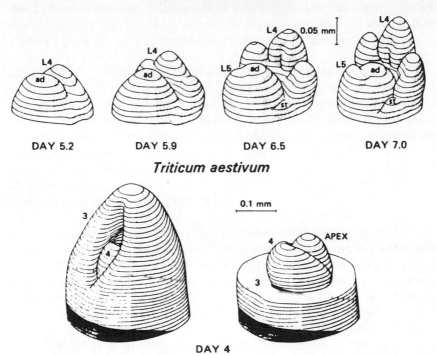

DAY 5.2   DAY 5.9   DAY 6.5   DAY 7.0

Triticum aestivum

DAY 4

Fig. 4.13. Three-dimensional drawings of the shoot apex of subterranean clover (above) showing changes during the plastochron between development of primordium of leaf four (L4) and leaf five (L5). Ages of the shoot apex are given to the nearest tenth of a day, ad = apical dome, st = stipule. Note the initiation and development of the lateral leaflets and stipules on the shoulders of the primordium of the terminal leaflet. Drawing of the shoot apex (below) shows leaf three (3) encircling the shoot apex and much of leaf four (4). Cutting away leaf three (right) exposes the apex where the primordium for leaf five will be initiated opposite leaf four. Contour lines are 0.01 mm apart. Reproduced with permission from Williams (1975).

and some dicots such as soybean can produce a trifoliate leaf every 3–4 days. Other environmental factors, such as soil fertility and light intensity, usually have much less dramatic influences on phyllochrons than they do on ultimate size and weight of the leaf.

LEAF GROWTH

During leaf development, a primordium on the terminal meristem begins to enlarge by cell division to form a cylindrical protuberance (Fig. 4.13). When this protuberance is about 1 mm long in species such as tobacco, it consists of the central axis (petiole and midrib) of the leaf. Then certain meristematic cells on the margins begin to grow, and the developing leaf takes on a more flattened appearance (Fig. 4.14). These active cells, and the edge they form, become the marginal meristem that will form the blade or lamina of the leaf. Tip growth of the central axis stops when the length approaches 3 mm, and all subsequent development is by the marginal meristem.

Marginal meristems consist of two types of initials or dividing cells. The marginal initials divide anticlinally (at right angles to the surface of the leaf) to form the surface layer or epidermis. Below the epidermal layer are a series of submarginal initials (Fig. 4.14) that form the inner tissues. Depending on the number of submarginal initials, a definite number of cell layers is formed in the leaf. Further, under any given set of environmental conditions, this number of layers will remain relatively constant. This occurs because the initials for each layer tend to divide anticlinally, so the area of the leaf, but not the number of cell layers making up the thickness, increases steadily.

While the previous discussion gives the mechanism for leaf growth, it does not allow for the development of different cells and the characteristic arrangement of cells in the leaf (Fig. 4.15). The arrangement of cells and air spaces results from meristems ceasing cell division at different stages. In the upper epidermis of tobacco leaves, cell production stops when the leaf is about 16–20% of its final size. The cells continue to enlarge, however, until the leaf reaches its final size. In contrast, palisade cells continue to divide and so give rise to many tightly packed cells in the developing leaf. They cease dividing and enlarging just before the leaf reaches its final size, so that in the end they are pulled apart slightly, giving rise to the intercellular spaces characteristic of mature palisade cells. Cells of the spongy mesophyll stop

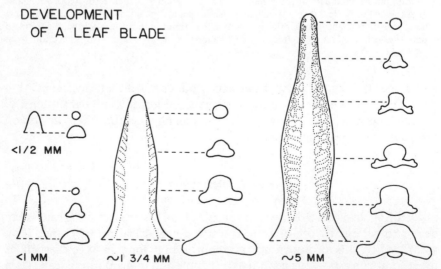

Fig. 4.14. The leaf primordium of tobacco first appears as a cone-shaped appendage from the shoot apex (upper left). Before the primordium is 1 mm long it begins to flatten, and before it is 2 mm long there is early development of main lateral veins. When about 5 mm long there is conspicuous development of the provascular system. Adapted with permission from Avery (1933).

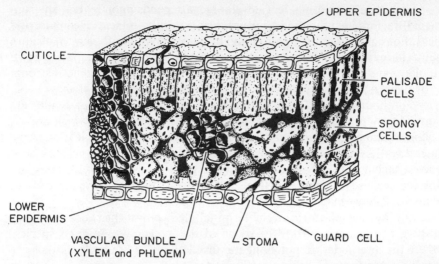

Fig. 4.15. Cross section of a leaf of a dicotyledonous plant showing component parts. Reprinted with permission from Mitchell (1970).

PRIMORDIA DEVELOPMENT

■ SUBAPICAL INITIAL

☐ PROCAMBIAL CELLS

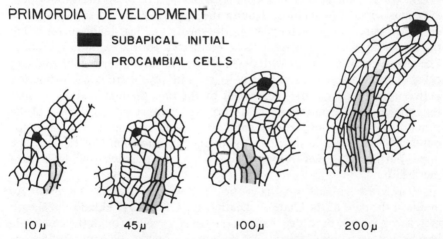

10 μ   45 μ   100 μ   200 μ

Fig. 4.16. Developmental stages of a leaf primordium of flax showing procambial cells and the subapical initial cell that develops cells of the primordium. Adapted with permission from Girolami (1954).

dividing and enlarging earlier than cells of the palisade layer, giving rise to their larger intercellular spaces and more irregular arrangement.

Vascular bundles are formed in the leaf from the procambium (Fig. 4.16), formed near the base of the primordium when it is very small. As the primordium develops the cylindrical shape, the procambium grows toward the tip of the new leaf to form the midvein and toward the vascular tissue of the stem to form a connection. The procambium further develops the distinguishable phloem and xylem tissues. Once xylem and phloem have differentiated and the marginal meristem becomes active to form the leaf lamina, lateral vein growth occurs toward the leaf margins (Fig. 4.14). Later, the fine network of veins appears.

Complete and normal leaf development depends upon several environmental factors including temperature, water supply, mineral elements, and light. For light, quantity, quality, and duration have an influence. An entire subject area, photomorphogenesis, has evolved that relates shape and form of plants and plant parts to light. Leaf growth tends to be influenced strongly by light. Auxin and cytokinin also play important roles in leaf growth.

Compound leaves, which have more than one lamina or blade, are formed very similarly to the way just described. In legumes, which are trifoliolate, the tip of the primordium develops into the central leaflet, while lateral leaflets develop from the shoulders on each side (Fig. 4.13). Lobed leaves, such as those of cotton or maple (e.g., *Acer saccharum* L.) trees, are formed because marginal meristems near the vascular tissue are more active than those between major veins.

During each plastochron the terminal meristem of the shoot continues to be active in laying down the tissue of the internodes. In grass species, where the internodes do not elongate until late in the life cycle, initials are left behind to form the intercalary meristems.

Leaf growth of grasses occurs in a manner different from dicots (Fig. 4.13). When primordia first appear as a protuberance on the shoot apex, they must first grow laterally around the apex or dome to completely encircle the axis. This growth results in the sheath and leaf blade surrounding the apex and causes the sheath to be connected entirely around the node. Following initiation, all growth becomes restricted to the zone of dividing cells, or intercalary meristem at the base. Initially, there is no distinction between the blade and the sheath, but by the time the leaf is about 10 mm long the ligule, a membranous projection at the inner side of the leaf at the junction of the blade and sheath, begins to form, and a layer of parenchymatous tissue develops to divide the intercalary meristem. The upper portion continues to form blade tissue, and the lower will later form the sheath.

The grass leaf matures from the tip downward, as the meristem remains at the base of the blade or sheath (Fig. 4.17). As the blade and sheath increase in length, the older tissue at the tip is forced through the coleoptile or whorl of developing leaves. Leaf growth stops when the ligule (collar) has emerged. Attempts to stimulate further growth of leaf meristems of various grass species by partially removing leaf blades have failed if the ligule had emerged. Growth of the leaf sheath, however, tends to be less dependent on ligule exposure and is apparently governed by another means.

It has been suggested that the change in the leaf environment from filtered radiation in the whorl, which is mostly far-red, to normal radiation upon exposure to direct sunlight may have a governing influence on growth of the blade and sheath. However, this hypothesis has not been firmly substantiated. Even so, length of a given grass leaf tends to be related to the length of the whorl through which it must grow. For that reason, leaf length of corn and other grain crops tends to increase during successive stages of seedling growth.

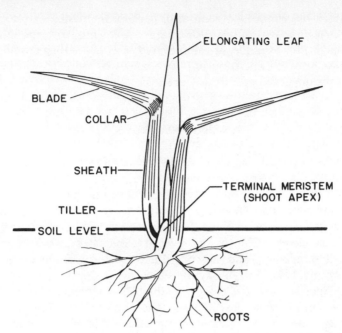

Fig. 4.17. Fully expanded leaves of a grass plant consist of the blade, collar, and sheath. A ligule and auricles (not shown) may also be present. The sheaths have been partially removed to show the base of the elongating leaf that is attached to the unelongated stem just below the shoot apex. A second elongating leaf is shown that would be encircled within the exposed elongating leaf, but attached to the stem one node above. An axillary bud is located at the base of each leaf and consists of meristematic cells that are capable of developing into tillers. Roots develop adventitiously from the bases of the lower, unelongated internodes.

Successively formed leaves of grasses are usually wider. This occurs because there are more columns of cells in the leaf and partially because of wider cells. Leaf width of grasses is usually much less influenced by environment and management practices than is leaf length. Therefore, large changes in leaf area are usually associated with length changes.

In crops such as wheat, barley, and oat, usually two leaves are already initiated in the ungerminated seed, and seven to nine leaves are formed during the life cycle. In spring, the plastochron of cereals is about 2–3 days, so that the flag leaf may be initiated about 15–20 days after germination of spring-sown cereals or after growth of autumn-sown cereals is reinitiated in the spring. Mature grain sorghum has 10–14, and mature corn 16–20 leaves. In corn, five or more leaves are already initiated in the seed, the sixth plastochron between successive primordia is less than 5 days, and successive ones become shorter. Those just preceding tassel initiation may be as short as 0.5 day. Obviously, temperature affects these rates; thus, plastochron changes will be markedly influenced by such management practices as time of year for establishment and depth of seed placement.

The time interval between appearance of fully developed leaves is usually considerably longer than the primordia plastochron. Even though all leaves start as primordia of similar or perhaps increasing size, successive leaves grow less rapidly, but develop a larger size. Thus, even though leaves of a corn plant, for example, are all initiated early in seedling development, the duration between initiation and attainment of full size will be much longer for the upper than for the lower leaves.

ENERGY SOURCES

Many biochemical changes occur as the seedling emerges and becomes autotrophic. This is particularly true of the photosynthetic and nitrogen metabolism systems. Chloroplast development begins early in the ontogeny of the primordium, and the plastids continue to grow and differentiate as the leaf grows. Increases in chloroplast number tend to follow increases in leaf area very closely. Chlorophyll level also increases as the chloroplasts are exposed to radiation, and this contributes to the increasing development of dark green foliage. Enzyme synthesis occurs within a few hours of exposure of light so that the leaf is capable of supporting photosynthesis. Ribulose 1,5-bisphosphate carboxylase, the primary fixation site for CO_2, reaches maximum activity at about the time the leaf is fully developed.

In germination and early stages of growth, all energy must come from the endosperm or cotyledon storage sites. Thence, an orderly transition occurs as the leaf tissue emerges and develops. During that time current photosynthate gradually provides energy for growth as seed storage becomes depleted. Cooper and MacDonald (1970) showed that even though corn plants had the third leaf emerging from the coleoptile on day 6 (Fig. 4.18), plants kept in the dark and those exposed to light showed no difference in weight of the embryo axis until after day 10. At that time the endosperm was nearly exhausted, as indicated by the dark-grown control. This suggests that even though leaves are exposed to light, the endosperm is the dominant energy source for growth, even after emergence. Another interesting relationship from this study was that both the growth of roots and shoots and the endosperm depletion were nearly linear from days 2–10. Over this range, about 0.65 mg of root and shoot growth occurred for each 1.00 mg of endosperm depletion. This compares very closely with the 70–75% efficiency that is considered to be the maximum for carbohydrate conversion to new growth. The remaining energy is utilized by respiration.

Photosynthetic rate per unit area of the developing leaves was about 650 mg CO_2 m^{-2} s^{-1} on day 9 when leaf area per seedling was 2.2 cm^2 (Fig. 4.18). By day 10 it increased to about 1220, and then leveled off at about 970 mg m^{-2} s^{-1} for days 11 to 13. This again illustrates the relatively rapid synthesis and activation of photosynthetic enzymes when exposed to light, and how quickly photosynthesis can become the dominant energy source. In an experiment comparing normal corn kernels with kernels having part of the endosperm excised, photosynthetic rate was not greatly affected, but growth rate was slower. A larger proportion of the photosynthate was

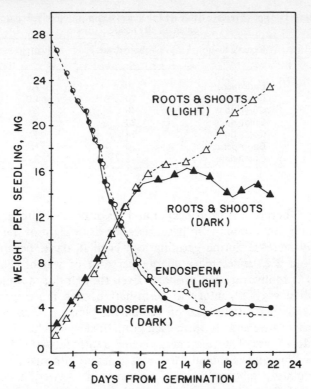

Fig. 4.18. A comparison of the weight decrease of endosperm and simultaneous increase of roots and shoots of a corn seedling grown in light and dark. Note that changes were similar for light and dark until day 10 even though the first true leaf emerged from the coleoptile on day 6. Adapted with permission from Cooper and MacDonald (1971).

translocated to the roots of plants established from seed with excised endosperm. Presumably, stored energy rather than current photosynthate was used preferentially for root growth in normal plants, and thus photosynthate was used for increasing leaf growth to allow greater energy capture. Other research shows that developing leaves use most of their photosynthate for their own growth. When leaves become fully formed, translocation to other growing areas becomes dominant.

In seedling growth, about 25–30% of the energy that is translocated to the meristem is lost as respiration for doing work, and the remaining 70–75% is incorporated into structural and metabolic tissue. It is well known that nitrogen and phosphorus influence seedling vigor. This response, however, does not appear to change the efficiency, but apparently alters the rate of breakdown, translocation, and incorporation of carbon into growth.

In contrast with plants that store mainly starch in the seed, plants that utilize stored oils and proteins require extensive interconversions of metabolites, since the chemical composition of new cells is markedly different from that of the storage material (Table 4.1). For example, castorbean stores 60–70% lipid in its seed, which must be converted to sucrose in the

Table 4.1. Type of reserve tissue and chemical composition of selected seeds on an air-dry basis.

Species	Reserve tissue	Carbohydrate	Protein	Lipid
		%		
Maize	Endosperm	70	10	5
Wheat	Endosperm	70	13	2
Pea	Cotyledons	40	20	2
Peanut	Cotyledons	20	25	45
Soybean	Cotyledons	14	37	17
Castorbean	Endosperm	0	18	64
Sunflower	Cotyledons	2	25	50

storage tissue before it is translocated and resynthesized into new tissue of the developing embryo axis. In these cases the dry weight of the total plant can actually increase during germination, even in dark, because the oils contain about 2.25 times more energy per unit of weight than the predominantly carbohydrate end product. Even though this storage form has an apparent energy advantage in germination, one must recognize that during development of the seed on the parent plant, it took considerably more carbohydrate energy to form each gram of seed weight.

Breakdown rate of storage energy is apparently mediated by growth of the new seedling and is thought to be associated with metabolism of the growth regulator, gibberellin. For example, if cotyledons of pea are removed from the growing embryo axis, enzyme activity of the cotyledons will not increase as it would if left intact. Moreover, senescence of the detached cotyledons is delayed markedly.

ENVIRONMENTAL INFLUENCES ON EMERGENCE

Seedlings are largely in the soil or close to the soil; thus, the soil environment has a major role in growth and development. Growth of tops, however, is also influenced by the aerial environment, including air temperature, relative humidity, water and nutrient supplies, and the gaseous environment. Root growth is influenced by similar factors, but is influenced more than the top growth by the gaseous environment. An additional factor influencing root growth is the physical constraint of the soil.

During initial development before the seedling has emerged, the soil environment is the dominant influence. Once the seed germinates and the radicle emerges to form the primary root, environmental forces tend to influence rate of development. Root growth depends on cell elongation, which depends on turgor pressure, but instead of the simple formula shown earlier for leaf or stem growth, a third factor must be added. For root cell elongation, the turgor pressure must be high enough to overcome the constraint of both wall pressure and resistance of the soil medium.

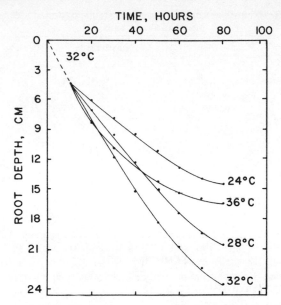

Fig. 4.19. Influence of time and soil temperature on growth of cotton root tips. Adapted with permission from Taylor et al. (1972).

The pressure needed to overcome the physical resistance of the soil is called root growth pressure. This pressure has been measured experimentally and found to be about 1.1–1.2 MPa for cotton and pea seedlings. However, when the gaseous environment of the soil was changed from 21 to 3% O_2, root growth pressure of cotton was reduced to about 0.5 MPa, illustrating that the environment has a dramatic effect on root penetration. Undoubtedly, O_2 influences on respiration were partially responsible for this response.

Each plant species has a minimum, optimum, and maximum temperature for root penetration and growth. Above the minimum, growth is almost linearly increased to the optimum for that species (Fig. 4.19). With further temperature increases, the rate decreases again. Since soil temperatures, at least in the depths where most seedlings are growing, are subject to diurnal variation, root growth probably also has a diurnal rhythm. Root growth of cotton was found to react in 15 min to changes in soil temperature, then tended to stay near the new rate for 40–60 hours (Fig. 4.19). These changes in growth rate are suspected to be related to metabolic activity of the root, which is mediated by enzymes.

Soil temperature is known to affect elongation rates of hypocotyls during emergence (Fig. 4.20). The 35°C optimum temperature for cotton is likely much higher than in cool-season crop species such as alfalfa or white clover, but the curve nevertheless suggests a mechanism for explaining rapid emergence in soils at optimum temperature. The 14°C minimum temperature for elongation of cotton hypocotyls is also higher than the 9–10°C re-

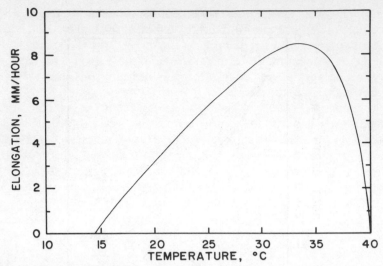

Fig. 4.20. Influence of temperature on hypocotyl elongation rate of cotton. Adapted with permission from Arndt (1945).

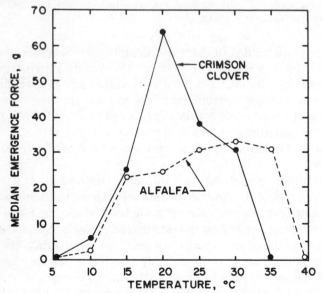

Fig. 4.21. Influence of temperature on force developed during emergence of seedlings of alfalfa and crimson clover. Adapted with permission from Williams (1963).

ported as minimum for corn coleoptiles or soybean hypocotyls. Most cool-season crops have minimum temperatures for emergence of 3–4°C, whereas they are 9–11°C for warm-season crops.

Emergence force exerted by the hypocotyl arch is also influenced strongly by temperature. Optimum temperature for developing emergence force for crimson clover was found to be 20°C, and the force was nearly double that of alfalfa at its optimum of 30°C (Fig. 4.21). At temperatures above or below optimum the emergence force for crimson clover decreased quickly. The exact mechanism responsible for temperature dependence of emergence force is not known, but it is suspected to be associated with cell wall metabolism and water uptake. In any event, this study suggests that seedlings at nonoptimum temperatures are more likely to have poor emergence because of crusted soil or highly resistant soils than they would be at optimum temperatures.

Available soil water is likely the major factor influencing seedling emergence and survival. An old generalization was that seeds should be planted deep enough to be in soil moisture. This rather crude understanding of seeding depth, nevertheless, did take into consideration the water environment. Seedlings most often fail due to lack of adequate water long enough for the seedlings to become established. Soil moisture tension has a direct effect on availability of water for uptake by plants.

Emergence of alfalfa is influenced strongly by soil moisture tension, which affects water availability. Triplett and Tesar (1960) reported that emergence was related inversely to tension, and that no emergence occurred at tensions greater than about 1.1 MPa. Pressing the soil around the seed to obtain better contact between soil and seedling improved emergence without irrigation (Fig. 4.22). Irrigation helped eliminate surface drying so that percent emergence and survival of seeds placed shallowly in the soil increased greatly.

Cultivars of several crops have been classified according to their ability to germinate and emerge at varying levels of soil moisture tension (Table 4.2). All cultivars of alfalfa germinated well at low osmotic pressures used to simulate soil moisture tension, but at higher pressures, and especially at

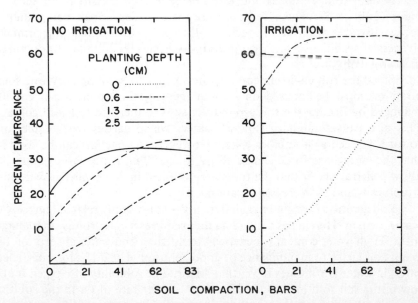

Fig. 4.22. Influence of planting depth, soil compaction, and irrigation on percent emergence of alfalfa. Note that firming the soil around the seed and irrigation had the most effect on seeds planted at shallow depths. Authors measured soil compaction in bars, a unit of force equal to 0.1 MPa. Adapted with permission from Triplett and Tesar (1960).

Table 4.2. Percent germination at 20°C for six alfalfa cultivars at three different osmotic
pressures. Adapted from Dotzenko and Dean (1959).

| Cultivar | Osmotic pressures in megapascals (MPa) | | |
	0	0.7	1.2
		%	
Buffalo	100	64	1
Vernal	93	58	2
Ranger	98	94	12
Lahontan	100	88	25
African	98	98	62
Caliverde	93	81	78
LSD 5%	ns	18	13

1.2 MPa, the cultivars differed. 'Buffalo', 'Vernal', and 'Ranger' are commonly grown in the Midwest, whereas 'Lahontan', 'African', and 'Caliverde' are adapted to the arid Southwest and germinated better at the greater stress level.

The soil environment influences mineral availability to germinating seedlings. Nitrogen is generally released rapidly from organic matter in well-oxygenated, warm soils, but it often must be applied as a chemical to get good vigor of young seedlings at low temperatures. Corn, for example, responds more to starter fertilizer applications in cold than in warm soils. Rhizobia require critical temperatures to infect root hairs, and nitrogen fixation rate is also temperature-dependent. These factors should be considered in the overall response of seedling vigor.

Several forage grasses and legumes have been classified for seedling vigor (Table 4.3). These responses become critical when compounding forage mixtures of different species. If the species differ greatly in competitiveness at seedling stages, the most vigorous species will tend to dominate in the final mixture.

Since the soil environment is so important to seedling survival, some attention must be focused on its management. Soil moisture is usually managed by firming the soil around the seed to increase soil-seed contact. This also causes capillary flow of water, which carries the soluble ions toward the seed as it imbibes water. However, compaction can be so great that the emergence force generated by the seedling is not large enough to allow penetration, or that the roots are inhibited in downward growth both structurally and by decreased aeration.

Soil aeration is related to soil porosity and is usually related inversely to water content. Herein lies a dilemma that requires a compromise in management. High water content is beneficial to nutrient and water absorption but is not conducive to maintenance of good aeration. The diffusion coefficient of O_2 in water is 10^4 times lower than in air. The solubility of oxygen is also low in the soil solution, so the effective transfer rate of O_2 in the solution phase is 10^5 times lower than through the air-filled pore spaces. In poorly aerated soils, ethylene and CO_2, both products of respiration, increase in

Table 4.3. Relative ranking of forage species from most competitive to least competitive during establishment. Adapted from Blaser et al. (1952).

Legumes	Grasses
red clover	perennial ryegrass
sweetclover	orchardgrass
alfalfa	tall fescue
alsike clover	bromegrass
ladino clover	meadow fescue (*Festuca pratensis*, Huds.)
birdsfoot trefoil (upright)	timothy
birdsfoot trefoil (prostrate)	redtop
white clover	Kentucky bluegrass

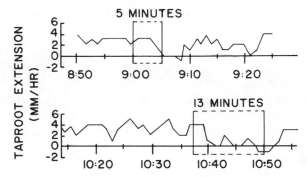

Fig. 4.23. Influence of O_2 deficiency on elongation rate of a cotton taproot. Conditions were imposed 2 days following initiation of germination by flushing the soil-root system with 100% nitrogen gas for the time duration noted. Adapted with permission from Huck (1970).

the gaseous environment. These gases tend to restrict root growth. Taylor and his co-workers (1972) reported that root growth of cotton seedlings through a low-density soil (low physical resistance to root growth) was not affected until O_2 concentration was decreased to less than 10%. In a high-density soil, root growth was reduced, but soil O_2 concentration had little effect between 21 and 1.2%. In the same soils, high CO_2 concentration inhibited growth in the low-, but not in the high-density soil. The authors concluded that soil resistance, and not the gaseous composition, was limiting root growth in the high-density soil.

Other research showed that the growth response of roots to change in O_2 concentration was rapid and reversible if duration was less than 30 min, after which root death in the region of elongation occurred (Fig. 4.23). When oxygen was restored, secondary root development above the dead portion led to continued root growth.

Selecting sites with good internal drainage and suitable soil structure to resist compaction and minimize surface crusting are probably the most important aeration management techniques. Placing the seed at reasonable depth and firming the soil around the seed to achieve good water relations without severely interfering with aeration are controllable factors that also markedly affect seedling survival.

The management of air and soil temperature has probably received more research attention than most other environmental factors. Organic mulching with crop residues, or living mulches like dormant crops can be used to lower soil temperature. Surface soil temperature in spring can be increased by ridging, or by tillage to reduce the mulch effect. Irrigation water can be used to alter soil temperature. Both cooling and heating can be accomplished, the latter taking on new significance with the advent of water-cooled nuclear reactors for power generation.

Managing seedlings for optimum emergence and survival to the autotrophic state depends greatly on the genetic potential of the species or variety. Once the genetic potential is understood, the management know-how of the agriculturalist helps to approximate the optimum environment and allow for maximum expression of that potential.

REFERENCES

Arndt, C. M. 1945. Temperature-growth relationships of the roots and hypocotyls of cotton seedlings. Plant Physiol. 20:200–220.

Avery, G. S., Jr. 1933. Structure and development of the tobacco leaf. Am. J. Bot. 20:565–592.

Blaser, R. E., W. H. Skrdla, and T. H. Taylor. 1952. Ecological and physiological factors in compounding forage seed mixtures. Adv. Agron. 4:179–216.

Burris, J. S., and W. R. Fehr. 1971. Methods for evaluation of soybean hypocotyl length. Crop Sci. 11:116–117.

Cooper, C. S., and P. W. MacDonald. 1970. Energetics of early seedling growth in corn (*Zea mays* L.). Crop Sci. 10:136–139.

Dotzenko, A. D., and J. G. Dean. 1959. Germination of six alfalfa varieties at three levels of osmotic pressure. Agron. J. 51:308–309.

Girolami, G. 1954. Leaf histogenesis in *Linum usitatissimum.* Am. J. Bot. 41:264–273.

Huck, M. G. 1970. Variation in taproot elongation rate as influenced by composition of the soil air. Agron. J. 62:815–818.

Larson, K. L., W. J. Russell, and C. J. Nelson. 1975. Agricultural plant science. Kendall/Hunt Publ. Co., Dubuque, Iowa.

Mitchell, R. L. 1970. Crop growth and culture. Iowa State Univ. Press, Ames.

Ray, P. M. 1972. The living plant. 2nd ed. Holt, Rinehart, and Winston, Inc., New York.

Roelefsen, P. A. 1965. Ultrastructure of the walls in growing cells and its relation to the direction of the growth. p. 67–149. *In* R. D. Preston (ed.) Advances in botanical research, Academic Press, Inc., New York.

Sunderman, D. W. 1964. Seedling emergence of winter wheats and its association with depth of sowing, coleoptile length under various conditions, and plant height. Agron. J. 56:23–25.

Tagawa, T., and J. Bonner. 1957. Mechanical properties of the *Avena* coleoptile as related to auxin and ionic concentrations. Plant Physiol. 32:207–212.

Taylor, H. M., M. G. Huck, and B. Klepper. 1972. Root development in relation to some soil physical properties. p. 57–77. *In* D. Hillel (ed.) Optimizing the soil physical environment toward greater crop yields. Academic Press, Inc., New York.

Triplett, G. B., and M. B. Tesar. 1960. Effects of compaction, depth of planting, and soil moisture tension on seedling emergence of alfalfa. Agron. J. 52:681–684.

Williams, R. F. 1975. The shoot apex and leaf growth. Cambridge Univ. Press, New York.

Williams, W. A. 1963. The emergence force of forage legume seedlings and their response to temperature. Crop Sci. 3:472–474.

SUGGESTED READING

Bannister, P. 1976. Introduction to physiological plant ecology. Halstead Press, John Wiley & sons, Inc., New York.

Bonner, J., and A. W. Galston. 1959. Principles of plant physiology. W. H. Freeman and Co., San Francisco, Calif.

Gemmell, A. R. 1969. Developmental plant anatomy. Edward Arnold Publ., Ltd., London.

Hillel, Daniel (ed.) 1972. Optimizing the soil physical environment toward greater crop yields. Academic Press, Inc., New York.

Maksymowych, Roman. 1973. Analysis of leaf development. Cambridge Univ. Press, New York.

Milthorpe, F. L., and J. D. Ivins. 1966. The growth of cereals and grasses. Butterworth, London.

Pierre, W. H., Don Kirkham, John Pesek, and Robert Shaw. 1966. Plant environment and efficient water use. Am. Soc. of Agron., Madison, Wis.

Russell, R. S. 1977. Plant root systems: their function and interaction with the soil. McGraw-Hill Book Co., New York.

Street, H. E., and Helgi Opik. 1970. The physiology of flowering plants: their growth and development. Am. Elsevier Publ. Co., New York.

Wareing, P. F., and I. D. J. Phillips. 1970. The control of growth and differentiation in plants. Pergamon Press, New York.

Wright, L. N. 1971. Drought influence on germination and seedling emergence. *In* K. L. Larson and J. D. Eastin (ed.) Drought injury and resistance in crops. Crop Sci. Soc. of Am. Spec. Pub. No. 2, Madison, Wis.

5

Photosynthesis, Respiration, and Photorespiration in Higher Plants

DALE N. MOSS
Department of Crop Science
Oregon State University
Corvallis, Oregon

Carbon dioxide (CO_2) from the air is the major raw material from which plants are made. Indeed, through the reduction of CO_2 and the formation of "energy compounds," the process of photosynthesis provides the organic raw materials and the energy for all the synthetic reactions in plants, which form useful food products such as carbohydrates, proteins and fats. Many factors that limit plant yield, such as lack of minerals or water, do so in part because they suppress plant photosynthesis.

Despite the obvious importance of photosynthesis to crop production, the relationship is not a direct one (Gifford and Evans, 1981). A crop with leaves having a high rate of photosynthesis will not necessarily have a high economic yield. On the other hand, a crop cannot yield well if it does not have adequate photosynthesis. In this chapter, some of the factors affecting the rate of photosynthesis in green plant tissues and in crop canopies will be discussed, as will the role of photosynthesis, respiration, and photorespiration in controlling primary productivity in plants.

Published in *Physiological Basis of Crop Growth and Development*, © American Society of Agronomy—Crop Science Society of America, 677 South Segoe Road, Madison, WI 53711, USA.

PHOTOSYNTHESIS

The process by which green plants are able to capture radiant energy and use it to produce sugar is extremely complex and, up to now, has not been reproducible in a test tube. Volumes have been written describing what is known about the process (for example, Rabinowitz, 1945, 1956), but consideration here must be limited. You can find extensive discussions of the various topics discussed here in the general references given at the end of this chapter (especially Burris and Black, 1976; Hatch et al., 1971; Johnson, 1981; Rabinowitch and Govindjee, 1969; Sestak et al., 1971; and Zelitch, 1971).

THE BIOCHEMICAL REACTIONS OF PHOTOSYNTHESIS

The chemistry of photosynthesis can be summarized as a process by which water and carbon dioxide (CO_2) react to form carbohydrates. This process takes place in the presence of chlorophyll, the green pigment in plant leaves, and requires light energy to drive the synthetic process. The energy-capturing steps are known as the light reactions or photoreactions of photosynthesis. The steps involved in reducing CO_2 and making sugar are referred to as the dark reactions, since they can proceed in darkness once the energy is captured by the light-harvesting molecular networks and stored in appropriate compounds.

The Photochemical Reactions

The capture of light requires the intimate association of many chlorophyll molecules with compounds that catalyze chemical reactions (Arnon, 1960; Govindjee, 1967; San Pietro and Black, 1965). The details of the process are not yet fully understood. It begins, however, by a packet of light energy[1] being absorbed by chlorophyll. When this happens, the chlorophyll molecule becomes excited. This excited chlorophyll molecule is capable of

---

[1] Radiation is the propagation of energy through space. For many applications, the electromagnetic spectrum of radiation can be described as waves that have a constant velocity but vary in length and frequency. The short wavelength of the spectrum is radiation described as gamma rays or the slightly longer x rays. Light is visible radiation. It has wavelengths betwen 400–700 nm. Each color has its specific wavelength.

For photochemical reactions, the wave property notation for radiation is inadequate. In chemical reactions driven by radiant energy, studies show that radiation consists of packets of energy called photons or quanta. The energy of a photon is proportional to the frequency of the radiation and, therefore, the wave and packet terminology can be interrelated. Photons are discrete units and cannot be divided. Only photons containing energy equal to or greater than that needed to drive a reaction can cause a chemical reaction to occur.

Wave properties of radiation are described by the equation $C = \lambda f$, where C = velocity and is constant for all wavelengths (3×10^8 m sec^{-1}), λ = wavelength, and f = frequency. The energy of a photon is described by the equation $e = h\nu$, where h = Planck's constant and ν = frequency.

doing work, in the chemical sense that it can pass on its energy to other compounds. In the chloroplast, chlorophyll molecules are arranged in linked networks, wherein some 200 molecules feed energy into a reactive center. Two such reactive centers, which must act sequentially, are involved in the process of capturing light energy and storing it in energy-rich molecules that can be used by the plant in synthetic reactions.

The photoreaction sequence of photosynthesis is illustrated in a simple diagram in Fig. 5.1. When a photon is absorbed by the network of chlorophyll molecules known as photosystem II and transmitted to a reactive site, a physical-chemical reaction takes place between the captured energy and the water molecule. The water molecule is cleaved, oxygen is released, and an electron is freed and raised to a higher energy state. This capture of energy by chlorophyll is illustrated in Fig. 5.1 as the vertical jump by the electron to a more negative electromotive force (emf) in the schematic electron pathway. The photon is symbolized as $h\nu$.

Visualize that light has been absorbed by photosystem II, an electron has been activated, and O_2 has been evolved. As the activated electron moves along the series of electron carriers between the photosystem II and photosystem I networks, it loses some of its chemical potential energy. In that process, however, as the electron "flows downhill" on the chemical potential energy scale, one of the excited proteins catalyzes the reaction of

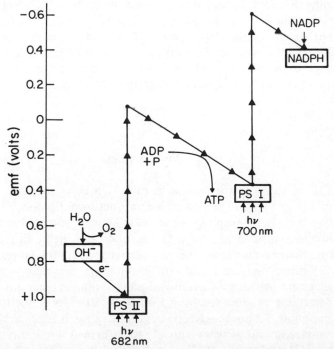

Fig. 5.1. A schematic representation of photosystem I (PSI) and II (PSII), as they function in photosynthetic energy capture. The number below each photosystem shows the maximum wavelength that will cause excitation of an electron by that photosystem.

adenosine diphosphate (ADP) with an inorganic phosphate ion to form adenosine triphosphate (ATP). Thus, part of the energy derived from the photon is transferred to a high-energy phosphate bond in ATP.

Subsequent to the formation of ATP, the electron passes to and activates photosystem I. A second packet of light energy is then absorbed, driving an electron to a still more negative emf. This electron then flows along a second series of carrier molecules that leads to the reduction of nicotinamide adenine dinucleotide phosphate (transformation of NADP to NADPH as shown in Fig. 5.1). The ATP and NADPH are the energy "currency" that drive the photosynthetic dark reactions.

The Dark Reactions

The chemical reaction by which CO_2 is reduced in photosynthesis involves a complex cycle, which is illustrated in Fig. 5.2. This is often referred to as the Calvin cycle, in deference to Dr. Melvin Calvin who first unraveled its intricacies (Bassham, 1962). In the Calvin cycle, NADPH and ATP are used to form the compound ribulose-1,5-bisphosphate (RUBP). In the presence of the enzyme ribulose bisphosphate carboxylase, RUBP reacts with CO_2 to form two molecules of 3-phosphoglyceric acid. Then, by a series of reactions involving compounds containing three, four, five, six, and seven carbon atoms, RUBP is reformed. Each time the cycle turns, a net amount of one carbon atom is incorporated into the plant.

Carbon moves into the Calvin cycle as CO_2 from the atmosphere and is drained off from the cycle by reactions of glucose and fructose to form sucrose. The sucrose is translocated from the chloroplast to cells throughout the plant and provides both energy and carbon for all other synthetic reactions within plants.

C_3 and C_4 Photosynthesis

The Calvin pathway for carbon in photosynthesis is found in green leaves of all higher plants. The first stable product from the fixation of CO_2 in the Calvin cycle is 3-phosphoglyceric acid. Thus, if you feed radioactive carbon dioxide to a plant leaf for a few seconds and then quickly kill the leaf to stop further reactions, the radioactivity is found largely in the compound 3-phosphoglyceric acid. In recent years, however, studies revealed that certain species have additional photosynthetic reactions in which the first detectable product resulting from CO_2 fixation is not the three-carbon compound, 3-phosphoglyceric acid, but rather a four-carbon compound, oxaloacetic acid, which is quickly transformed into either malic or aspartic acid (Hatch et al., 1971). To distinguish between plants whose leaves have this adaptation and those that do not, it is customary to refer to

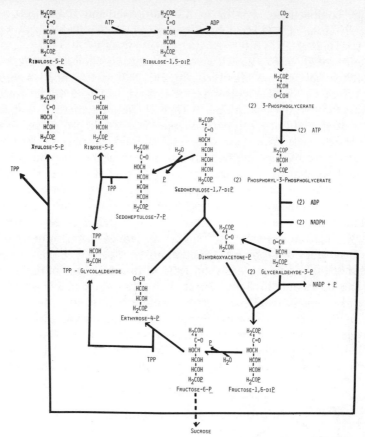

Fig. 5.2. Pathway of carbon in photosynthesis as derived by research of Calvin and his co-workers. Abbreviations: ADP, adenosine diphosphate; ATP, adenosine triphosphate; NADP or NADPH, oxidized or reduced nicotinamide adenine dinucleotide phosphate; TPP, thyamine pyrophosphate. The dashed line leading to sucrose is used to indicate that several reactions are involved in its synthesis.

species in which the Calvin cycle alone functions in photosynthesis as C_3 species. Likewise, the symbol C_4 is applied to species in which a four-carbon acid is the first stable product of CO_2 reduction.

Some crop species are C_3 types; other are C_4 (see Chapter 1). In optimum environments, the C_4 species such as corn, sorghum, and sugarcane are among the most highly productive crops.[2] Indeed, it was agricultural research workers, seeking to understand the high productivity of these crops, who discovered the C_4 photosynthetic pathway.

The C_4 photosynthetic system appears to be an evolutionary modification for increasing the CO_2 concentration at the functional sites of the carboxylating enzyme, RUBP carboxylase. To accomplish this concentration of CO_2, C_4 species have evolved a complicated CO_2 pump that features

---

[2] Scientific names of important crop plants are given in Table 1.1, Chapter 1.

a unique compartmentalization within the leaves and a division of labor among different cells. A schematic diagram of the biochemical reactions of the C_4 system is given in Fig. 5.3.

Leaves of C_4 species have a unique structural arrangement that permits the biochemical scheme illustrated in Fig. 5.3 to function. Anatomical structures of C_3 and C_4 species are compared in Fig. 5.4. This structure consists of numerous leaf veins, with each vein encased in a sheath or cylinder of specialized green parenchyma cells. These prominent, vascular bundle sheath cells have thick walls and contain numerous, large chloroplasts. It was recognized many years ago that these vascular bundle sheath cells had a peculiar physiological role when it was discovered that the only chloroplasts in corn leaves normally forming starch were in the bundle sheath cells. In the late 1960s, researchers at the Hawaiian Sugar Planters Association Experiment Station reported that if they fed CO_2 labeled with radioactive carbon to illuminated sugarcane leaves and then quickly killed the leaf, the labeled carbon was found almost exclusively in the four-carbon compound, malic acid. It was soon learned that CO_2 was initially fixed in sugarcane leaves only in the leaf mesophyll cells, which are attached to and radiate like fingers from the vascular bundle sheath into the internal air spaces of the leaf. The enzyme responsible for the initial CO_2 fixation in sugarcane was found to be phosphoenolpyruvate carboxylase (PEP carboxylase). This enzyme has a high affinity for CO_2 and is efficient at capturing it, even at the low CO_2 concentration found in the atmosphere.

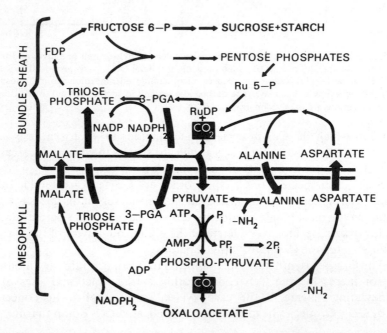

Fig. 5.3. A schematic representation of the CO_2 pump in C_4 photosynthesis. After Hatch et al. (1971). Used by permission.

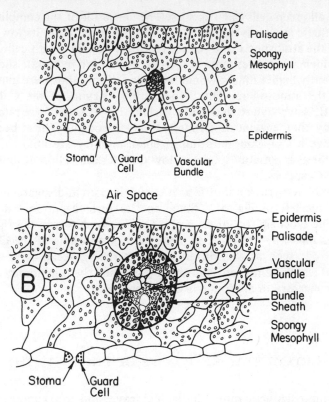

Fig. 5.4. Typical leaf cross-section of C_3 (A) and C_4 (B) species. The chloroplasts containing mesophyll cells have no particular orientation to the vascular bundles in C_3 species. In contrast, in C_4 species the vascular bundle is surrounded by a prominent sheath or collar of cells containing many large chloroplasts.

Nearly all the PEP carboxylase in sugarcane leaves is found in the mesophyll cells. These fingerlike cells present a large surface area to the air inside the leaf and thus present the smallest possible physical barrier to the entrance of CO_2. After the CO_2 is captured by the PEP carboxylase system, malic acid is formed (Fig. 5.3) and quickly transported to the bundle sheath cells. This transport apparently occurs by a special conduction mechanism, featuring numerous pit fields in the walls of adjacent mesophyll and bundle sheath cells through which the cytoplasms of the two cell types are connected.

Once the malic acid is inside the bundle sheath cells, it is decarboxylated by an enzyme called malic enzyme, and CO_2 is released. The remainder of the malate molecule is in the form of pyruvate, which is then shuttled back to the mesophyll cells. There it is phosphorylated to begin the cycle again.

The CO_2 released inside the bundle sheath cell is held within that cell by some unknown mechanism. All of the enzymes of the Calvin cycle are found within the bundle sheath cells and photosynthesis occurs there just as it does in all green cells in C_3 species. The net result of the complicated C_4 photosynthetic system is that CO_2 has been captured from its low concentration in the atmosphere and released into the bundle sheath cells. Experimental evidence indicates that the CO_2 concentration in bundle sheath cells is significantly higher than it would be if the cell solutions were in equilibrium with the atmosphere. The carboxylating enzyme of the Calvin (C_3) cycle, RuBP carboxylase, has a low affinity for CO_2, and the rate of CO_2 fixation by that enzyme is strongly CO_2 dependent. Thus, because C_4 species have a CO_2 concentrating mechanism, they are able to perform photosynthesis in an atmosphere of low CO_2 concentration at much faster rates than C_3 species.

The malate-pyruvate shuttle as it occurs in corn and sugarcane is illustrated to the left in Fig. 5.3. Some species form aspartate rather than malate. The biochemistry of aspartate formers is shown in the right-hand portion of Fig. 5.3. Although some slight modifications of the C_4 system have evolved, such as this difference in carrier molecules, the essential features of leaf anatomy, enzyme distribution, and the functioning of a CO_2 pump or concentrator are found in all C_4 species.

ENVIRONMENTAL RESPONSES OF PHOTOSYNTHESIS

The photosynthetic rate of individual leaves and crop canopies in both C_3 and C_4 species strongly depends on environmental conditions. Some of the more important factors are light, CO_2 concentration, temperature, mineral nutrients, and water.

Light

The light response curves of photosynthesis of some individual leaves are shown in Fig. 5.5. Note that, under high irradiance, corn leaves have an appreciably greater capacity to fix CO_2 than do orchardgrass leaves and that rates in orchardgrass are greater than in white oak (*Quercus alba* L.) or maple (*Acer saccharum* Marsh.) leaves. The orchardgrass response (Fig. 5.5) is typical of temperate forage grass species and such crops as wheat, barley, or alfalfa. These crops all have a higher capacity for photosynthesis than many woodland species from shady habitats; however, their leaf photosynthetic rates are significantly less than rates found in many grass species of tropical origin such as corn, sugarcane, or sorghum.

Both orchardgrass and oak have the C_3 system for fixation of CO_2, but the orchardgrass response is typical of C_3 field crop species. Likewise, the corn response shown in Fig. 5.5 is typical of the light response of all crop species having the C_4 system for CO_2 fixation.

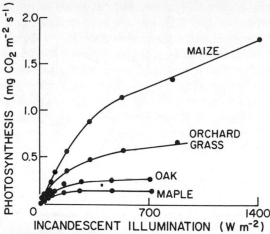

Fig. 5.5. Photosynthetic light response curves for leaves of several species. These measurements were made in a laboratory using incadescent lamps as an energy source. Adapted from Hesketh (1963).

Another important characteristic of the light response curves shown in Fig. 5.5 is the shape of the curve. As the intensity of illumination of an orchardgrass leaf increases from zero, the response evoked by a unit of light becomes progressively less until additional light evokes no additional uptake of CO_2. The leaf is then said to be light saturated. Out-of-doors, this condition of light saturation for a fully exposed leaf of a C_3 species occurs at about one-fourth of midday sunlight intensity. In striking contrast to this response by C_3 crop species, the corn photosynthetic rate continues to respond to increasing light up to the intensity of full midday sun, although the response is not linear. Investigation of this difference in response to light among crop species led to the discovery of the C_4 system for CO_2 fixation.

Although the capacity for photosynthesis of individual leaves is important for crop productivity, we must remember that crop growth is determined by many factors affecting plant photosynthesis. Obviously a plant with large leaf area may have the potential for greater growth than a plant with a smaller leaf area, even if, per unit leaf area, the second plant has the greater photosynthetic capacity. Likewise, any plant growing in the shade of another plant or any leaf in the shade of other leaves cannot fix large quantities of CO_2, regardless of its internal biochemistry, because it lacks the energy to drive the process. Thus, the rate of photosynthesis of all the leaves in a crop canopy is a complex function of the biochemistry and the availability of energy to drive the biochemical processes (see also Chapter 6).

In all crop canopies many leaves are fully or partially shaded or are at oblique angles to the sun. Photosynthesis in these shaded leaves will increase as the irradiation increases even though other leaves may be light saturated. Therefore, photosynthesis of all crop canopies is closely correlated with the intensity of solar radiation. This is illustrated for corn in

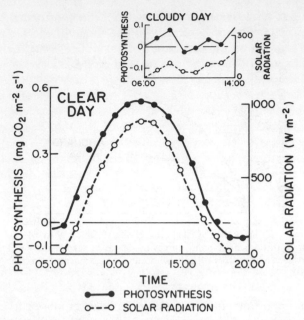

Fig. 5.6. Time course over a day for photosynthesis of corn for a clear day and, in the inset, for a cloudy day. Adapted from Moss et al. (1961).

Fig. 5.6, where the rate of photosynthesis of a crop canopy and the light intensity are shown for a clear and cloudy day.

Wheat, rice, alfalfa, sugar beet, and others have crop canopy light response curves similar in shape to Fig. 5.6. The magnitude of the response and the area beneath the curve differ for these crops, however. Likewise the curve differs for a given crop in different environments. The magnitude of photosynthesis of the crop canopy may depend on temperature, on the nutrient status of the plants, or on leaf area, for example. For each crop, you can visualize a family of possible daily photosynthetic curves, the height of each member of the family being controlled by the environment. And, of course, the area under the curves, which is proportional to the total daily photosynthesis, is likewise a function of the environment. Thus, although many factors influence the height of the midday peak, the crop canopy's response to light will still be a curve shaped like Fig. 5.6. Inadequate light restricts productivity in a nutrient deficient crop just as it does in a crop growing under optimal conditions, even though the maximum rate of photosynthesis may be dramatically different for the two crops.

Efficiency of Use of Light

Although light provides the energy to drive the photosynthetic reactions, a plant under nutrient stress may intercept and absorb the same amount of light as a plant under no stress. Yet the stressed plant may have appreciably less photosynthesis. In that case, the nonstressed plant has a higher efficiency of use of light. Since light is one of the factors limiting crop yields, it is important that crop systems utilize light as efficiently as possible. Thus, one of the questions researchers ask is, "What is the maximum efficiency possible for the photosynthetic process?"

As indicated in footnote 1, light consists of packets of energy called quanta or photons. The energy content of a photon is directly related to the frequency and inversely related to the wavelength of the radiation. Thus, a photon of short wavelength radiation contains more energy than one of long-wave radiation. Photosystem II requires radiation of wavelength 682 nm or shorter, while photosystem I will operate with radiation of wavelength 700 nm or less (Fig. 5.1). Any photochemical reaction requires the absorption of at least one photon. Photons are physical entitles and cannot be divided. As shown in Fig. 5.1, when a photon is absorbed by photosystem II and an electron ejected and transported to photosystem I, ATP is formed in the process. Then, a second photon must be absorbed for an electron to be raised to an energy content where NADPH can be formed. Thus, one photon is required for each ATP and each NADPH.

The photosynthetic carbon reduction cycle illustrated in Fig. 5.2 shows that the incorporation of one molecule of CO_2 into the plant as part of a sucrose molecule requires two NADPH and three ATP. Sucrose may be considered the end product of photosynthesis because it is the compound most commonly transported from leaves. Since a photon is required for the formation of each ATP and each NADPH, the minimum number of photons required to incorporate one molecule of CO_2 into the plant in the form of a simple sugar is five.

Plants do not consist only of sugars, however. Additional energy is required to transform glucose or sucrose into other compounds that make up the plant. These additional synthetic reactions require further ATP or NADPH molecules. Thus, the overall quantum requirement to incorporate 1 carbon atom into a plant is estimated to be 10.

Plants can be burned in a calorimeter and the amount of energy released in this combustion measured. The end products of combustion are CO_2 and water, the starting materials of photosynthesis. Calorimetry measurements show that, for the average reduction state of carbon found in plant materials, 4.4×10^5 J are released for each mole of carbon atoms combusted.

The energy content of a single photon is extremely small. For calculation purposes in photochemistry, the einstein, which is the energy of a photon multiplied by the Avagadro number (i.e., a "mole" of photons), is used. Ten einsteins of 500 nm wavelength radiation (the center of the visible spectrum) contain 22.8×10^5 J. Thus, if 10 photons are required to reduce 1 molecule of CO_2 to the average reduction state of plant material (or for calculations, 10 einsteins/mole), then we can see that the energy required for the process of photosynthesis and other associated synthetic reactions, compared to the energy stored in carbon bonds in plant material, is 4.4×10^5 J/22.8×10^5 J. Thus, the theoretical maximum efficiency of photosynthesis is about 20%. The actual efficiency of field crops is far from this theoretical maximum, however. Most crops store 1–2% of the solar radiation incident upon the crop during the growing season. There are a few instances of crops fixing as much as 10% of the incident solar radiation for short periods. This happens only during the phase of maximum growth of a crop or at low light intensities.

A hectare of corn in Iowa that produces 9.5×10^3 kg grain ha^{-1} (a good yield, but one attained by many farmers) will produce another 9.5×10^3 kg of stalk and leaves and about 3.8×10^3 kg of roots for a total production of 2.3×10^4 kg of dry matter per hectare. Each kg of dry matter contains approximately 17.6×10^6 J of energy. Thus, a hectare of corn fixes $(17.6 \times 10^6) \times (2.3 \times 10^4) = 4.0 \times 10^{11}$ J ha^{-1} during a single growing season. During that season (1 May to 15 October) for an average year, the sun's energy incident on an hectare in central Iowa is about 3.1×10^{13} J. Thus, an Iowa corn-field stores $(4.0 \times 10^{11})/(3.1 \times 10^{13})$ or approximately 1.3% of the solar energy incident on the field during the growing season. However, only about 45% of the solar energy is in the visible range (400–700 nm) effective in photosynthesis. Thus, a highly productive corn crop captures and stores in dry matter about 3% of the visible radiation incident upon the crop during the growing season. In addition to the energy stored, of course, an amount equivalent to about 25% of the crop is lost through crop respiration during the crop growth. Thus, perhaps 4% of the visible radiation is fixed in the photosynthetic process.

Obviously, even our most efficient crops utilize little of the energy available to them. Therefore, researchers are interested in increasing the efficiency of energy capture. Attempts have been made to select genetic lines with higher rates of photosynthesis, but little progress has been made in that effort. A number of agronomic practices have increased the efficiency of capture of solar energy, however. Varieties have been developed that utilize more of the growing season, thereby "keeping the factory going" for a longer time. Where the season permits, double cropping has become a standard practice because it is impractical to utilize a long growing season efficiently with a single crop. Early planting in the spring and use of narrow rows and high plant populations are practices that provide a maximum ground cover and result in light being intercepted by leaves instead of being wasted on bare soil. The result of such practices is to increase photosynthesis per unit of ground area.

Crop plants fail to function at maximum efficiency for many other reasons besides failure to intercept light. Crops may suffer from diseases or shortages of water or nutrients, insects may affect the fixation of CO_2, or light may be absorbed by nongreen tissues. Thus, although the potential to increase crop yield by increasing the efficiency of light utilization seems great, much research is needed to determine how greater efficiency can be achieved.

Carbon Dioxide

As discussed earlier, CO_2 is the major reactant in photosynthesis and the enzyme that catalyzes the reaction of CO_2 with RUBP has a relatively low affinity for CO_2. The rate of photosynthesis, therefore, is strongly dependent on the concentration of CO_2 (Fig. 5.7). The CO_2 concentration in the atmosphere is only 0.03%; therefore, in a closed space or in still air, an illuminated leaf uses the CO_2 in its immediate neighborhood rather quickly. Diffusion of CO_2 in still air is a slow process. It can be greatly facilitated by slight air movements. Thus, under certain experimental conditions, the rate of photosynthesis is associated with wind velocity. Under normal field conditions, however, leaves use CO_2 rapidly, only in intense light, and under intense irradiation, the leaf absorbs large quantities of energy and is warmed in the process. The warm leaf, in turn, warms the air in contact with it, causing it to rise and be replaced by cooler, heavier air, a process called convection or turbulence. This brings in new CO_2 by mass flow, a much more rapid process than diffusion. Thus, only on rare occasions is the rate of photosynthesis substantially aided by increased wind, because natural convection replenishes CO_2 rather quickly when the sun is shining intensely.

Crop productivity would be low without this turbulence found under field conditions during periods of intense radiation, which renews the CO_2 supply to leaves. Although convective processes maintain the CO_2 concentration in the air next to the leaf surface at a concentration close to that of the bulk atmosphere, the bulk atmosphere concentration of CO_2 is very low. Photosynthesis in intense light would be more rapid if the CO_2 concentration of the atmosphere were higher.

Although turbulence helps supply atmospheric CO_2 to the leaves, the normally occurring turbulence also precludes adding CO_2 to open field crops to get higher yields. The same turbulence that brings CO_2 to a crop

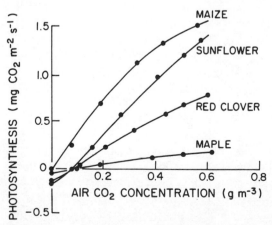

Fig. 5.7. Response of leaf photosynthetic rates of different species to CO_2. From Hesketh (1963).

surface under normal conditions will carry CO_2 away if it is fed to a field at higher than atmospheric concentration. Thus, there is no known way to take advantage in the field of the dramatic response of the photosynthetic mechanism to CO_2. In greenhouses, however, the atmosphere is confined, and its CO_2 concentration can be controlled. Enriching greenhouses with CO_2 is a common practice and several-fold increases in yield have been obtained in numerous crop plants such as tomato (*Lycopersicon esculentum* L.), lettuce (*Lactuca sativa* L.), cucumber (*Cucumis sativis* L.), or carnation (*Dianthus caryophyllus* L.).

Greenhouse crops are usually grown in the winter months, when light intensity is low. Even more dramatic effects from CO_2 enrichment would be expected under high light intensity, especially if crop varieties could be developed that take special advantage of the additional photosynthate resulting from the CO_2 feeding. Thus, it appears that additional photosynthesis could markedly enhance yield. Crop physiologists and plant breeders are interested in the possibility of breeding varieties that have a greater capacity for photosynthesis (see Chapters 6 and 11).

Temperature

Most crop species have rather broad temperature optima for photosynthesis, with the rate decreasing at temperatures either warmer or cooler than the optimum (Fig. 5.8). Corn and soybean have rather high temperature optima (30 to 35°C) and, in temperature climate zones, they grow well only in midsummer. In contrast, small grains and many forage grasses have temperature optima near 20°C, and are most productive in the spring. During midsummer, the temperate grasses become dormant. High temperatures early in the season result in low yields in crops like wheat. Thus, the correlation of photosynthesis and yield is high in these examples.

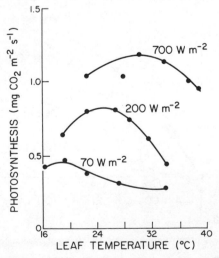

Fig. 5.8. Temperature response of photosynthesis of corn leaves at different light intensities. From Moss, D. N. (unpublished data).

Mineral Nutrients (see also Chapter 7)

One of the major effects of deficiencies of any essential mineral element is a sharp reduction in the rate of photosynthesis. Some elements, such as iron (Fe) and magnesium (Mg), are essential components of chlorophyll. An example of the curvilinear relationship of photosynthesis and Mg content of corn leaves is shown in Fig. 5.9. Evidence that this relationship is due to the effect of Mg on chlorophyll is shown in Fig. 5.10 (data from the same leaves), where the rate of photosynthesis is directly proportional to the chlorophyll concentration. Similar results have been found for nitrogen (N), which is required for protein synthesis and thereby affects production of chlorophyll and RUBP carboxylase.

Phosphorus (P) deficiency does not cause chlorosis, but P is an essential element in the biochemical steps of photosynthesis, and a lack of it results in severe inhibition of photosynthesis.

Potassium (K) is a key cofactor in many biochemical reactions. It is also an essential element in the osmotic process by which stomata open. In K-deficient leaves, the stomata fail to open, and CO_2 cannot get into the leaf. The effect of K on photosynthesis is shown in Fig. 5.11. The critical level for K is much higher than for Mg (compare Fig. 5.9). Potassium-deficient leaves may not be chlorotic and, in severely deficient plants, leaves that appear healthy may have no photosynthesis at all. Thus, a plant's appearance does not always indicate the severity of the effect of a mineral nutrient deficiency on physiological processes or on potential crop yield.

Plants deficient in minerals usually have small leaves, and the older leaves die more rapidly. Thus, nutrient-deficient crops have low leaf area indices (LAI) and reduced photosynthesis. The lack of minerals also affects yield in other ways, but the low rate of photosynthesis in deficient plants is certainly one of the major reasons for their low yield.

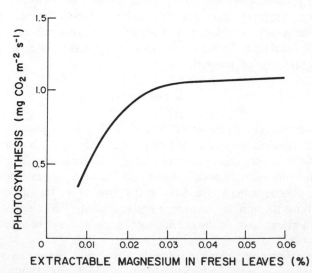

Fig. 5.9. Response of photosynthesis of corn leaves to leaf magnesium content. Adapted from Peaslee and Moss (1966).

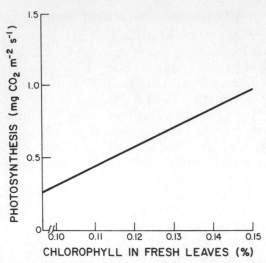

Fig. 5.10. Relationship of photosynthesis of corn leaves to chlorophyll concentration for leaves differing in magnesium content. Adapted from Peaslee and Moss (1966).

Water

From a worldwide perspective, perhaps no other factor causes greater yield losses in crops than lack of water. Severe water stress results in death of plants, but low yields in situations where plants are stressed but not killed are common in most environments (see also Chapter 8). The most common effect of water stress is stunted plants. Stunted plants have a severely reduced leaf area, greatly reducing the size of the photosynthetic factory. In addition to reduced size, moreover, water stress causes closure of stomata which, in turn, sharply depresses CO_2 uptake. Stressed plants produce less biomass because they are smaller and because the supply of raw material for synthesis is restricted. Thus, water stress suppresses yield, in large part, through its effect on photosynthesis.

Summary

From the foregoing discussion, it is clear that many factors controlling yield do so by controlling the rate of photosynthesis. It is also clear that photosynthesis is a complex process that depends on many different factors to function at its maximum rate. From studies of how environmental factors affect photosynthesis, we often learn how these factors affect yield. Thus, learning to optimize photosynthesis in controlled environments can tell us much about how we should manage crops in the field to maintain optimum growth.

As scientists understand more fully how plants grow, they will learn more about how energy can be utilized more efficiently. Using this knowl-

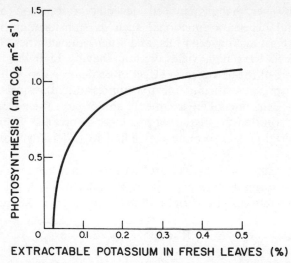

Fig. 5.11. Response of photosynthesis of corn leaves to leaf potassium content. Adapted from Peaslee and Moss (1966).

edge, it is likely that more efficient cultivars will be produced, and management practices will be devised to increase the efficiency of carbohydrate production and utilization.

RESPIRATION

Respiration is the process by which plants and animals utilize various carbon-containing compounds to convert energy to useful forms, such as the chemical potential energy ATP and NADH. Interestingly, the process is similar in plants and animals. The biochemical steps are complex and differ in number and reactions for carbohydrates and more reduced compounds such as fats or proteins. Many of the biochemical reactions in respiration were discovered in the 1930s by Krebs and his coworkers; the reactions by which sugars are oxidized to yield CO_2 and water often are referred to as the Krebs cycle or the tricarboxylic acid cycle. You may read in detail about these reactions in any biochemistry text. For our purposes, it is sufficient to think of respiration as the process in which plants recover the light energy fixed in carbon compounds during photosynthesis in a usable form when and where it is needed. In general, the process is:

$$CH_2O + O_2 \rightarrow CO_2 + H_2O + energy$$

This summary statement of respiration appears to be the inverse of photosynthesis. Sometimes this leads us into the trap of thinking that, if respiration could be suppressed, less stored photosynthate would be used; therefore, suppression of respiration should lead to higher yields. We must remember, however, that all plant cells depend on respiration for the energy to perform all necessary functions, such as translocating sugars, synthesizing proteins, making cell walls, and other vital activities. Therefore, it is impossible to have high yields without having high respiration. Undoubtedly, part of the detrimental effect of cool air or cool soils on growth during early spring is attributable to suppression of respiration by low temperatures. Cells in cool environments do not have the necessary energy available to permit energy-requiring processes to proceed. Thus, adequate rates of respiration are absolutely essential in order for plants to grow rapidly.

In recent years, many agricultural scientists have been interested in the relationship of respiration to plant growth. McCree (1970, 1982) evaluated respiration of white clover and fit his data to the linear equation:

$$R = kP + cW$$

where R is the rate of dark respiration of the plant, P is the rate of photosynthesis summed over the day (a function of the amount of light received by the plant and the size of the plant), W is the plant dry weight and k and c are constants. The constant k is dimensionless, but c has the dimensions time^{-1}. The best values for the constants were found to be 0.25 and 0.15 per day respectively when R and P were expressed as g (CO_2) day^{-1} m^{-2} and W in g (CO_2 equivalents of dry weight) m^{-2}. McCree assumed that the term kP represented the carbon loss associated with growth processes occurring in the dark, while the term cW represented the carbon loss to the plant associated with maintenance of the plant. Analysis of respiration in this way is sometimes referred to as quantitative biochemistry (Penning de Vries, 1975). It shifts attention from rate of respiration to efficiency of utilization of respiratory substrates. Perhaps when the requirements of plant growth for respiration are more fully understood, scientists may find ways to develop cultivars of crops that utilize respiratory substrates more efficiently. If so, it may be possible to breed crop cultivars adapted to adverse environments such as high or low temperatures (see Chapter 11).

The rate of respiration of plants increases rapidly as the temperature increases, approximately doubling for each 10°C increase. After the temperature rises to the point that the basic energy needs of the plant are being met, it is generally assumed that further temperature increases cause wasteful respiration and are harmful to the plant. Many crops do well at relatively cool night temperatures, and yields may be suppressed by extended periods of high nighttime temperatures, perhaps because of excessive loss of dry matter due to rapid respiration.

Because there is some evidence that respiration can be wasteful, some crop scientists speculate that yields might be increased by breeding varieties with low respiration rates. However, since genotypes of crop species differing in respiration rates have not been identified, this concept remains in the realm of speculation. A counter argument can be made, however, by comparing the respiration rates of different species. Many desert and shade plants have low rates of respiration. This adaptation permits them to survive in adverse environments lethal to most crop species because respiration occurs slowly, even at high temperatures. Therefore, sugar and other compounds are not burned up quickly, as would occur in most crop plants at high temperatures. These slow-respiring species characteristically have slow growth rates also. It appears that it may be necessary to have the ability to respire rapidly to have sufficient energy for rapid growth.

Another factor to consider in evaluating the role of respiration in plant productivity is that many different chemical compounds are formed during the process. These compounds are not only important links in the respiratory chain, but many of them serve as essential building blocks for other vital compounds in the plant. Thus, the respiratory process is much more complex than being merely the reverse process of photosynthesis.

PHOTORESPIRATION

At about the same time that high photosynthetic rates in corn and sugarcane leaves were discovered, biochemists were reporting that illuminated tobacco leaves synthesized a small, two-carbon compound, glycolic acid, in relatively large quantities. Experiments using radioactive CO_2 at low concentrations such as found in the atmosphere and at temperatures near 30°C showed that much of the CO_2 taken up by a tobacco leaf passes through the compound glycolic acid. Since glycolic acid was not an intermediate of the Calvin cycle (Fig. 5.2), it was believed that, in some plants, there might be another pathway for photosynthesis that had not yet been discovered, in which glycolic acid was a major product.

As biochemists studied these new discoveries; it became apparent that glycolic acid had a unique property—it could be oxidized extremely rapidly in plant tissue with at least half of the carbon converted to CO_2. Since glycolic acid was formed in some unknown way early in this photosynthetic process, it began to appear that certain plants had a "leaky system" for photosynthesis, in which part of the CO_2 fixed was rapidly converted again to CO_2, glycolic acid being an intermediate compound.

Since glycolic acid can be oxidized quickly by leaves in light or darkness, yielding CO_2, the biochemical oxidation of glycolic acid was termed respiration. Because significant quantities of glycolic acid could be found in leaf tissue only when a leaf was illuminated, the term *photorespiration* was applied to the generation of CO_2 from glycolic acid.

Photorespiration was found initially in tobacco, but appeared to be totally absent in corn. Another trait of green leaves that became important in unraveling the mystery of photorespiration was that the CO_2 compensation point—that concentration of CO_2 at which the CO_2 taken up in photosynthesis by an illuminated leaf just balances the CO_2 released in respiration—was low or near zero in all plant species having the C_4 pathway for photosynthesis. In contrast, species with the C_3 pathway had CO_2 compensation points near 50 ppm (v/v). It soon became apparent that C_3 species have photorespiration while C_4 species do not, and much of the early information classifying species into C_3 or C_4 photosynthetic types was gained by measuring the CO_2 compensation point for leaves of the species.

Photorespiration appears to be wasteful respiration. There does not appear to be any net gain in usable energy for the plant from the reactions leading to the release of CO_2. From that standpoint, plant scientists have been very interested in finding a way to diminish or inhibit photorespiration. Much of the research has been directed at determining how glycolic acid is formed initially in the plant. Some of that research by Ogren and his co-workers led to the discovery that the enzyme RUBP carboxylase, which catalyzes the reaction of CO_2 and RUBP, also catalyzes the reaction of RUBP with oxygen (Ogren and Bowes, 1971). When that happens, rather than CO_2 being fixed and two molecules of 3-phosphoglyceric acid being formed (Fig. 5.2), a single molecule of 3-phosphoglyceric acid and one molecule of glycolic acid are formed.

The discovery of the dual nature of the enzyme involved in CO_2 fixation in photosynthesis and in photorespiration explained many of the mysterious results of experiments on photorespiration and on the metabolism of glycolic acid. Since that discovery, scientists have made an intensive effort to find a chemical means of favoring the CO_2 fixation process and to prevent the formation of glycolic acid. Others have searched for genetic differences seeking a plant having a genetic aberration that could prevent the formation of glycolic acid. None of these efforts has been successful, and it now appears that there is no way to manage or diminish the loss of carbon from the plant through photorespiration.

CONCLUDING STATEMENT

This brief discussion of photosynthesis is intended to touch upon several aspects of photosynthesis and respiration that are important in producing crops and in crop science research. The literature on this broad topic is massive, and this presentation is hardly more than an outline of the subject matter. Students wishing to pursue any particular topic under this broad subject are referred to the references. Extended discussions are given in books, and specific topics, with citations to original literature, are covered in review articles.

Managing crops for maximum yields of necessity involves managing the photosynthetic process so that it functions as rapidly as possible. Early planting, use of cultivars adapted to local climatic conditions, and multiple cropping in areas with long growing seasons are all practices designed to make best use of the photosynthetic factory. Also, many other practices, such as optimal fertilization and good water management, affect yield by altering the rate and duration of photosynthesis. Thus, much of crop management is photosynthetic management. A better understanding of how the photosynthetic process functions and which environmental factors place constraints on its rate has helped crop management specialists devise management schemes that have increased yields dramatically. Scientists are now evaluating the potential of further yield advances by genetically or chemically enhancing the capability of leaves to fix CO_2. Success in these research endeavors may lead to continued increases in yield potential in future crop cultivars.

REFERENCES

Arnon, D. I. 1960. The role of light in photosynthesis. Sci. Am. 203:104–118.

Bassham, J. A. 1962. The path of carbon in photosynthesis. Sci. Am. 206:88–100.

Burris, R. H., and C. C. Black (ed.) 1976. CO_2 metabolism and plant productivity. University Park Press, Baltimore, Md.

Gifford, R. M., and L. T. Evans. 1981. Photosynthesis, carbon partitioning, and yield. Annu. Rev. Plant Physiol. 32:485–509.

Govindjee. 1967. Transformation of light energy into chemical energy: Photochemical aspects of photosynthesis. Crop Sci. 7:511–560.

Hatch, M. D., C. B. Osmond, and R. O. Slatyer (ed.) 1971. Photosynthesis and photorespiration. Wiley Interscience, New York.

Hesketh, J. D. 1963. Limitations to photosynthesis responsible for differences among species. Crop Sci. 3:493–496.

Johnson, C. B. 1981. Physiological processes limiting plant productivity. Butterworth, London.

McCree, K. J. 1970. An equation for the rate of respiration of white clover plants grown under controlled conditions. p. 221–229. In I. Setlik (ed.) Prediction and measurement of photosynthetic productivity. PuDoc, Wageningen, The Netherlands.

————. 1982. The role of respiration in crop production. Iowa State J. Res. 56:291–306.

Moss, D. N., and R. B. Musgrave. 1971. Photosynthesis and crop production. Adv. Agron. 23:317–336.

————, R. B. Musgrave, and E. R. Lemon. 1961. Photosynthesis under field conditions. III. Some effects of light, carbon dioxide, temperature, and soil moisture on photosynthesis, respiration and transpiration of corn. Crop Sci. 1:83–87.

Ogren, W. L., and G. Bowes. 1971. Ribulose diphosphate carboxylase regulates soybean photorspiration. Nature (New Biol.) 230:159–160.

Peaslee, D. E., and D. N. Moss. 1966. Photosynthesis in K- and Mg-deficient maize (*Zea mays* L.) leaves. Soil Sci. Soc. Am. Proc. 30:220–223.

Penning de Vries, F. W. T. 1975. The cost of maintenance processes in plant cells. Ann. Bot. 39:77–92.

Rabinowitch, E. 1945–1956. Photosynthesis and related processes. Vol. I, 1945; Vol. II, Part I, 1951; Vol. II, Part II, 1956. Wiley Interscience, New York.

————, and Govindjee. 1969. Photosynthesis. Wiley Interscience, New York.

San Pietro, A., and C.C. Black. 1965. Enzymology of energy conversion in photosynthesis. Annu. Rev. Plant Physiol. 16:155–174.

Sestak, Z., J. Catsky, and P. G. Jarvis (ed.) 1971. Plant photosynthetic production: Manual of methods. W. Junk N. V., The Hague.

Zelitch, Israel. 1971. Photosynthesis, photorespiration, and plant productivity. Academic Press, New York.

6 Growth of the Green Plant

R. H. BROWN
Department of Agronomy
University of Georgia
Athens, Georgia

Growth of plants may be expressed in a number of ways. Increase in height is the most obvious manifestation of growth, but it may not be of great importance. Increase in dry weight may be the most important aspect of growth in forage and pasture crops, but in grain crops dry weight increases in nongrain components are important only in producing a plant capable of high grain yield. Growth may also be expressed as the advance of a plant from one development stage to another. A seed germinates, the seedling develops roots, leaves, and stem, and the plant grows to maturity. The plant flowers and produces seed to complete its life cycle.

On a more basic level, growth may be expressed as the division of a cell to form two cells and the enlargement of the newly divided cells. The division and enlargement of cells is a complicated process involving synthesis of many organic compounds such as protein, cellulose, and nucleic acids and requiring physical forces that cause cell enlargement. The growth of a plant is much more complicated because it involves division and expansion of many different cell types. Cells in root tips, leaf buds, and developing seeds divide and grow in basically the same way, but they serve vastly different functions for the plant.

This chapter will deal with the descriptive aspects of plant growth and factors influencing that growth, focusing on growth at the organ (leaf, fruit, root, etc.) or whole plant level rather than at the cellular level. It will include some of the mathematical models for describing plant growth. Except for Fig. 6.4 and 6.8, graphs represent generalized patterns and not specific crops.

Published in *Physiological Basis of Crop Growth and Development* © American Society of Agronomy—Crop Science Society of America, 677 South Segoe Road, Madison, WI 53711, USA.

PATTERNS OF GROWTH

The growth of most plants follows a similar pattern, generally sigmoid in shape. A relatively slow growth rate characterizes early or seedling growth. The rate of growth increases as the plant becomes larger, is greatest just before or in early stages of flowering, and then decreases as the plant matures.

Dry Matter Accumulation

The dry weight accumulation pattern of plants in the early stages has been compared to compound interest earnings of a bank account. The weight of the plant at any given time is likened to the size of the bank account and the increase in weight equated with the interest. In the early stages when plants are very small, the actual dry weight increase per day (interest) is small (Fig. 6.1, first 4 weeks), but as the plant becomes larger, its weight gain per day increases (Fig. 6.1, first 10 weeks). This logarithmic increase in growth cannot continue indefinitely, of course. All plants mature and growth ceases (approximately 18 weeks in Fig. 6.1) because of completion of the life cycle or the onset of dormancy.

The slow growth rate of seedlings relative to large plants may be attributed to the relatively small number of cells that can divide, the small leaf area available for light interception and photosynthesis, and perhaps to the relatively large percentage of photosynthate (photosynthetic products) going to roots. As plants increase in size, larger meristematic regions (areas of cell division) per plant exist, and more leaves are present to act as photosynthetic energy sources. Thus each plant is capable of capturing and utilizing more energy as it increases in size.

The decline in growth rate near maturity may be attributed to several factors. Many species have determinate growth characteristics; that is, vegetative growth ceases prior to or concurrent with reproductive growth. In these species, such as corn[1] or small grains, the formation of fruit signals an end to growth of other plant parts. Even in species such as cotton where vegetative and reproductive growth occur simultaneously, the onset of reproduction reduces vegetative growth rates.

As plants become older a large portion of the plant structure becomes inactive. Lower leaves may become heavily shaded or nonphotosynthetic due to senescence. Leaves may fall from the plant, representing loss of dry weight. Large portions of the plant may be comprised of stem or other tissue relatively low in metabolic activity and therefore not contributing to growth. In addition, competition from neighboring plants for water, nutrients, and light may cause a reduction in growth rates when plants in crop stands become large.

---

[1] Scientific names of important crop plants are given in Table 1.1, Chapter 1.

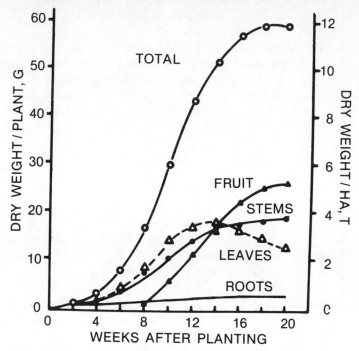

Fig. 6.1. Dry weight accumulation patterns for a hypothetical crop plant and its parts. The dry weights per unit of land area are for the same plants as the individual plant weights, but plant population is assumed to be 200 000 plants/ha. Plants mature at 18 to 20 weeks after planting.

Leaf Area Expansion

Leaf growth is included in Fig. 6.1 as a part of the total dry matter accumulation. Leaves serve a much more important function, however, than just being a part of the dry weight of the plant. Leaf surfaces intercept light and absorb carbon dioxide in the process of photosynthesis. Photosynthesis and dry matter production of a plant are proportional to the amount of leaf area on the plant as long as some leaves are not heavily shaded by others.

Leaf weight (Fig. 6.1) and leaf area (Fig. 6.2) follow patterns similar to that of total dry weight during the first half of the growing season. The leaf area index (LAI) shown in Fig. 6.2 is the ratio of leaf area (one side of leaf only) to soil area it occupies. The rate of increase of leaf area is important, because it determines the rate of increase in the photosynthetic capacity of the plant. The growth rate of plants during the seedling and up to mid-season stages may depend on the expansion of new leaves. Plants that put a large proportion of their photosynthetic products into leaf production during early growth may subsequently grow at a faster rate.

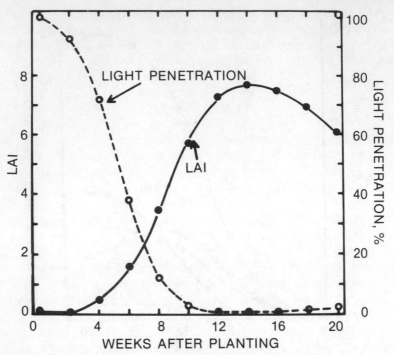

Fig. 6.2. Increase in leaf area index (LAI) and decrease in light penetration to the soil surface during growth of the crop represented in Fig. 6.1. Note that the percentage of light penetrating the stand decreases to 5% when LAI reaches a value of about 5.

The pattern of leaf area and weight accumulation may differ drastically from the total dry matter accumulation late in the growth cycle of the plant. Figures 6.1 and 6.2 show that leaf weight and area peak at about 15 weeks and then decline. This decline in amount of leaf is due to the death and abscission of leaves at a faster rate than new leaves are formed. In completely determinate species, no new leaves are initiated during the reproductive portion of the growth cycle. Indeterminate species continue to form leaves during fruiting, although usually at a slower rate.

Plant Height

As mentioned earlier, increase in height is usually the most obvious change in growth of most plants. The increase in height may give plants an advantage in competing with other plants in a community, but otherwise may not be an important characteristic. A consequence of the increase in plant height and the formation of new leaves at the top is that younger, more efficient leaves are usually above older ones and receive more sunlight. This important characteristic benefits the plant by having the most efficient leaves in the most favorable position for photosynthesis.

The pattern of increase in plant height with age may be very similar to the dry weight increase in some species, but not in others. Corn, for example, reaches its maximum height before the fruit starts to develop, but in other plants, height continues to increase during fruiting.

Root Growth

Growth of roots is obvious only to those interested in studying them. This is, nevertheless, a very important component of growth in all plants. In Fig 6.1 root weight is presented as a very small proportion (5%) of total plant weight during late stages of growth. This is probably near the minimum proportion of total weight that plants put into roots. In some plants, such as potato and sugar beet, the proportion of weight in root systems can be 50% or more.

The function of roots is more important than their weight. Roots anchor the plant and support its vertical stance. Nearly all water and mineral nutrients used by plants must pass through the root system. Therefore, the extent of the root system is important, if not necessarily its weight. In many cases, movement of nutrients within the soil, by diffusion or mass flow, is not sufficient to supply the needs of plants. When a plant is small, its need for minerals and water does not require a large root system, but as it grows, the root system must grow to exploit larger volumes of soil for nutrients and water. The fact that the youngest growing portions of the roots are most active in mineral nutrient uptake is another reason that the root system must continually grow to provide adequate nutrition for the plant.

GROWTH ANALYSIS

Crop growth in the field has been characterized by a system of growth analysis based mostly on dry matter accumulation rates. Since crop yields are based on land area, the most meaningful analysis of crop growth is on a land area rather than an individual plant basis. Development of methods to analyze crop growth has provided a better understanding of growth processes and yield limitations.

Crop Growth Rate

The dry matter accumulation rate per unit of land area is referred to as crop growth rate (CGR), normally expressed as g (m of land area)$^{-2}$ day^{-1}. Crop growth rate is measured by harvesting plants at frequent intervals and

Table 6.1. Maximum crop growth rates reported for some crop plants.

Crop	Maximum CGR†	Reference
	g m^{-2} day^{-1}	
Corn	51	Williams et al. (1965)
Sorghum	51	Loomis and Williams (1963)
Millet	54	Begg (1965)
Sugarcane	42	Monteith (1965)
Rice	27	Chandler (1969)
Soybean	23	Koller et al. (1970)
Peanut	28	Williams et al. (1975)
Alfalfa	20	Nelson and Smith (1968)
Sugarbeet	32	Watson (1956)

† These values are not valid for exact comparisons among species, since climatic conditions were quite different for the crops listed.

calculating the increase in dry weight from one harvest to the next. Roots are excluded in many studies of CGR. For a given time interval,

$$CGR = \frac{W_2 - W_1}{SA\,(t_2 - t_1)}$$

where W_1 and W_2 are crop dry weight at beginning and end of the interval, t_1 and t_2 are the corresponding days, and SA is the soil area occupied by the plants at each sampling. The most accurate measure of CGR throughout a growing season is obtained by sampling at frequent intervals.

Growth on a soil area basis follows a pattern identical to the growth pattern of individual plants (Fig. 6.1) and for the same reasons. The maximum CGR (steepest slope of the curve in Fig. 6.1) occurs when plants are large enough or dense enough to exploit all the environmental factors to the greatest degree. In favorable environments, this maximum CGR occurs when leaf cover is complete and may represent the maximum potential dry matter production and solar energy conversion rate for a given species. Values in Table 6.1 represent some of the maximum CGR reported for crop species. Tropical or warm-season grass crops such as corn, sorghum, millet, and sugarcane have higher CGR than the other crop species. The reason for this difference, at least in part, is the more efficient photosynthesis of most tropical grasses with the C_4 photosynthetic pathway. Rice is a tropical grass but lacks C_4 photosynthesis. Differences in species with C_3 and C_4 photosynthesis are discussed in Chapter 5.

Growth rate is low early in the growth of the crop because of incomplete cover and the low percentage of sunlight intercepted. A rapid increase in CGR occurs as the crop develops, the leaf area expands, and less light penetrates through the crop to the soil surface (Fig. 6.2). The maximum CGR generally coincides with the early fruiting stage and decreases as the plant matures because of cessation of vegetative growth, loss of leaves, and senescence (Fig. 6.3).

Studies of CGR are important in the interpretation of yield differences among crop varieties and cultural practices. Since crop yields may be affected by many factors during the growing season, analysis of growth throughout the growth cycle may be helpful in explaining yield differences.

Relative Growth Rate

Relative growth rate (RGR) is the increase in plant weight per unit of weight already present and may be expressed as g (g of dry weight)$^{-1}$ day^{-1}. At any given instant during growth,

$$RGR = \frac{1}{w} \cdot \frac{\delta w}{\delta t}$$

where w is the plant dry weight and $\delta w/\delta t$ is the change in dry weight per unit time. Just as income from a large savings account is greater than from a small one, potential growth of a large plant is likely to be greater than that of a small plant.

Relative growth rate decreases with plant age. This decrease is due to the fact that an increasing part of the plant is structural rather than metabolically active tissue and as such does not contribute to growth. The decrease is also due inpart to shading and increased age of lower leaves.

Net Assimilation Rate

Since leaf surfaces intercept sunlight and absorb CO_2 in photosynthesis, it is desirable in some cases to express growth on the basis of leaf area. The dry matter accumulation rate per unit of leaf area is termed net assimilation rate (NAR). At any instant during crop growth,

$$NAR = \frac{1}{A} \cdot \frac{\delta w}{\delta t}$$

where A is the leaf area and $\delta w/\delta t$ is the change in plant dry weight per unit time. The NAR is expressed as g (m of leaf area)$^{-2}$ day^{-1}.

Net assimilation rate is a measure of the average efficiency of leaves on a plant or in a crop stand. It is highest when all leaves are exposed to full sunlight and therefore highest when plants are small and leaves are few enough that none are shaded by others. As plants grow, more and more leaves are fully or partially shaded; thus, NAR decreases during the growing season (Fig. 6.3). The decrease in NAR with plant age may also be partially due to the older average leaf age and resulting lower photosynthetic efficiency.

A disadvantage of using leaf area as a basis for growth expression is that only the average efficiency of leaves in producing dry matter is known.

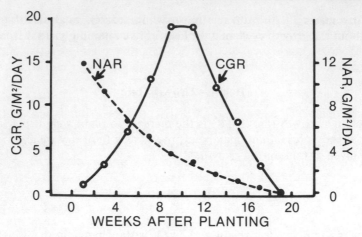

Fig. 6.3. Change in crop growth rate (CGR) and net assimilation rate (NAR) during growth of the crop represented in Fig. 6.1. Each point on the graph is for a 2-week period.

Older leaves at the bottom of the plant or crop stand may be contributing nothing to growth, while young, upper leaves may have an efficiency much greater than indicated by the NAR. A second disadvantage is that leaf petioles (usually not included in leaf area) and stems may carry on photosynthesis, and in some crops the inflorescence may contribute substantially to fruit formation. In cereals such as barley the inflorescence may contribute up to 50% of the dry weight of the grain.

PARTITIONING OF PHOTOSYNTHETIC PRODUCTS

Most crop plants are characterized by a period of vegetative growth (no flowers or fruit produced) followed by reproductive growth and then death or dormancy. During growth, the dry matter accumulated is partitioned into the various plant parts shown in Fig. 6.1. The final yield of an economically important product is of most concern to the producer, but much of the material produced by the plant is waste or by-product after harvest. In the case of a few crops, such as forages, producers are interested principally in the vegetative growth. The manner in which plants partition products of photosynthesis into various plant parts is important in determining growth rate and yield.

Vegetative Growth

From the time a plant emerges from the soil until it starts to flower, we say it grows vegetatively. For many crops, the importance of the vegetative growth is simply to produce a large enough photosynthetic factory to obtain maximum yields and a sturdy plant to support the fruit. The amount of vegetative growth is, however, important. Rapid and extensive vegetative growth can aid in weed control by competing for sunlight and other factors. In some cases, large accumulation of vegetative plant material may help provide needed nitrogenous compounds during a very rapid seed production period.

As seedlings grow, the plant has basically three components to which photosynthetic products can be partitioned: leaves, stems, and roots. The growth rate of the young plant may be increased by investing a greater percentage of the increased weight in leaves, since the faster the leaf area expands, the greater the increase in light interception and photosynthesis. Root growth and penetration is also important for increased nutrient and water uptake capacity. Some crop plants partition very little of their early growth into stems. Fall-planted cereals, for example, produce primarily leaves and roots during fall, and stems do not develop until spring.

Leaves, stems, and roots may compete for available photosynthetic products. The partitioning of products may be controlled by plant hormones or by environmental factors. If soil temperature is more favorable for root growth than air temperature is for top growth, then a greater percentage of dry matter may be diverted to root growth. A large supply of nitrogen usually stimulates growth of leaves and stems more than roots; thus, at high nitrogen fertilization rates, the percentage of the plant weight in roots decreases. Water stress generally reduces leaf and stem growth more than root growth.

Reproductive Growth

Since reproductive growth is usually most important in crop production, the diversion of photosynthetic products to fruit is the objective of plant breeding and cultural practices. In plants of indeterminate growth habit, in which reproductive and vegetative growth occur simultaneously, the two components may compete for photosynthetic products. Such competition for photosynthetic products by vegetative growth may reduce the yield of fruit. This competition is, of course, not a problem in plants that stop vegetative growth when fruiting starts.

Evidence of competition between vegetative and reproductive growth has been demonstrated in some crop species. For example, if fruiting is prevented in peanut and cotton by removing flowers, vegetative growth proceeds rapidly, but if a heavy fruit load is formed, vegetative growth is greatly reduced.

The partitioning of dry matter to vegetative and reproductive growth is important even in cases where the two do not compete. Determinate plants, like corn and small grains, divert photosynthetic products to vegetative growth early in the season and to grain later. If vegetative growth is completed before plants are of sufficient size for maximum photosynthesis, then

Table 6.2. Percentage of total above-ground dry weight in grain for old and
new cultivars of rice and wheat.

Crop	Cultivars	Dry weight in grain
		% total dry wt
Rice†	Newer cultivars	
	IR8	53
	Taichung Native 1	54
	Tainan 3	55
	Older cultivars	
	Hung	37
	Peta	37
	Nang Mong S₄	33
Wheat‡	Newer cultivars	
	Bluebird #6	44
	Bbxlnia	44
	Tordo	40
	Older cultivars	
	Gabo	34
	Nainari 60	34
	Napo 63	34

† From Chandler (1969).
‡ CIMMYT Annual Report (1972).

grain yield will be reduced. Small plants will not necessarily produce lower
grain yield, however, if the plant population is increased enough to compensate for the smaller plants.

The yield potential of a crop may be viewed as a product of the photosynthate produced and the fraction of photosynthate diverted to the desired
plant part (usually fruit). If wheat cultivar A has an equal photosynthetic
potential to cultivar B but diverts 60% of its total weight to grain rather
than 50% as does cultivar B, grain yield is improved in cultivar A without
actually increasing photosynthetic capacity. Increased yields of modern
grain varieties appear to be due in part to diversion of more of the dry
weight to grain (Table 6.2). Grain yield may also be increased in some crop
varieties by increased photosynthesis rates.

Storage

Storage is represented by accumulation of dry weight in a form reusable in plant metabolism. Cell walls and other structural materials, including some structural protein, are forms of organic materials not reusable
by plants. Most of the stored material is in the form of carbohydrate
(starch, fructosan, or sugars) or lipid, although some protein may be stored.
Reproductive growth discussed in the preceding section involves mostly a
specialized form of storage of photosynthetic products. The filling of grain
is growth in terms of dry matter accumulation, but very little new structural
tissue is formed.

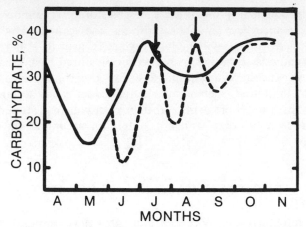

Fig. 6.4. Trend of stored carbohydrate percentage (of dry weight) in alfalfa roots from April to November. The crop represented by the solid line was not mowed; that represented by the broken line was mowed three times as indicated by the arrows. The decrease in stored carbohydrate percentage following mowing is attributable to its utilization in the regrowth of the plant tops. From Smith (1962).

Although temporary storage may occur in leaves, large quantities of materials in most species are stored in seeds, stem bases, or underground organs (roots, rhizomes, tubers, corms, etc.). Storage in seeds is, of course, to provide for the metabolic needs of seedlings. Other methods of storage serve such purposes as regrowth of hay following mowing and early growth of perennial plants in spring after winter dormancy, as shown by the following two examples.

After harvest of alfalfa for hay, carbohydrates stored in a large tap root are translocated to the new growing points (Fig. 6.4). This translocation is necessary because the mowed alfalfa plants have few or no leaves remaining and no other energy source for regrowth. As soon as sufficient leaves are produced to supply the dry matter and energy for growth, carbohydrate storage in the tap root begins anew. This cycle of storage and reutilization is repeated as many times during the season as plants are mowed. If the alfalfa is mowed too often, storage in the tap root is not replenished, and the plants are weakened and may die.

In autumn when perennial, herbaceous plants approach dormancy, carbohydrate is stored. Low temperatures and short days in autumn apparently reduce vegetative growth more than photosynthesis, so excess photosynthetic products may be stored. This storage is concentrated in stem bases and stolons or underground organs of crop species that survive the winter. Storage of organic compounds is necessary for winter survival and early spring growth.

The partitioning of photosynthetic products to storage may reflect the balance between the photosynthetic capacity of the plant and the need for the photosynthetic products in growth. When photosynthesis exceeds the need for its products in growth, the excess is stored. A number of factors affects this balance between photosynthesis and growth. High levels of nitrogen applied to grasses stimulate vegetative growth apparently more than photosynthesis, resulting in reduction of stored carbohydrate. Low temperatures reduce growth and increase stored carbohydrate. Some synthetic growth inhibitors, such as maleic hydrazide, cause accumulation of storage carbohydrate. Storage in reproductive organs (e.g., tubers, seeds) is an active process, however, and not simply an accumulation of surplus carbohydrate.

PHOTOSYNTHATE TRANSLOCATION

Photosynthetic rates may control plant growth rates, but there are indications that, in some instances, translocation and utilization of photosynthate (photosynthetic products) in growth may limit photosynthesis. During a day of high photosynthesis, leaves in full sunlight may increase their dry weight by 20–30%. Most of the weight increase is starch. This results from translocation and use of photosynthate being slower than its synthesis. The accumulated starch is usually broken down and exported from the leaf at night.

Rate of Translocation

The rate at which photosynthate is exported from leaves is important in linking photosynthesis to growth. Though relatively little is known about the mechanism of movement, many studies have been made of the rate of transport. One of the most common ways to estimate rate of translocation is to allow leaves to assimilate CO_2 containing radioactive carbon (^{14}C) and then measuring the quantity of ^{14}C transported from the leaf. The linear rate of movement of photosynthate may be as high as 200–300 cm hr^{-1}.

Perhaps more important than linear rate of translocation is the quantity transported per unit of time. Leaves may export in 6 hours as much as 70 to 80% of the photosynthate produced in a short time interval. The percentage may, however, be much lower depending on the plant species, environmental conditions, and plant vigor. Factors known to reduce plant growth, such as water stress, low temperatures, and mineral nutrient deficiencies, also reduce translocation.

Species are known to have different translocation capacities. Leaves of plants exhibiting C_4 photosynthesis export a greater percentage of their photosynthate in short time intervals (6 hours or less) than do C_3 species. This more efficient translocation may be related to the specialized leaf anatomy of C_4 species (see Chapter 5) and to a greater amount of phloem in their leaves. The greater movement of photosynthate into ears of modern wheat species was found by Evans et al. (1970) to be related to the larger phloem cross section in the stem of these species compared to more primitive ones.

Patterns of Translocation

The movement of photosynthate from leaves serves two purposes: 1) to nourish nonphotosynthetic plant parts; and 2) to store photosynthate in storage organs for later use. Photosynthate moves to these plant parts in relation to demand. In a plant growing vegetatively, photosynthate from a leaf may move to the roots, growing points of stems, and young developing leaves. The relative amounts of photosynthate going to these various plant parts from a given leaf depends on the distance from the leaf to the other plant parts. Lower leaves export a larger proportion of their photosynthate to roots than do upper leaves. Conversely, upper leaves export a larger percentage of their products to the upper portion of the plant.

When a plant develops a heavy fruit load, the fruit seem to have priority for the photosynthate from most leaves. In plants having many fruits dispersed on the main stem or branches, the photosynthate from a leaf goes primarily to the nearest developing fruit. Only small amounts of photosynthate are translocated from one heavily fruited branch to another.

Leaves act as importers of photosynthate during their early growth and as exporters later. Immature leaves grow at rates faster than can be supplied by their limited photosynthesis. Later, however, they produce much more photosynthate than they need. Photosynthesis in leaves of some species increases from the time they appear until they reach full size. After reaching full size, leaves have much less need for photosynthate. Prior to full expansion, leaves reach a point when they export more photosynthate than they import.

Source-sink Concept

Photosynthate moves from the source (usually leaves) to the point of utilization (the sink). The relationship between the source and sink has been the object of much crop physiology research. Plant yield has been considered a result of photosynthetic (source) capacities on the one hand and capacity to utilize photosynthate (sink activity) on the other. The limitation of yield due to one of these factors rather than the other is very difficult to prove; both are undoubtedly important.

FIELD ASPECTS OF GROWTH

While most of the growth aspects discussed previously in this chapter apply to individual plants or plant communities, certain topics relating to field crop production need further elaboration. Most crops are grown in stands of sufficient density to create almost complete cover of the soil; therefore, they are studied as plant communities rather than individual plants. Yields are computed on a land area basis and, to maximize yields, practices must be used that make full use of the land area.

Solar Radiation Interception

Solar radiation must be intercepted by leaves if it is to be used in dry matter production. Any radiation not impinging on plants must be dissipated by the soil either through reflection, evaporation of soil water, or heating of the soil. Radiation intercepted by the soil contributes nothing to the current growth of the crop unless soil temperature is below optimum for plant growth and sunlight striking the soil may warm it.

Only about 70–80% of all solar radiation striking an individual leaf is absorbed. Of the remainder, about 10–15% penetrates the leaf and 10–15% is reflected. The leaf does not absorb all wavelengths of radiation equally. The green color of leaves results from somewhat less absorption in the green region than in the blue (470 nm) and red (660 nm) regions. Approximately 80% of the visible radiation (400–700 nm) is absorbed compared to less than 50% of the infrared radiation (700–3000 nm). The reduced absorption of infrared radiation helps prevent overheating of leaves or excessive water loss through transpiration. In the remainder of this chapter solar radiation will be referred to as "light" and only the 400–700 nm wavelengths are included.

The percentage of light penetrating through a crop stand decreases as leaf area increases (Fig. 6.5). Light penetration can be related to LAI as follows:

$$\log_e I/I_o = -k \,(LAI)$$

where I is the irradiance under the crop canopy, I_o the irradiance above the crop, and k is a constant or extinction coefficient, which is the fraction of the light intercepted per unit LAI. In Fig. 6.5, two crop canopies are represented, and the slopes of the lines are shown as k-values of 1.0 and 0.6. It may be seen (indicated by arrows) that a stand with k = 0.6 requires an LAI of 5 to reduce light penetration to 5%, whereas a stand with k = 1.0 requires only about 3 LAI units. The LAI required to intercept 95% of sunlight (light penetration = 5%) has been termed critical LAI (Brougham, 1956).

It is important that leaf area expand rapidly and a critical LAI is attained so that photosynthesis and growth rate reach a maximum early in the life of a crop. This helps to insure high yields and has the added advantage of reducing competition from weeds. In Fig. 6.2, 6.3 and 6.5, the crop attains a higher LAI than needed to intercept 95% of the light (LAI of about

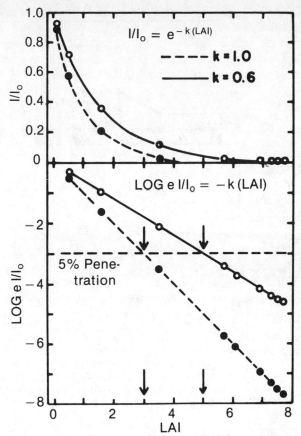

Fig. 6.5. Light penetration through a plant stand (I) expressed as a fraction of light at the top of the stand (I_o) and plotted against LAI. The Naperian logarithm of the light penetrating the stand is plotted in the lower part of the graph. The solid lines represent data taken from the curves in Fig. 6.2 and the broken lines are for the same LAI values, but a k-value of 1.0 is assumed. The arrows indicate critical LAI, where 95% of the light is intercepted and only 5% penetrates the canopy.

5) and produce maximum CGR. The extra LAI does not appear in most cases to be detrimental to growth or yield.

Canopy Architecture

The difference in slope of the two curves in Fig. 6.5 reflects properties of the plant canopies involved in light interception. The percentage of sunlight penetrating through a plant canopy with a given LAI depends mostly on the pattern of leaf display. Two main characteristics, leaf angle and grouping of leaves, influence the k-value.

It is easy to visualize that less leaf area is needed to intercept a given percentage of sunlight if the leaves have their broad surfaces rather than

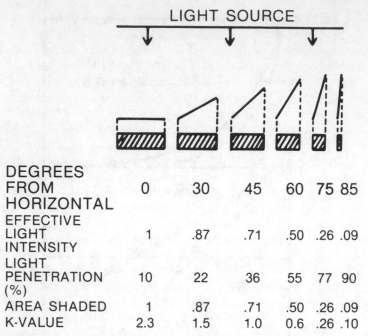

DEGREES FROM HORIZONTAL	0	30	45	60	75	85
EFFECTIVE LIGHT INTENSITY	1	.87	.71	.50	.26	.09
LIGHT PENETRATION (%)	10	22	36	55	77	90
AREA SHADED	1	.87	.71	.50	.26	.09
K-VALUE	2.3	1.5	1.0	0.6	.26	.10

Fig. 6.6. Representation of light penetration through plant stands with leaves at various angles. The LAI is assumed to be 1 and light penetrating through an individual leaf is taken as 10%. For a given soil area leaves with more vertical angles intercept less sunlight as can be seen by the smaller shadow they cast (shaded area). Effective irradiance (light intensity) is the irradiance at the leaf surface relative to a value of 1 for horizontal leaves. The area shaded is also relative to a value of 1 for horizontal leaves. The k-value is a calculated extinction coefficient.

their edges pointed toward the sun. Figure 6.6 shows that plants with an LAI of one can intercept nearly all of the sunlight if the sun is directly overhead and the leaf layer is displayed horizontally in the same plane with no overlapping. But even horizontal leaves are displayed somewhat randomly so that an LAI of one does not intercept all of the sunlight. Nevertheless, it is clear that horizontal orientation of leaves is more effective in light interception per unit LAI. If we assume the amount of light penetrating through individual leaves is 10%, then that amount (10%) penetrates the horizontal plane covered by one horizontal leaf, but 77% penetrates the same horizontal plane if the leaf is 75° from horizontal.

In thick crop stands, more nearly vertical leaves may be an advantage. If leaves are 85° from horizontal (Fig. 6.6), ten leaves could occupy the same space as one horizontal leaf without shading each other. While leaves at this angle in the field would not be so evenly spaced as shown in Fig. 6.6, vertical leaves would produce much less mutual shading than horizontal ones. At an LAI of 10, there would be 9 layers of horizontal leaves shaded.

Effective irradiances (light intensity) at leaf surfaces, relative to horizontal leaves, are given in Fig. 6.6. As leaves form greater angles with the horizontal, irradiance at the leaf surface decreases. Assuming vertical light rays, the light at the leaf surface is the product of the irradiance measured in a horizontal plane and the cosine of the angle of the leaf from horizontal. Photosynthesis of the tilted leaf may not be decreased if irradiance is still great enough for saturation. The advantage of more vertical leaves comes from the fact that several leaves receive enough light for reasonable photosynthetic rates, rather than fewer horizontal leaves receiving more light than they can use (above light saturation, see Chapter 5) while shading lower leaves. The effects of leaf angle may diminish toward sunset and dawn when the sun is low on the horizon and at latitudes greater than 23.5°. In some species leaves move throughout the day, however, in a tendency to "face" the sun.

Another aspect of canopy architecture that influences sunlight interception is the dispersal of leaves. If leaves tend to be grouped together [lettuce (*Lactuca sativa* L.) and cabbage (*Brassica oleracea capitata* L.) are extreme examples] light interception will not be near complete even at a high LAI. Corn may be planted in rows wide enough apart that an LAI of 6 would not intercept 95% of sunlight, but at closer spacing an LAI of 4 may be sufficient. The most efficient spacing pattern of plants for light interception is equidistance among plants in all directions.

Mutual Shading

The ideal illumination of leaves for maximum yields per plant is when every leaf receives as much light as it can use. In the effort to maximize yields per unit land area, such an ideal is not feasible. Productivity per plant must be reduced in plant communities to obtain maximum yield of the community. Maximum yield of a plant community is not attained until plant populations are large enough to produce competition. A major part of this competition may be for sunlight, and the result is the shading of leaves by other leaves (mutual shading).

When LAI is low, NAR (leaf efficiency) is high (Fig. 6.7). As LAI increases, NAR decreases even though CGR is increasing during the same period. The decrease in NAR results in part from progressive increases in mututal shading of leaves.

In 1947, Watson first related CGR to LAI and NAR as follows:

$$CGR = (LAI)(NAR)$$

He reasoned that CGR would increase as LAI increased because of greater light interception and, thus, more photosynthesis. Crop growth rate should increase with added LAI up to the point that bottom leaves receive just enough light for photosynthesis to offset respiration. If more leaves are added, bottom leaves do not receive enough light for net photosynthesis, and they therefore lose more CO_2 than they gain. According to Watson's

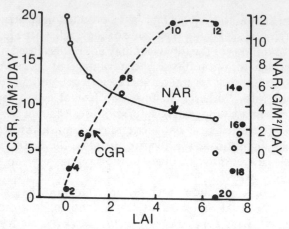

Fig. 6.7. Patterns of change in CGR and NAR as LAI increases during the growing season for
the crop represented in Fig. 6.1. Numbers beside closed circles indicate weeks after planting.
Data for the crop beyond 12 weeks do not follow the pattern of earlier growth because of
loss of leaves and maturity of the plants.

original proposal, these leaves respire more CO_2 than they assimilate (they
are "parasitic"), and at a LAI higher than the optimum, CGR decreases
rather than increasing. If LAI accumulates to a high enough level, respiration of lower leaves would balance photosynthesis of upper leaves, and
NAR and CGR would drop to zero. Most experiments since Watson's early
work show that CGR does not drop greatly at LAI values above the critical
or optimum LAI, probably because lower leaves respire at only a fraction of
the rate of upper, younger leaves. Most species do not export photosynthetic products to old leaves, so lower leaves, which are usually the oldest
ones, are not very parasitic. In addition, specific leaf weight (leaf
weight/leaf area) decreases as LAI increases, which tends to reduce respiration per unit of leaf area.

As LAI and light interception increase, it follows that CGR will increase until light interception is nearly complete. That is, the optimum LAI
for CGR should be similar to the critical LAI. Figure 6.7 indicates an optimum LAI of about 4 to 6 for the hypothetical crop examined in previous
figures. CGR data for the latter part of the growth cycle (beyond 12 weeks)
do not fit on the curve because factors other than leaf area are controlling
growth rate.

Leaf and Plant Age

As leaves become older, their photosynthetic capacity decreases. In
dense plant canopies, leaves of crop plants may live only for a month or
two. In growth of most plants, new leaves are initiated at nodes higher on a
stem than preceding leaves. This places older leaves lower in the canopy and

in a more shaded environment than younger leaves. Since younger leaves have higher photosynthesis rates, it is an advantage to the plant communities to have older rather than younger leaves in shaded positions.

Aging of leaves may be detrimental to plant growth in some cases. After the transition from vegetative to reproductive growth, if new leaves are not produced, dry matter production in fruit may be reduced by leaf aging. It has been observed that photosynthesis of the flag (upper) leaf of wheat decreases during grain filling. This decrease may or may not limit wheat yield, depending on whether the demand for photosynthetic products by the grain is met.

Most crop species have a definite cycle of growth, especially in temperate regions, during which vegetative and reproductive growth occur. The aging of the plant then follows a fairly set pattern. Maturity or senescence may be hastened or retarded to some degree by climatic factors. The time required to reach maturity varies among cultivars that have been developed for different regions or cropping systems.

The pattern of development makes some crops susceptible to stress by various factors. If drought occurs when corn is in the flowering stage, yield is reduced because of poor pollination. Because of its aging pattern, corn cannot recover from drought damage at that stage. Some indeterminate species have the ability to continue to flower after drought so that fruit yields may be affected much less.

Competition

Plants compete with each other when placed in close proximity. The competition may be for mineral elements or water in the soil, or for light. The competition may be among plants of the same species, in which case weight per plant is reduced. Plant height may actually increase in competition for light, because of the etiolation effect of heavy shade.

Competition among plants is manifested in components of yield. Yield per plant is reduced when plants are close enough to produce competition. In soybean this reduction is caused by a decrease in the number of seeds per plant (Fig. 6.8). In this case, total seed yield was not affected as plant population was increased from 64 000 to 510 000 plants/ha although seed per plant decreased from 308 to 45. Therefore, competition among plants was such that reduced yield per plant almost exactly offset increased plant numbers. Weight per seed was not affected by population. In the case of corn and many other species, high plant populations increase the percentage of plants bearing no fruit at all.

Competition among plants of different species in a mixture is usually expressed by increased number and size of plants of the dominant species. In pasture mixtures one species tends to dominate under conditions that favor growth of that species. In the example presented in Table 6.3, weights

Table 6.3. Dry weights and numbers of alfalfa and orchardgrass seedlings
planted in March and August.†

Seeding rate		Dry weight				No. seedlings			
		Alfalfa		Orchardgrass		Alfalfa		Orchardgrass	
Alfalfa	Orchardgrass	March	August	March	August	March	August	March	August
kg ha⁻¹				mg plant⁻¹				Seedlings m⁻²	
11.2	0	66	152	--	--	330	290	--	--
11.2	10	59	130	44	34	280	310	400	310
22.4	0	64	144	--	--	440	600	--	--
22.4	10	56	117	17	18	550	480	530	300

† Seedlings counted and sampled 67 days after the March seeding and 51 days after the August seeding. From Blaser et al. (1956).

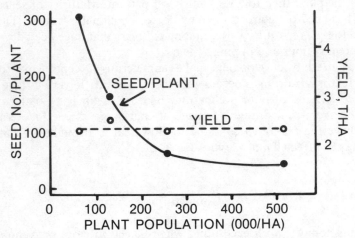

Fig. 6.8. Effects of soybean plant population on yield and the number of seed per plant. The increase in plant population is offset by a decrease in seeds per plant so that yield is not affected. Plants were grown in rows 1.02 m apart. Drawn from data of Weber, Shibles, and Byth (1966).

of alfalfa and orchardgrass seedlings show that alfalfa dominates the mixture at 50 to 70 days after seeding. The alfalfa is more dominant in August than in March plantings, probably because fall temperatures are more favorable for alfalfa. Increasing the seeding rate of alfalfa from 11.2 to 22.4 kg/ha decreased the weight of the alfalfa seedlings, but more drastically reduced weight of orchardgrass seedlings. This shows that alfalfa seedlings are more competitive with orchardgrass than with other alfalfa seedlings.

The degree of dominance a given species exerts in a mixture depends on how well the environment suits that species. High nitrogen rates stimulate growth of grasses more than legumes. When mixture of subterranean clover (*Trifolium subterraneaum* L.) and ryegrass (*Lolium rigidum* Gaud.) is not fertilized with nitrogen, the clover dominates because it can assimilate atmospheric nitrogen (Table 6.4). When 225 kg of nitrogen are applied per

Table 6.4. Weights of ryegrass (*Lolium rigidum* Gaud.) and subterranean clover (*Trifolium subterraneum* L.) seedlings grown in a mixture with and without nitrogen applied.†

Days from planting	Weight of seedlings			
	No nitrogen		Nitrogen at 225 kg ha⁻¹	
	Clover	Grass	Clover	Grass
	g/m²			
69	1.3	0.2	1.4	0.8
99	5.5	0.5	1.8	5.1
133	12.8	1.2	0.8	12.9

† From Stern and Donald (1962).

hectare, the grass dominates. With added nitrogen, the clover actually decreased in weight beyond 99 days after planting.

An important aspect of competition in crop production is weed control. Most weed control practices succeed because they favor the crop species. Seldom are weeds eliminated, even by chemical weed control. Weed populations are reduced by herbicides to the point that crop species dominate, sometimes rather completely. Few herbicides applied at recommended rates control weeds completely in the absence of a competing crop.

REFERENCES

Begg, J. E. 1965. High photosynthetic efficiency in a low-latitude environment. Nature (London) 205:1025–1026.

Blaser, R. E., W. L. Griffith, and T. H. Taylor. 1956. Seedling competition in compounding forage seed mixtures. Agron. J. 48:118–123.

Brougham, R. W. 1956. Effect of intensity of defoliation on regrowth of pasture. Aust. J. Agric. Res. 7:377–387.

Chandler, R. F. 1969. Plant morphology and stand geometry in relation to nitrogen. p. 265–289. *In* J. D. Eastin, F. A. Haskins, C. Y. Sullivan, and C. H. M. Van Bavel (ed.) Physiological Aspects of Crop Yield. Am. Soc. of Agron., Madison, Wis.

CIMMYT (Centro International de Mejoramiento de Maiz y Trigo). 1972. Annu. Rep. Londres, Mexico. p. 38.

Evans, L. T., R. L. Dunstone, H. M. Rawson, and R. F. Williams. 1970. The phloem of the wheat stem in relation to requirements for assimilate by the ear. Aust. J. Biol. Sci. 23:743–752.

Koller, H. R., W. E. Nyquist, and I. S. Chorush. 1970. Growth analysis of the soybean community. Crop Sci. 10:407–412.

Loomis, R. S., and W. A. Williams. 1963. Maximum crop productivity: an estimate. Crop Sci. 3:67–72.

Monteith, J. L. 1965. Light distribution and photosynthesis in field crops. Ann. Bot. 29:17–38.

Nelson, C. J., and D. Smith. 1968. Growth of birdsfoot trefoil and alfalfa. II. Morphological development and dry matter distribution. Crop Sci. 8:21–28.

Smith, D. 1962. Carbohydrate root reserves in alfalfa, red clover and birdsfoot trefoil under several management schedules. Crop Sci. 2:75–78.

Stern, W. R., and C. M. Donald. 1962. Light relationships in grass-clover swards. Aust. J. Agric. Res. 13:599–614.

Watson, D. J. 1947. Comparative physiological studies on the growth of field crops. I. Variation in net assimilation rate and leaf areas between species, and within the between years. Ann. Bot. 11:41–76.

——. 1956. Leaf growth in relation to crop yield. p. 178–191. *In* F. L. Milthorpe (ed.) The Growth of Leaves. Proc. 3rd Easter School of Agric. Sci. Butterworth Scientific Publications, London.

Weber, C. R., R. M. Shibles, and D. E. Byth. 1966. Effect of plant population and row spacing on soybean development and production. Agron. J. 58:99–102.

Williams, J. H., H. H. Wilson, and G. C. Bate. 1975. The growth of groundnuts (*Arachis hypogaea* L. cv. Makula red) at three altitudes in Rhodesia. Rhod. J. Agric. Res. 13: 33–43.

Williams, W. A., R. S. Loomis, and C. R. Lepley. 1965. Vegetative growth of corn as affected by population density. II. Components of growth, net assimilation rate and leaf-area index. Crop Sci. 5:215–219.

7 Nitrogen and Minerals

J. G. STREETER AND A. L. BARTA
Ohio State University/Ohio Agricultural
Research and Development Center
Wooster, Ohio

The importance of mineral elements to the growth and productivity of plants was demonstrated several centuries ago. The mineral nutrition of plants is still one of the most important factors determining ultimate crop productivity, especially in underdeveloped countries where chemical fertilizer is not readily available. The increasing demand for food production combined with a steady loss of prime agricultural land to development will result in more crops being planted on marginal soils. Thus, while there continues to be increasing emphasis on increasing crop productivity via increasing photosynthesis, symbiotic N_2 fixation, and other factors, we must not underrate the importance of plant nutrition.

In this chapter we describe how crop plants obtain nutrients and how these nutrients are used by plants to maximize growth and reproduction. Emphasis is on the plant; for a more detailed discussion of mineral nutrition as it relates to soil chemistry and fertility, you should consult other sources. Most of the biochemical details relating to the uptake and utilization of nutrients are treated in a simplified manner. For a more detailed but general treatment of these subjects, the books by Aldrich (1980), Hardy and Gibson (1977), Hewitt and Cutting (1979), Miflin (1980), Epstein (1972), Martin-Prevel (1978), Gauch (1972), and Quispel (1974) should be consulted. For entry into the technical literature, chapters in the *Annual Review of Plant Physiology* are recommended.

Published in *Physiological Basis of Crop Growth and Development,* © American Society of Agronomy—Crop Science Society of America, 677 South Segoe Road, Madison, WI 53711, USA.

NITROGEN

Nitrogen is the element that most often limits crop yields. Our planet contains a massive quantity of nitrogen, estimated at 100 to 200 $\times$ 10^{15} t. About 98% is a constituent of primary rocks and most of the other 2% is present as gases (mostly N_2) in the earth's atmosphere. Very small amounts, relative to the total, are present in land and sea plants and animals.

We can see immediately, then, that the limitation of crop yield by N is not related to overall amount. Rather, the limitation depends on the ability of crops to use only specific forms of N. For example, N_2 gas, which consists of two N atoms joined by a triple bond (N $\equiv$ N), is one of the most inert of all chemical compounds. This vast supply of N_2 is useless to plants, with the exception of N_2 utilized via N_2 fixation.

Most of the N in plants is in organic forms, i.e., the N is incorporated in carbon-containing compounds. These include nucleic acids, some vitamins (e.g., riboflavin), hormones (e.g., indole acetic acid), membrane components (e.g., phosphatidylcholine), coenzymes (e.g., nicotinamide derivatives), and pigments (e.g., chlorophyll). Thus, the structure and, especially, the function of all plant cells is intimately linked to the chemistry of N.

As interesting as these N-containing compounds are, we must focus our attention on proteins. Proteins deserve emphasis because of the large quantities of N required for their synthesis and because of their great significance in animal nutrition. Although there are a few exceptions, at least 90% of the N in plants is in the form of protein. The synthesis, storage, and utilization of protein are processes central to the survival of all living organisms.

Protein Composition and Quality

Proteins are composed of compounds known as amino acids, so called because they include an amino group ($-NH_2$) and a carboxyl (organic acid) group ($-COOH$). While several hundred amino acids are known to occur in plants, only 20 are utilized in the synthesis of essentially all proteins (Table 7.1).

To form a protein, the carboxyl group of one amino acid is linked to the amino group of a second amino acid by a peptide bond. Long chains of amino acids linked through peptide bonds give the protein its primary structure (Fig. 7.1). The chain may be from 10–1000 or more amino acids long.

The actual structure of a protein differs from Fig. 7.1 because in reality, the chain of amino acids forms a helical structure. In addition, the helical chain folds in various directions to form a complex three-dimensional structure. Plants contain many hundreds of different proteins, each with unique chemical properties. These unique properties are based on the

Table 7.1. Alpha amino acids involved in protein synthesis.

Name	Number of N atoms	% N	Unique chemical group
Glycine	1	18.7	none
Alanine	1	15.7	α-methyl
Serine	1	13.3	hydroxyl
Cysteine	1	11.6	sulfhydryl
Methionine†	1	9.4	sulfide
Valine†	1	12.0	⎤
Leucine†	1	10.7	⎥ aliphatic carbon chains
Isoleucine†	1	10.7	⎦
Threonine†	1	11.8	hydroxyl
Aspartic acid	1	10.5	2nd carboxyl
Asparagine	2	21.2	amide
Glutamic acid	1	9.5	2nd carboxyl
Glutamine	2	19.2	amide
Phenylalanine†	1	8.5	phenyl
Tyrosine	1	7.7	hydroxyphenyl
Tryptophan†	2	13.7	indole
Histidine	3	27.1	imidazole
Proline	1	12.2	secondary amino
Arginine	4	32.2	guanidino
Lysine†	2	19.2	ε-amino

† Essential amino acid; see text.

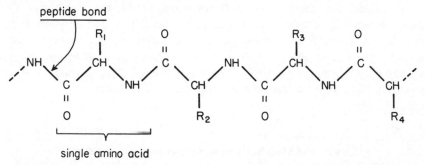

Fig. 7.1. Primary structure of proteins. R_1, R_2, etc. indicate various chemical groups of amino acids (Table 7.1).

quantities and the sequential order of the amino acids in the chain and on differences in three-dimensional structure that tend to expose or hide chemical groups on various amino acids.

Eight of the amino acids found in proteins cannot be synthesized by higher animals, including humans. These amino acids (Table 7.1) are usually termed essential amino acids because their inclusion in the diet is essential for the survival of these animals. (Ruminant animals are an exception to this rule. Bacteria living in the digestive tract of ruminants synthesize the essential amino acids and supply them to the animal.)

Proteins vary widely in their nutritional quality. Since the essential amino acids cannot be synthesized by animals, the relative amount of these eight amino acids usually governs protein quality. As a rule, plant proteins are of lower quality than animal proteins (meat, milk, eggs) for human nutrition. The amino acids most commonly deficient in plant proteins are methionine and lysine.

Protein content of plant tissues is usually estimated by analyzing total N concentration and multiplying by 6.25. This approach is useful because it is easier and less expensive to analyze total N than the actual proteins present. Adjustments in the multiplication factor, 6.25, are made where it is known that the N concentration of the protein in that tissue deviates significantly from 16% N. Such deviations do occur occasionally because the N concentration in amino acids varies from 9.4–32.2% (Table 7.1).

One other important property of proteins should be mentioned before concluding this section. Proteins serve not only as storage forms of N but also as catalysts. These protein catalysts, called enzymes, are present in all living cells where they regulate and increase the rate of many hundreds of chemical reactions.

As the amino acid chains in a protein are folded to form the three-dimensional structure, various amino acid groups (Table 7.1) are brought into precise configurations. These configurations function in the binding of particular compounds (substrates) at the active site of the enzyme and in chemical rearrangement of the substrate to form the product. We will discuss some important enzyme-catalyzed reactions in the following sections.

Finally, some proteins serve both for storage of N and as catalysts. The best example is the enzyme ribulose bisphosphate carboxylase, which functions in the fixation of carbon dioxide (Chapter 5). This enzyme may comprise 50% or more of the total protein content of a green leaf. When demand for N exceeds supply, most of the enzyme protein will be hydrolyzed to supply amino acids to other parts of the plant.

Uptake and Metabolism of Nitrate and Ammonium

As noted previously, massive amounts of N are present in and around the earth, but this N is essentially unavailable for direct use by plants. We now need to consider in detail the uptake and utilization of forms of N, namely nitrate and ammonium, that can be exploited by plants for growth and reproduction. The relationship of nitrate and ammonium to other important forms of nitrogen is illustrated in Fig. 7.2, which introduces a number of terms discussed in the following sections.

Nitrogen Transformations in Soils

Over 90% of the N in most soils is in the form of organic matter. This organic matter fraction comprises 1–5% of the dry weight of most cultivated mineral soils. It is an extremely complex mixture of organic compounds including carbohydrates, proteins, lignin, and amino acids. It

Table 7.2. Approximate nitrogen requirements of several crops.

Crop	Product	Yield	N requirement
		Mg ha^{-1}	kg ha^{-1}
Alfalfa	forage, dry matter	18.0	600
Corn	seed	12.5	300
Soybean	seed	4.0	360
Cotton	lint	1.3	160
Wheat	seed	5.4	150

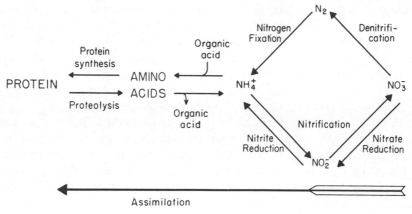

Fig. 7.2. The biological nitrogen cycle.

represents the dead remains of plants, bacteria, fungi, and microscopic animals in various stages of decomposition.

The carbon to nitrogen ratio of soil organic matter is relatively constant at 10:1. While the organic matter content of soils varies widely, we can make a rough estimate that most cultivated soils contain 3000 kg N ha^{-1} or more in the plow layer. This is more than enough N to produce high crop yields (Table 7.2)—if the N in organic matter were available for absorption by plant roots.

Actually, N from organic matter becomes available to plants very slowly. Only 2–3% of the N in organic matter is converted to available forms in a year. Thus, a soil containing 3000 kg N ha^{-1} in organic matter can be expected to provide only 60 kg N ha^{-1} during a cropping season. The conversion of organic matter N to available forms is controlled by the activity of soil microbes and varies little from year to year.

In the final analysis, we find that N must be supplied to most crops to achieve optimum yields. Nitrogen fertilizers most widely used include anhydrous ammonia (NH_3), urea (NH_2CONH_2), ammonium nitrate (NH_4NO_3), and ammonium sulfate [$(NH_4)_2SO_4$]. The latter three forms are often supplied in solutions of varying composition. Once it is dissolved in the soil solution, urea is hydrolyzed by soil microbes to ammonium plus carbon dioxide in a few days. Thus, we need to consider only the fate of ammonium and nitrate.

Most of the soluble nitrogen in well-drained, cultivated soils is present in the nitrate form. This will be the case even after the application of ammonium fertilizer if the soil is well drained and neutral to slightly acidic. The predominance of nitrate is the result of a very important biological process known as nitrification. Nitrification is carried out by several species of soil bacteria that derive energy from the oxidation of ammonium to nitrate (Fig. 7.2).

Nitrification greatly complicates our efforts to manage the supply of available N. This is because ammonium (NH_4^+) bears a positive charge and is bound to negatively charged clay minerals, while nitrate (NO_3^-), with its negative charge, is not bound. Thus, significant quantities of nitrate may be lost through leaching or, to a lesser degree, through surface runoff. Not only is this N wasted from the standpoint of crop production, but also it is considered a pollutant when it enters our supplies of drinking water. Chemicals that inhibit microbial nitrification are now commercially available and may reduce the loss of fertilizer N applied to soil.

A second microbial process, denitrification, may also result in loss of soil nitrogen under special circumstances, namely, when the soil is waterlogged. Denitrification, an anaerobic process, results in the reduction of nitrate to nitric oxide gas (NO), nitrous oxide gas (N_2O), and N_2 gas (Fig. 7.2). These gases are lost to the atmosphere where they are, of course, not available to support plant growth.

Uptake of Nitrogen from the Soil Solution

Essentially all of the N absorbed from the soil by plant roots is in the form of nitrate and ammonium. Because of nitrification, most of the soil N available to plants is in the form of nitrate.

Nitrate is absorbed by the root when ions in the solution around and within root cell walls cross the plasma membrane of the root cell. Nitrate uptake is an active process; i.e., it requires expenditure of energy by root cells. The uptake of ammonium ions by plant roots also probably involves an active mechanism. Uptake of both nitrate and ammonium may be influenced by temperature, pH, and oxygen and carbohydrate supply to the roots.

Once inside the root cells, ammonium and nitrate are converted to other compounds or loaded into the xylem of vascular bundles where they are transported to other parts of the plant in the transpiration stream. It is important to note that nitrate is not toxic to plant cells and may accumulate to high concentrations. In contrast, ammonium is toxic to plant cells, and its concentration in plant tissues is very low. Plants maintain low concentrations by incorporating ammonium into organic compounds near the site of absorption, before loading into the xylem.

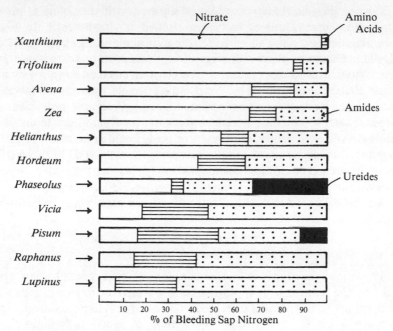

Fig. 7.3. Composition of stem exudate from plants supplied with 140 mg NO_3-N L^{-1} of nutrient solution. The amides were asparagine and glutamine. The structures of the ureides are shown in Fig. 7.6. From Pate (1973). Used by permission.

One of the most definitive ways of determining the fate of N entering the root is to analyze the contents of the xylem just above the soil surface. This can be done with most plants by severing the stem and collecting the stem exudate (or "bleeding sap"). The composition of stem exudate varies widely among plant species even when they are grown under identical conditions (Fig. 7.3). The variation in composition may reflect the presence or absence of root nodules and differences among plant species in nitrogen metabolism in roots.

These differences will be discussed in more detail in later sections. For now, Fig. 7.3 serves to emphasize two points: 1) plants supplied with abundant nitrate may metabolize nearly all of the nitrate (*Raphanus*) or almost none of the nitrate (*Xanthium*) in their roots; and 2) virtually no ammonium is translocated to shoots, indicating that essentially all ammonium formed in or absorbed by roots is assimilated before transport.

Assimilation of Nitrate

The first step in nitrate assimilation (incorporation into organic compounds) is the reduction of nitrate to nitrite, a reaction catalyzed by the enzyme nitrate reductase (Fig. 7.2). Nitrate reductase is found in bacteria, fungi, cyanobacteria (blue-green algae), algae, and higher plants, and the enzyme from many sources has been studied in great detail. In higher plants, reducing power is supplied by the pyridine nucleotides NADH and NADPH. [The concept of reducing power is important to this and later topics. When a substrate is reduced it receives an electron supplied by a reductant. (When a substrate is oxidized, it loses an electron.) In metabolism, the passing of electrons may be mediated by one of several molecules. The foremost examples are the pyridine nucleotides nicotinamide adenine dinucleotide (NAD) and nicotinamide adenine dinucleotide phosphate (NADP). The reduced forms of these compounds are written NADH and NADPH. In cases where either of these compounds may serve as reductant, depending on various conditions, the reductant is written NAD(P)H.]

A simplified version of the nitrate reduction reaction may be written:

$$NO_3^- + NAD(P)H \rightarrow NO_2^- + NADP$$

The amount of nitrate reductase in plant cells is very small when nitrate uptake by the plant is low; rapid formation of active enzyme occurs when nitrate supply to the plant is increased. The enzyme is localized in the cytoplasm of plant cells. In leaves, enzyme activity is significantly higher in the light than in the dark, a response at least partly due to generation of reducing power in the light. Each molecule of nitrate reductase contains one atom of molybdenum (Mo). The Mo atom is located at the active site of the enzyme and participates in the transfer of electrons to nitrate.

The second step in nitrate assimilation is the reduction of nitrite to ammonium, a reaction catalyzed by the enzyme nitrite reductase (Fig. 7.2). In leaves, the reductant is ferridoxin, a small protein that is also involved in the transfer of electrons in the reactions of photosynthesis. Since the reductant is ferridoxin, it is not surprising that nitrite reductase in leaves is localized in chloroplasts. Ferridoxin has not been detected in roots, and the reductant involved in nitrite reduction in roots is not known. Nitrite reductase contains three atoms of iron (Fe) per molecule, and at least one of these Fe atoms participates in the transfer of electrons. A simplified version of the reaction may be written:

$$NO_2^- + \text{Ferridoxin (reduced)} \rightarrow NH_4^+ + \text{Feridoxin (oxidized)}$$

Plants generally have much more nitrite reductase than nitrate reductase. Thus, the conversion of nitrate to nitrite is thought to be the limiting step in nitrate assimilation, and some effort has been devoted to selecting crop plants having high nitrate reductase activity. Unfortunately, the correlation between nitrate reductase activity and yield is not especially high and is not consistent among plant species. No new crop cultivar has yet been selected on the basis of nitrate reductase activity.

Assimilation of Ammonium

Ammonium ions absorbed from the soil solution or generated by nitrite reductase are rapidly incorporated into amino acids. This is usually accomplished in a reaction catalyzed by glutamine synthetase. This reaction requires the expenditure of energy in the form of ATP. [Adenosine triphosphate (ATP) is often involved in energy-requiring reactions. The energy in this molecule is present in the phosphate bonds. Hydrolysis of these bonds generates energy, and the synthesis of ATP from adenosine diphosphate (ADP) requires energy. The use of energy from ATP is mediated via the enzyme involved in the reaction. For example, the active site of an enzyme may be altered slightly during ATP hydrolysis, allowing the binding and/or alteration of the substrate to occur.]

The reaction results in the conversion of one glutamate to glutamine:

$$\text{glutamate} + NH_4^+ + ATP \rightarrow \text{glutamine} + ADP + PO_4^{3-}$$

Note that the assimilation of ammonium in this reaction results in the formation of an amide ($-\overset{\overset{\displaystyle O}{\|}}{C}-NH_2$) group and not a new α-amino ($-\overset{\overset{\displaystyle NH_2}{|}}{C}-COOH$) group. To complete the synthesis of a new α-amino group, the amide N of glutamine is transferred to an α-keto acid to form an α-amino acid. The reaction is catalyzed by the enzyme glutamate synthase, and the reaction requires a reductant. The use of NAD(P)H is illustrated here, but ferridoxin may serve as the reductant in green leaves:

$$\text{glutamine} + \alpha\text{-ketoglutarate} + NAD(P)H \rightarrow 2\,\text{glutamate} + NAD(P)$$

If we sum the two reactions catalyzed by glutamine synthetase and glutamate synthase, the net result is the following:

$$NH_4^+ + ATP + NAD(P)H + \alpha\text{-ketoglutarate}$$

$$\rightarrow \text{glutamate} + ADP + NAD(P) + PO_4^{3-}$$

Another enzyme, glutamate dehydrogenase, is found in essentially all plant cells and is capable of catalyzing a reaction very similar to the one above, namely:

$$NH_4^+ + NAD(P)H + \alpha\text{-ketoglutarate} \leftrightarrow \text{glutamate} + NAD(P)$$

Although ammonium could be assimilated by this reaction, it is now widely accepted that, in higher plants, ammonium is assimilated via glutamine synthetase and glutamate synthase. This conclusion is based on the incorporation of ^{15}N-labeled NH_4^+, the effect of compounds that inhibit specific reactions, and the properties of the enzymes.

Glutamate dehydrogenase probably functions in the conversion of glutamate to α-ketoglutarate plus ammonium. This would allow for the synthesis of glutamine and asparagine from nitrogen in glutamate. This remobilization of nitrogen could be important during the senescence of leaves, for example.

The α-amino N of glutamate is utilized for the synthesis of other amino acids in reactions known as amino transferase reactions. These reactions have the general form:

$$\alpha\text{-keto acid} + \text{glutamate} \leftrightarrow \alpha\text{-amino acid} + \alpha\text{-ketoglutarate}$$

Amino transferase reactions are reversible and do not require the expenditure of energy or reducing power. Synthesis of nearly all of the amino acids in Table 7.1 involves an amino transferase reaction. Two of these, the synthesis of aspartate and alanine, are illustrated in Fig. 7.4. Synthesis of asparagine, an important translocation form of N in many plants, involves the formation of an amide group, and glutamine serves as the amide donor (Fig. 7.4).

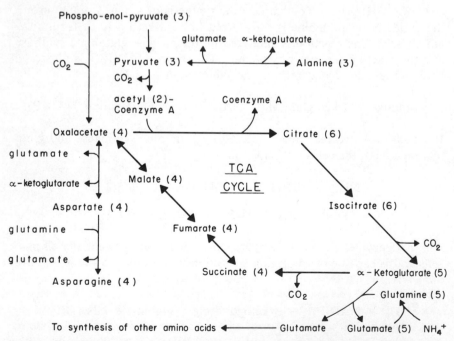

Fig. 7.4. The tricarboxylic acid (TCA) cycle, the assimilation of ammonium, the synthesis of a few amino acids, and the replenishment of the TCA cycle via phospho-enol-pyruvate carboxylase. The number of carbon atoms in intermediates is indicated in parenthesis.

Note in Fig. 7.4 that there is no net generation of organic acids during the operation of the tricarboxylic acid (TCA) cycle; for the input of each acetyl group (two carbons), two CO_2 molecules are evolved. Thus, as α-ketoglutarate is withdrawn from the cycle for the synthesis of glutamate, additional carbon must enter the cycle. This is accomplished by the carboxylation of phospho-enol pyruvate to form oxalacetate. As we would predict from these relationships, plant tissues having high rates of ammonium assimilation also have high rates of CO_2 uptake.

Symbiotic Nitrogen Fixation

Nitrogen fixation involves the synthesis of ammonium from nitrogen gas (Fig. 7.2). This reaction requires large amounts of energy and reducing power to break the $N \equiv N$ triple bond and reduce the N to ammonium.

Many microorganisms are capable of N_2 fixation. Some are free-living organisms such as bacteria (several genera and many species) and cyanobacteria. In some habitats these organisms may constitute a significant source of ammonium for plant growth, but in most cultivated soils these organisms generally fix less than 10 to 20 kg of N ha^{-1} per year. Slightly more N (15–30 kg N ha^{-1}) may be provided by cyanobacteria growing in flooded soils used in the production of rice.

In addition to the free-living organisms, many symbiotic associations capable of fixing N are also known. A symbiotic association is defined as an association where two organisms live together for the benefit of both. Some of the simpler associations include the lichens, which involve fungi and cyanobacteria, and the association between the water fern *Azolla* and the cyanobacterium *Anabaena*. It is hoped that an understanding of these associations will help to unravel the mysteries of more complex associations.

One such complex association involves certain actinomycetes. The association with alder (*Alnus*) trees has been most widely studied although these associations are known to occur among several genera of woody shrubs and trees. Large, perennial nodules are formed on roots and significant quantities of nitrogen are fixed by the actinomycetes that inhabit them. These associations are important to forestry and forest management. However, their importance to field crop production is presently minimal.

In recent years, another association has attracted the attention of crop physiologists. This association involves tropical grasses and N_2-fixing bacteria of the genus *Azospirillum*. It is thought that the bacterium does not actually infect (enter) the cells of the plant root. Instead, the bacteria proliferate around the root and may inhabit spaces between cells. A significant portion of the plant's N requirement may be supplied through N_2 fixation, but this is highly variable among plant species and, especially, among environments. While the impact of this association on crop production is presently minimal, there is hope that this association can be exploited in the decades ahead.

Of all the symbiotic associations known, those involving plants belonging to the family *Leguminosae* and the bacterial genus *Rhizobium* are the most important to crop production. The *Leguminosae* is one of the largest of all plant families; it is comprised of over 14 000 species. The species infected by *Rhizobium* bacteria include such major crops as bean, pea, soybean, alfalfa, clover, cowpea, and lentil. Because of the great importance of these leguminous crops, we will focus our attention on the symbiotic associations of the legumes.

Nodule Formation and Structure

Rhizobia are widely dispersed in the soils of the earth. However, where a legume is to be grown for the first time, the correct species of bacterium should be introduced at the time of seeding. This is accomplished through commercially available *Rhizobium* inoculants. Once introduced, the bacteria survive in the soil for many years. However, it is common practice to resupply the bacteria at every seeding to insure an adequate rhizobial population in the soil.

Rhizobium bacteria enter the legume root via root hairs. A root hair may be receptive to infection for only a short time, but infections occur over periods of weeks or months. The first nodules are formed on the primary root near the soil surface, and later infections occur on secondary roots. The invading bacterium induces the formation of a structure known as an infection thread that grows through the root hair cell and into the cells of the root cortex. The infection thread is comprised of cell wall and membrane materials supplied by the plant. As the thread grows through root cortex cells, bacteria are released from the thread and are enclosed by a portion of the plasma membrane surrounding the thread.

The structure of mature nodules varies among legumes. For example, some nodules are spherical (e.g., soybean, Fig. 7.5), while others are ovoid (e.g., alfalfa). All nodules have in common the presence of an infected region located in the central portion of the nodule (Fig. 7.5). In this region, the plant cells have enlarged and are filled with thousands of bacteria (Fig. 7.5).

During nodule development, *Rhizobium* bacteria undergo morphological and physiological changes. Instead of the typical rod shapes found in the free-living bacteria, the bacteria in nodules assume elongated, twisted, and even branched shapes (Fig. 7.5). The principal physiological change is the expression of N_2-fixing activity, which is usually not expressed by free-living rhizobia. The bacteria in nodules are so unlike their free-living progenitors that they are referred to as bacteroids.

Some nodules undergo a single cycle of development, maturation, and senescence (e.g., soybean). Other nodules retain meristematic regions and have the capability for growth and development over longer periods of time

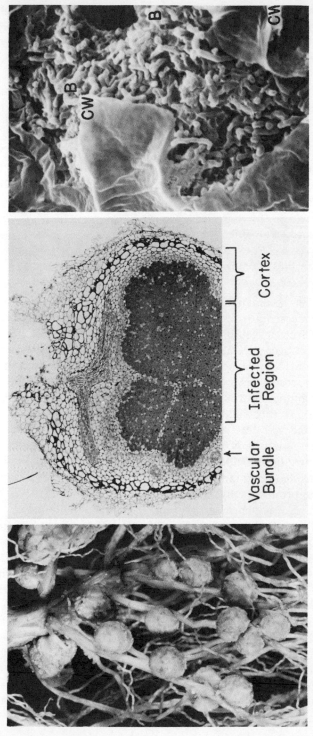

Fig. 7.5. *Left:* Nodulated soybean root, approximately actual size. *Center:* Light micrograph of a cross section of a soybean nodule, magnification about 25 ×. The cross section was taken at the point of attachment of the nodule to the root, and a major branch point of the connecting vascular bundle can be seen at the top of the picture. *Right:* Transmission electron micrograph of an infected cell from a soybean nodule, magnification about 1000 ×. CW = cell wall, B = bacteroid.

(e.g., alfalfa). Microscopic examination of these nodules may reveal regions of new growth, new infection, current N_2-fixing activity, and senescence, all in the same nodule.

Nitrogen and Carbon Metabolism

The reduction of N_2 to ammonium is catalyzed by a very large and very complex enzyme known as nitrogenase. Actually, the enzyme consists of two proteins, a Mo- and Fe-containing protein that catalyzes the $N_2 \rightarrow NH_4^+$ reaction, and a second Fe-containing protein that serves as a reductant for the Mo-Fe protein. The transfer of electrons to N is not completely understood but probably involves several steps:

$$
\begin{array}{c}
\text{Ferridoxin} \\
\quad e^- \quad\quad\text{or}\quad e^- \quad\quad\quad e^- \quad\quad\quad\quad\quad e^- \\
\text{NAD(P)H} \rightarrow \text{Flavodoxin} \rightarrow \text{Fe protein} \rightarrow \text{Mo-Fe Protein} \rightarrow N_2
\end{array}
$$

Nitrogenase has several important properties:

1) The enzyme requires essentially anaerobic conditions. Thus, the concentration of free oxygen in infected cells must be maintained at very low levels. This is accomplished in legume nodules by the presence of leghemoglobin. This red hemoprotein, which closely resembles animal hemoglobin in structure, has a high affinity for O_2. Thus, low O_2 concentration can be maintained around the bacteroids while a supply of O_2 to support respiration is still available.

2) The enzyme will catalyze the reduction of several alternate substrates such as cyanide ($C \equiv N^-$) and actylene ($HC \equiv CH$). These alternate substrates have no physiological significance. However, the reduction of acetylene has provided a convenient and sensitive assay for the enzyme activity because the product, ethylene, is released by bacteroids or nodules and can be analyzed by gas chromatography.

3) The enzyme will also catalyze the reduction of protons to hydrogen gas (H_2). Hydrogen gas may be discharged by bacteroids and nodules, resulting in the wasteful loss of reducing power. Recent studies have found that many bacteroids are capable of recapturing the reducing power in H_2 through the activity of another enzyme termed hydrogenase. There is presently considerable concern about insuring that all rhizobia used in inoculants possess hydrogenase.

Rhizobium bacteria are capable of synthesizing all of the protein amino acids. However, in bacteroids, amino acid synthesis is repressed, and bacteroids excrete the fixed N as ammonium. The assimilation of NH_4^+ takes place in the plant cell cytoplasm. Analysis of nodule exudate and stem exudate has shown that nodules of different legumes export various N-rich compounds. The N exported from legume nodules is generally in the form of amide amino acids (glutamine, asparagine) or ureides (allantoin, allantoic acid).

Synthesis of glutamine and asparagine have already been described briefly and are illustrated in Fig. 7.4. Synthesis of ureides is via a pathway that is also involved in the synthesis and degradation of purines (adenine, guanine). A detailed description of the numerous reactions required for the synthesis of allantoin from NH_4^+ is beyond the scope of this discussion. The structure of the allantoin and allantoic acid, the products of the pathway, are shown in Fig. 7.6. Note that the C to N ratio in these compounds is 1:1.

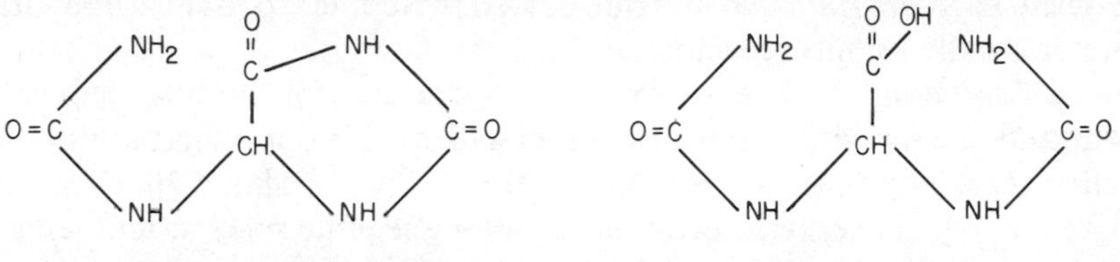

Allantoin   Allantoic Acid

Fig. 7.6. Structures of the ureides allantoin and allantoic acid.

The operation of nitrogenase requires large quantities of reducing power and ATP. The energy and reducing power to support symbiotic nitrogen fixation are derived from organic acids that, in turn, are derived from carbohydrates supplied by the shoot. Sucrose is the compound generally supplied by shoots, and positive correlations between sucrose content of nodules and their N_2-fixing activity have been observed.

Factors Affecting Nodule Activity

The seasonal amount of N fixed by a legume crop varies widely depending on plant species, the soil and atmospheric environments, and crop management practices. Most estimates of N supplied to legumes via symbiotic fixation range from 100–200 kg N ha^{-1}. Quantities in this range are large enough to be of great significance in crop production, but, in general, are not large enough to supply all of the N required for a maximum yield.

The balance of the N required by legume crops is obtained by absorption of nitrate and ammonium from the soil. In most cultivated soils, the amount of available soil N is usually sufficient to supplement N fixed by nodules, and the application of N fertilizer is not recommended. This is especially true where legumes are grown in rotations with nonleguminous crops that have been fertilized with adequate N.

Before concluding our discussion of legume nodules, we should give some attention to factors that may influence their activity.

1) *Mineral Nutrition and pH.* Under specialized circumstances, any one of the major or minor nutrients (see later sections) may limit nodule growth and activity. However, in productive, cultivated soils, this is rarely a

problem. Limitations by soil acidity or alkalinity are more common problems, and soil pH between 6.0–6.5 is required for optimum nodulation.

2) *Photosynthesis.* Because large amounts of reducing power and energy are consumed by nodules, it is not surprising that factors influencing carbohydrate supply affect N_2-fixing activity. Disruptions of photosynthesis by partial defoliation or shading reduce nodule activity. Conversely, increasing the CO_2 concentration supplied to shoots will increase photosynthetic activity of shoots and the total N_2-fixation activity of nodulated roots. However, in the absence of experimental manipulations like these, sufficient carbohydrate may be transported to nodules so that carbohydrate is not normally a limiting factor.

3) *Temperature.* The optimum temperature for nodule activity is around 25°C. Activity declines on either side of this optimum but does not decline markedly until temperatures below 20°C or above 30°C prevail. There are, of course, large differences among legume species, and temperate legumes have lower temperature optima than tropical legumes.

4. *Combined Nitrogen.* Combined N (ammonium, nitrate) may adversely affect infection of root hairs (number of nodules/plant), nodule growth (mass/nodule), and nodule activity (N_2-fixation/nodule mass). The impact of combined N on these three parameters is directly proportional to the amount of N supplied. Nodule formation may be completely eliminated if enough N is supplied. If high rates of N are supplied to legume plants that are already nodulated, N_2-fixing activity of nodules can be reduced to zero. The mechanisms underlying these effects are not known.

5. *Water.* Insufficient water supply can depress nodule activity in two ways. Water stress is known to reduce photosynthetic activity of leaves because of closing of stomata and concomitant decline in diffusion of CO_2 into the leaf. If this condition prevails long enough, the supply of carbohydrate to nodules is reduced. Secondly, water stress probably has a direct effect on nodule activity resulting from a decline in the diffusion of oxygen into the nodule and a concomitant decline in respiration. An oversupply of water (flooding) drastically reduces nodule activity because of restricted oxygen supply.

OTHER NUTRIENTS

There are two separate but equally important aspects of mineral nutrition: 1) uptake of nutrients from the soil and 2) utilization of the element within the plant. Plant species and genotypes within species differ in nutrient uptake and/or utilization in the plant. These differences play a major role in determining not only a plant's ability to survive in the ecosystem but also its potential productivity. This section describes how nutrients are used by plants and some of the ways in which plants increase the absorption of nutrients that are in deficient supply in the soil.

Roots generally explore and contact only 1–2% of the total soil volume. Therefore, nutrient absorption by direct contact with soil particles is minimal, and the root must obtain nutrients by other means. The soil solution surrounding the root is the major source of nutrients, but the concentration of nutrients in soil solution near the root is rapidly depleted by a rapidly growing plant. Ion concentration near the root is replenished by deadsorption from soil particles, diffusion, and mass flow of water to the root (as the root takes up water, water in the soil moves toward the root carrying with it ions from soil farther away from the root.)

Phosphate (PO_4^{3-}) is relatively immobile in the soil and can only diffuse a fraction of a millimeter away from the soil colloids. Thus, PO_4^{3-} uptake is often a limiting factor in crop productivity. Plant species adapted to low P soils usually have larger, more fibrous root systems, which allow for more efficient mining of the soil volume. Other genetic factors that allow plants to survive on low P or other soils with nutrient problems will be discussed later in this chapter.

Nutrient Deficiency

A plant's growth rate may be controlled by the supply of nutrients in the plant as illustrated in Fig. 7.7. For example, if the uptake of a nutrient such as N is reduced because of reduced availability in the soil, the content

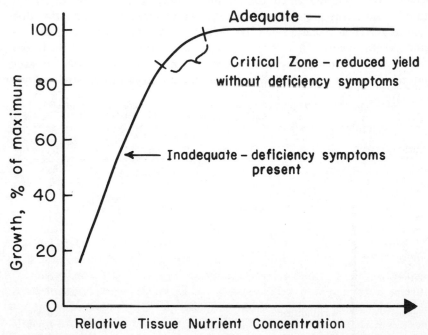

Fig. 7.7. Generalized description of crop growth as a function of relative tissue nutrient concentration.

of N falls in the plant. The plant responds to this imbalance by lowering its rate of N utilization, resulting in a reduced rate of growth and dry matter accumulation. If the deficiency becomes more severe, further reductions in plant growth may not be sufficient to maintain internal plant N content, and the plant begins to break down proteins and other N-containing compounds in older tissue for use in young leaves or developing fruits. These drastic events lead to visual nutrient deficiency symptoms characteristic for each element. The plant is "programmed" to complete the reproductive process and will readily sacrifice old leaves or other noncritical tissue to remobilize sufficient N to continue growth.

The important point to remember is that visual deficiency symptoms represent the most severe aberrations in plant growth and metabolism. Significant reductions in plant productivity can occur with no visual symptoms and can only be detected by tissue analysis. Tissue samples are analyzed for mineral concentration and the results compared to normal element concentration for that crop at a specified stage of growth.

Metabolic Role of Nutrients

The essential mineral elements required for plant growth and reproduction are shown in Table 7.3. In addition, silicon (Si), cobalt (Co), sodium (Na), and vanadium (V) are required for normal growth by certain plant species. The essential elements are commonly grouped into macronutrients and micronutrients as shown in Table 7.3. However, this classification does not imply that any one element or class of elements is more essential for plant growth. Deprivation of any essential element inhibits growth. Classification as a macronutrient and micronutrient only refers to the relative amount of the element required by the plant for maximum growth.

Table 7.3. Elements required for growth of most higher plants and typical levels found in dry matter.

Element		Chemical symbol	Percentage in dry matter
Oxygen		O	45
Carbon		C	45
Hydrogen		H	6
Nitrogen		N	1.5
Potassium		K	1
Calcium	Macronutrients	Ca	0.5
Magnesium		Mg	0.2
Phosphorus		P	0.2
Sulfur		S	0.1
Chlorine		Cl	0.01
Boron		B	0.002
Iron		Fe	0.01
Manganese	Micronutrients	Mn	0.005
Zinc		Zn	0.002
Copper		Cu	0.0005
Molybdenum		Mo	0.00001

While we know a great deal about the role of N and P in plant metabolism, we know much less about the function of other nutrients, especially micronutrients that are required in only trace amounts.

The role of nutrient elements in plants can be generalized into the following categories:

1) *Building Materials.* Besides the elements carbon, hydrogen and oxygen, some mineral elements such as nitrogen, phosphorus, sulfur, magnesium, and calcium form essential, structural parts of molecules.

2) *Control of Cell Osmotic Potential.* Regulation of cell water content and water potential is essential for plant metabolism. The high concentration of cations in plants contributes to and aids in regulating cell osmotic potential.

3) *Regulation of pH and Maintenance of Electrostatic Neutrality.* Elements are absorbed in the ionic form. Plants minimize changes in pH and maintain neutrality by altering the ratio of cations and anions present in the tissue.

4) *Regulation of Membrane Permeability and Hydration.* Proper functioning of membranes in the cell is essential for normal plant metabolism. The relative balance of monovalent (K^+ and Na^+) and divalent (Ca^{2+} and Mg^{2+}) cations affects hydration, permeability, and metabolic activity of membranes.

5) *Catalytic Activity.* Minerals affect the activity of enzymes by determining the three-dimensional structure of the enzyme. Elements may also be required at the active site and participate directly in the enzymatic reaction.

We will now describe some of the major roles of the macronutrients in plant growth and metabolism.

Phosphorus

Much of the P in a normal plant is found in the vacuole where it is not available for use in cellular activities. However, if P becomes deficient, it moves out of the vacuole into the cytoplasm and becomes metabolically active. Phosphorus is usually absorbed as the $H_2PO_4^-$ anion and is not reduced in the plant as are the other major anions, NO_3^- and SO_4^{2-}.

Phosphate is an essential component in the structure of many plant compounds. One of the most important groups of phosphate-containing compounds are the nucleotides, the basic structural unit of nucleic acids (Fig. 7.8). Individual nucleotides, in particular adenosine triphosphate, described elsewhere in this chapter, are extremely important in the transfer of energy in biochemical reactions.

Another structural role of P in cells is its function in phospholipids. The highly polar phosphate anion is attached to one end of a nonpolar long-chain lipid molecule. The layered arrangement of phospholipids contributes

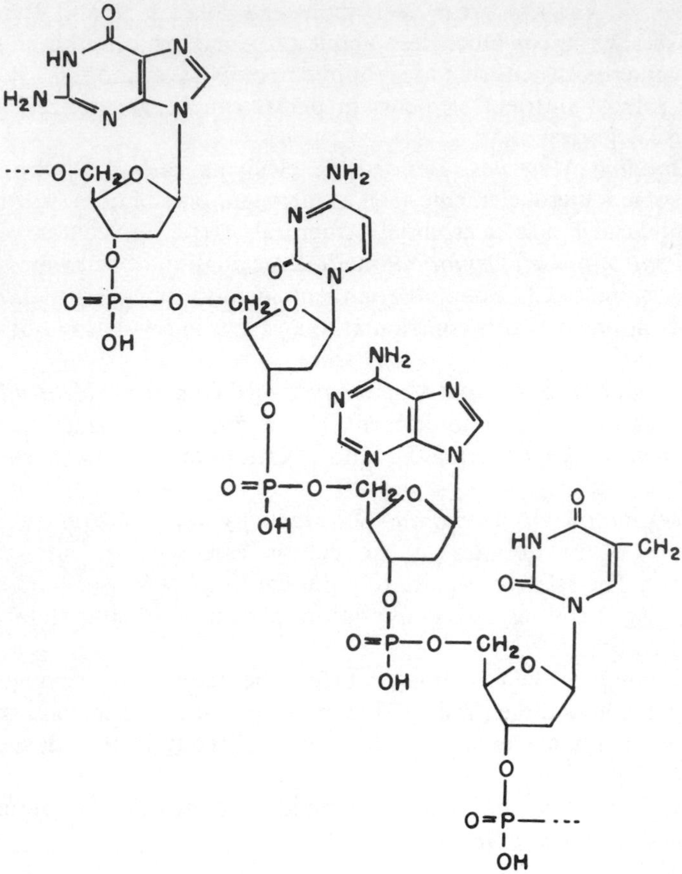

Fig. 7.8. Portion of DNA chain. Note that each nucleotide unit is connected via phosphate linkage.

to the unique characteristics of membranes, particularly their semiperme-
able nature. Large quantities of P are required in young cells, where the rate
of metabolism is high and cell division (membrane synthesis) is rapid. Phosphorous is a major component in starter fertilizer because of its importance
in promoting seedling growth.

Sulfur

Although the concentration of S in plants is similar to P concentration,
the importance of S in plant metabolism is not often stressed. There are
several explanations. First, S is not often deficient in the soil and is not a
major factor limiting yield compared to N, P, and potassium (K). Second, S
is not as widely utilized in plant organic compounds as N or P and thus may
be considered less important in plant metabolism.

Sulfur is taken up as the sulfate ion (SO_4^{2-}), and most is reduced to the sulfhydryl form (SH) before it is incorporated into organic matter. Sulfur is an important component of two amino acids, cysteine and methionine (Table 7.1). Since animals cannot reduce sulfate, plants play a vital role in supplying essential S-containing amino acids to them.

The SH group of cysteine is very important because it can be oxidized to form a disulfide (S–S) bond when the SH groups from two cysteine molecules interact. The disulfide bond is an important feature in stabilizing the three-dimensional structure of many enzymes since it can link chains of amino acids together to form the three-dimensional structure necessary for enzyme activity.

Potassium

While Ca is usually the predominant cation in the soil, K is usually the predominant cation in the plant. Plants require large quantities of K for growth, but the reasons for these high requirements are still not fully understood. However, the chemical properties of K, especially its chemical activity and its hydrated ionic radius, confer special properties to this element. The shell of water surrounding each K ion has the smallest radius of any cation (Na^+, Ca^{2+} or Mg^{2+}). Because of its size and mobility, K acts as a moisturizing agent to maintain hydration of cellular microstructures, including membranes. In this role K acts as a metabolic activator and promotes metabolism.

Some of the other major roles of K in the plant are provided below.

1) *Neutralization of Acids Produced During Metabolism.* During the metabolism of carbohydrates, organic acids are synthesized and accumulate in the cell. These anions must be neutralized with a cation, and, because of the abundance of K, organic acids are usually present as potassium salts.

2) *Effects Movement of Leaf Stomata.* Potassium is intimately involved in the opening and closing of stomata. Cells surrounding the guard cells can actively secrete K into the guard cells. The concentration of K in the guard cells is raised to high levels ($\geq$ 300 mM), which decreases the water potential in the guard cells. The decreased water potential promotes water movement into the guard cells, and the increasing turgor pressure results in the opening of the stomata. When the stomata close, the process is reversed, and K is actively excreted from the guard cells.

3) *Activation of Enzymes.* A large number of enzymes (at least 45–50) have been reported to be activated by K. Although K has not been shown to be required in any enzyme structure, its presence is required for the reaction to occur.

4) *Stimulation of Transport Processes Within the Plant.* Potassium has been shown to promote translocation of photosynthate from the leaves.

Rapid removal of photosynthate from leaves is necessary if high rates of photosynthesis are to be maintained. The increased translocation of photosynthate to roots and nodules may be important in the higher rate of symbiotic N_2 fixation observed under high K fertility.

Calcium

Calcium rarely limits the growth or yield of major crops with the possible exception of peanut. The concentration of Ca^{2+} in healthy tissue ranges from about 0.2% (dry weight basis) to several percent.

The bulk of the Ca^{2+} in the plant is found as the salts of anions, including organic acids, sulfates, and phosphates. These salts are found mainly in the vacuole. Calcium can render some anions insoluble and thus reduce their activity. For example, oxalic acid is produced in potentially toxic quantities in some plants, but it is detoxified by the formation of insoluble calcium oxalate.

Calcium also serves to counteract the effect of nutrient imbalances in the cell. An excess concentration of K or even Na may injure the plant by promoting too much membrane permeability. Calcium serves to reverse the effects of K and Na and thus limits their potentially injurious effects.

Calcium also has a major role in membrane structure and plasticity. Calcium combines with pectins (long-chain polymers of D-galacturonic acid, a carbohydrate derivative) and, when incorporated into the cell wall with protein, serves as a "cement" to strengthen the structure and give it flexibility. Once membranes have been formed, the presence of Ca^{2+} is necesary for continued maintenance of membrane integrity and function.

Lastly, in contrast to the other major cations K^+ and Mg^{2+}, Ca^{2+} appears to have limited roles as an enzyme cofactor.

Magnesium

The most commonly recognized role of Mg^{2+} in plants is in the structure of chlorophyll (Fig. 7.9). In fact, one of the first effects of Mg deficiency is the localized destruction of chlorophyll and chlorosis of leaf tissue.

Magnesium has much greater chemical activity and mobility than its sister divalent cation, Ca^{2+}. This is probably why Mg^{2+} is found as a cofactor in a number of enzymatic reactions. Some of the more important enzymes requiring Mg^{2+} are involved in energy transfer. Most enzymatic reactions involving the transfer of PO_4^{3-} require the presence of Mg^{2+}. Magnesium is also required for ribosome integrity and contributes to the structural stability of nucleic acids.

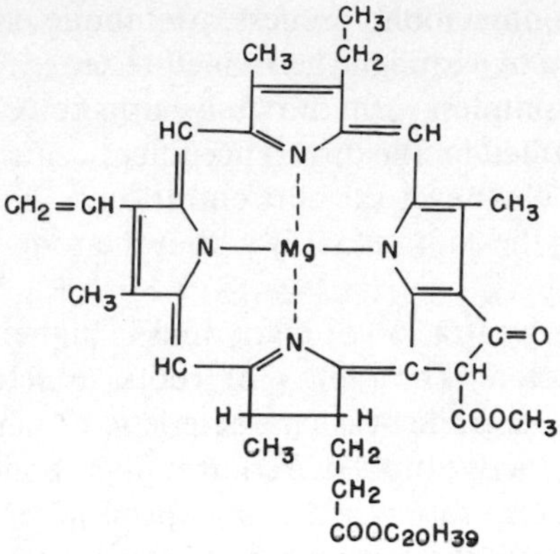

Fig. 7.9. Structure of chlorophyll a. Four nitrogen-containing rings, called pyrrole rings, are linked to a central Mg ion.

Micronutrients

The metallic trace elements are Fe, Mn, Zn, Cu, and Mo. Their concentrations in plants varies from 2 to 3 mmoles/kg for Fe down to around 0.01 mmole/kg or less for Mo. The roles of these elements are essentially enzymatic.

The role of Fe in plants has been studied more than the role of any of the other trace metals. Most of the Fe in the plant is involved in oxidation-reduction reactions. Iron facilitates electron transfer because of its variable redox potential ($Fe^{2+} \leftrightarrow Fe^{3+}$). Much of the iron in the plant is found in the mitochondria and chloroplasts, where these oxidation-reduction reactions are very prominent. Other trace metal elements can also have multiple valences and serve an oxidation-reduction role in enzymatic reactions. For instance, Cu is necessary for electron transfer in the oxidative degradation of phenolic substances, and Mo is involved in electron transfer in nitrate reduction.

In addition to the metallic trace elements, other elements that appear essential for growth of certain plants have been identified. These include Na, Si, Cl, B, V, and Co. Boron appears to be required by all plant species while the essentiality of the others is highly variable among species.

Plant Factors that Influence the Uptake and Utilization of Minerals

Efficiency of Uptake

Mineral nutrients are absorbed into the root in ionic form as either cations or anions. At the root surface or free space within the root, absorption can occur by either passive or active uptake mechanisms. Passive uptake is a physical, nonmetabolic process. Metabolic energy is not required and ions are absorbed by virtue of their kinetic energy.

Diffusion is the simplest form of passive uptake. The rate of ion uptake by diffusion is controlled by the difference in concentration of ions outside and inside the root. Whenever the concentration of an ion is higher in soil solution surrounding the root than it is within the root, net migration of the ion into the root occurs.

However, the concentration of many ions is higher inside the root than it is in the soil solution. The ability of roots to selectively absorb ions against a concentration gradient is characteristic of active ion uptake. Unlike passive uptake, active uptake requires metabolic energy. Anything interfering with root respiration and energy production (e.g., anaerobiosis) not only reduces ion uptake but also can increase loss of ions previously acquired and retained by the root.

The fact that an element is generally deficient in the soil does not necessarily mean that a plant growing on that soil will also be deficient in that element. Various plant species have evolved mechanisms allowing them to increase absorption of nutrients from a soil deficient in one or more nutrients. Roots of many plants develop associations with fungi called mycorrhizae (root fungi). These fungi obtain organic nutrients from the plant for their growth. In return the hyphae of the fungi, which permeate the soil surrounding the root, effectively increase the absorbing surface of the root. The hyphae absorb minerals (especially PO_4^{3-}) and move them to the root surface where the plant can absorb them. Several studies have shown that the efficiency of root P uptake is related to the population of mycorrhizae around the root. As we mentioned earlier, species with increased numbers of fine roots and roots with more root hairs also display increased absorption of nutrients because of their greater root surface area.

Roots of certain plant species are also able to increase P uptake by releasing an enzyme (acid phosphatase) that speeds the release of PO_4^{3-} from organic compounds close to the root. Some plants are also able to excrete protons and thus acidify the soil around the root. This increased acidity is believed to speed release of PO_4^{3-} from minerals in P-deficient soils.

Soybean cultivars capable of increasing Fe uptake under Fe-deficiency conditions are called Fe-efficient types. The controlling factor appears to be the ability of Fe-efficient roots to reduce ferric iron (Fe^{3+}), which they cannot absorb, to ferrous iron (Fe^{2+}), which they can absorb. It is obvious that plant species and genotypes with these mechanisms for increasing Fe uptake have a distinct advantage and are much more vigorous and productive on soils where Fe is limiting.

We have been discussing some of the ways plants increase uptake of specific nutrients present in the soil at deficient levels. Many soils have the opposite problem in that they possess toxic levels of nutrients such as Mn and Al because of low pH. Species able to survive under these conditions have evolved systems to modify their root environment or have other physiological mechanisms to avoid the toxic condition. Roots may excrete hydroxyl ions and make the root environment more alkaline, thus reducing the concentration of Al and Mn near the root surface. Recent studies on Al tolerance in wheat cultivars show vast differences in cultivar response to Al and in productivity of cultivars grown on low-pH soils.

Efficiency of Utilization

Genotypes that are efficient with respect to mineral nutrition produce more dry matter from a given amount of absorbed nutrient than do inefficient genotypes. For example warm-season grasses such as Caucasian bluestem and switchgrass can produce up to twice as much dry matter per unit P absorbed than cool-season grasses such as tall fescue and orchardgrass. Both cool- and warm-season species absorb similar amounts of total P, but because of the higher yield of warm-season grasses, the concentration of P in dry matter is about half that found in the cool-season grasses. The P-efficient warm-season grasses appear to have a lower internal P requirement for growth than cool-season grasses. It has been suggested that efficient species or genotypes are able to maintain a higher nutrient concentration at critical sites in the cell where the nutrient is required.

Efficiency is also related to the distribution of nutrients within the plant. Inefficient genotypes may exhibit adequate rates of nutrient uptake but fail to transport an adequate amount of the nutrient from the root to the shoot where it is needed. There may also be differences in the distribution of a mineral nutrient within the shoot or an organ, which also result in differential efficiency of utilization.

Efficiency of nutrient utilization is under genetic control. For instance, efficiency of K utilization in bean appears to be controlled by a recessive pair of alleles. It is now realized that some very valuable characteristics related to efficient nutrient utilization may have been repressed in the process of selecting for agronomic characters on nutrient-rich soils. Our approach to increasing yield has been to modify the soil (fertilizer, lime, etc.) to meet crop needs. We are now beginning to realize that significant improvements in crop productivity can also be achieved if we select genotypes adapted to the soil.

REFERENCES

Aldrich, S. R. 1980. Nitrogen—In relation to food, environment, and energy. Illinois Agric. Exp. Stn., Urbana, Ill.

Epstein, E. 1972. Mineral nutrition of plants: principles and perspectives. John Wiley & Sons, Inc., New York.

Gauch, H. G. 1972. Inorganic plant nutrition. Dowden, Hutchinson, and Ross, Stroudsburg, Pa.

Hardy, R. W. F., and A. H. Gibson (ed.). 1977. A treatise on dinitrogen fixation. Section III: biology; and Section IV: Agronomy and ecology. John Wiley and Sons, Inc., New York.

Hewitt, E. J., and C. V. Cutting (ed.). 1979. Nitrogen assimilation of plants. Academic Press, New York.

Martin-Prevel, P. 1978. The role of nutrient elements in plants. Fruits 33(7–8):521–529.

Miflin, B. J. (ed.). 1980. Amino acids and derivatives. *In* P. K. Stumpf and E. E. Conn (ed.) The Biochemistry of plants, Vol. 5. Academic Press, New York.

Pate, J. S. 1973. Uptake, assimilation and transport of nitrogen compounds by plants. Soil Biol. Biochem. 5:109–119.

Quispel, A. 1974. The biology of nitrogen fixation. Am. Elsevier Publishing Co., New York.

SUGGESTED READING

Bonner, J., and J. E. Varner (ed.). 1976. Plant biochemistry. 3rd ed. Academic Press, New York.

Clarkson, D. T., and J. B. Hanson. 1980. The mineral nutrition of higher plants. Annu. Rev. Plant Physiol. 31:239–298.

Guerrero, M. G., J. M. Vega, and M. Losada. 1981. The assimilatory nitrate-reducing system and its regulation. Annu. Rev. Plant Physiol. 32:169–204.

Lafever, H. N. 1981. Genetic differences in plant response to soil nutrient stress. J. Plant Nutr. 4(2):89–109.

Miflin, B. J., and P. J. Lea. 1977. Amino acid metabolism. Annu. Rev. Plant Physiol. 28:299–329.

Pate, J. S. 1980. Transport and partitioning of nitrogenous solutes. Annu. Rev. Plant Physiol. 31:313–340.

Phillips, D. A. 1980. Efficiency of symbiotic nitrogen fixation in legumes. Annu. Rev. Plant Physiol. 31:29–49.

Stevenson, F. J. (ed.). 1982. Nitrogen in agricultural soils. Am. Soc. of Agron., Madison, Wis.

8 Environmental Stress Influences on Plant Persistence, Physiology, and Production

JERRY D. EASTIN AND
CHARLES Y. SULLIVAN
Department of Agronomy
University of Nebraska
Lincoln, Nebraska

The beginning student may observe a confusing array of plant responses when a complex combination of stresses is imposed on a plant system. The effects of stresses such as temperature or water extremes are difficult to analyze individually because one stress seldom occurs in the absence of some other stress. For example, either very cold or hot temperatures alter the activity of water in plants. We shall consider the separate effects of individual stresses on plants and the interaction of stresses, namely water and temperature extremes, in terms of plant reaction. This will provide a reasonable background to consider questions on specific crops under specific environmental circumstances.

TERMINOLOGY

Types of Stress

Physical stress (Levitt, 1956, 1972) is measured in terms of force per unit area (pressure or suction/unit area), while the resultant strain is measured in terms of dimensional changes (shape, volume, etc.). Biological

Published in *Physiological Basis of Crop Growth and Development,* © American Society of Agronomy—Crop Science Society of America, 677 South Segoe Road, Madison, WI 53711, USA.

stresses differ from mechanical stresses in at least two important ways. First, plants may erect barriers to an environmental stress such as the closing of stomata (the barrier) to reduce water loss in a very dry environment (the stress). The component of biological interest in this system is water. The stress placed on the water would ordinarily be measured in terms of energy rather than force per unit area as in a purely physical system. The work capacity or chemical potential (activity) of the water relates to its functional capacity in the biological system. This point will be elaborated later. Second, stresses in biological systems sometimes cause permanent injury to the system. The strain effect might be loss of photosynthetic capacity (chemical), cytoplasmic streaming (chemical and physical), or inhibition of cell size expansion by imposing rigidity in cell walls (physical). Chemical and physical aspects of stress effects are usually difficult to separate and are rather arbitrary. However, strains induced in biological systems do not always result in dimensional changes in the system (as in purely physical systems) but may produce more subtle chemical and/or physical changes.

Plants vary in their abilities to resist damage due to various stresses. These resistances may be categorized according to the ability to avoid the stress or tolerate it. Hence, a freeze-resistant plant might avoid the stress by lowering freezing point below the lowest exposure temperature and thereby evade the stress. Alternatively, the plant might freeze without suffering damage (stress tolerance).

These concepts as outlined primarily by Levitt (1956, 1972) are generally useful in orienting a student's thinking about the implications of stress on plant development and yield.

Economic Perspective

Economic yield of food products ordinarily depends on 1) crop production efficiency and 2) length of the production period.

Production Efficiency

Crop production efficiency is the dry matter accumulation rate considered in terms of units of production input (quantities of light, nutrients, water, etc.). As detailed earlier, dry matter production depends on light energy transformation to chemical energy through photosynthesis. Simple chemical products of photosynthesis are then translocated to the sites of utilization and are elaborated into the storage components of interest (grain, melons, nuts, etc.). Efficiency of light energy conversion to economic product depends on a host of factors such as carbon dioxide supply, light intensity, light interception by the crop, soil fertility level, water availability, temperature, other environmental factors, and the plant itself.

Production Time

Total production is the product of efficiency or dry matter accumulation rate and time. For example, cereal grain yields equal grain accumulation rate × days in grain filling. Some of our research shows that under ideal conditions corn outyields grain sorghum by 30%, largely because its grain fill period is 35% longer. Photosynthesis rates are similar as are daily production rates. Frequently, analysis of limitations imposed by environmental stresses can be productively considered in terms of these efficiency and time factors. However, such an analytical approach may not be pertinent where the economic value of the products depends heavily on special quality factors, aesthetic factors (flower size or color), or recreational factors (golf greens, etc.).

STRESS INTERACTIONS

Some stresses, such as minerals or salts, are specific to limited locations whereas variations in temperature and water are universal. As mentioned earlier, water stress rarely occurs in nature in the absence of a temperature influence and vice versa. For example, field drought is generally accompanied by high temperatures. Likewise, cellular desiccation can be a direct result of freezing temperatures. The critical role of temperature–water interactions in natural production systems is obvious, and the need to measure these two variables is clear.

Temperature is commonly measured with thermistors, thermocouples, and infrared thermometry. Other techniques are available but receive less use. In general, the measurement is relatively simple compared to the difficulty in evaluating plant and soil water status.

Kramer and Brix (1965) listed desirable characteristics for measuring plant stress in general and water status in particular. Of chief concern was that the stress measurements correlate well with changes in physiologic process rates. The two most popular measurements currently are relative water content and water potential. Osmotic potential is growing in popularity.

Relative water content is a measure of the amount of water in the plant at the time of sampling relative to the amount at full turgidity. Relative water content (RWC) or relative turgidity (RT), as introduced by Weatherley (1950), is defined as:

$$100 \left(\frac{W_f - W_d}{W_t - W_d} \right)$$

where W_f is fresh weight, W_d is dry weight, and W_t is the turgid weight obtained by floating the tissue on water. Sullivan (1971) reviewed some of the difficulties arising with this method. Cell growth can occur during the water infiltration period, water infiltrates cut surfaces and intercellular spaces, and respiratory weight losses occur.

Currently the most popular index of plant water status is probably water potential. A limited derivation of water potential is given in the appendix. Only an intuitive view of water potential follows in the text. Basically, water potential is a thermodynamic term relating to the capacity of water to function in work processes in a system. Work capacity of water in a plant system depends on the chemical activity of the water. Chemical activity is generally quantitated or defined in terms of the mole fraction of water in the system. That is, the level of activity relates to the quantity of water in the system. The activity of water or water potential is related to that of pure water, which arbitrarily is assigned a value of zero. Thus, any factor reducing the activity of water in the system results in a negative value for water potential.

For example, plant leaves are cooled by transpiration. Under well-watered conditions, water is readily available (high water activity) and transpiration is rapid. Conversely, under drought conditions, the mole fraction and activity of water are lowered, which contributes to reduced transpirational cooling and elevated leaf temperature. The plant characteristics influencing water potential that relate to transpiration are interrelated, numerous, and complicated. Examples are root ability to grow and scavenge for water, internal plant resistance to water flow, and stomatal closure. A general water potential equation follows:

$$\psi = \psi_s + \psi_p + \psi_m$$

where ψ_s is the potential due to solute effects; ψ_p is the pressure potential created when cell solutes cause water to osmose into the cell and exert pressure (turgor) against the cell wall thereby increasing water activity; and ψ_m is the matrix potential resulting from adsorption of water to cell wall surfaces and contents.

How do these different potentials influence water potential? The ψ_s component reduces water potential (activity) because the solutes reduce the mole fraction of water in the system. Also, weak bonds between the solutes and water may reduce the activity of the water. ψ_p increases cell water activity by virtue of the limited cell wall expansion and resultant pressure imposed as water osmoses into the cell. The ψ_p has a positive value when the cell is turgid. It may rarely be negative when the plant is extremely water stressed. ψ_m restricts water activity through adsorption of water to cell walls and components. As mentioned earlier, the activity or water potential of pure water is set at zero. Changes in plant ψ caused by ψ_s, ψ_p and ψ_m result in negative values. ψ is expressed in joule m^{-3}, MPa (megapascals), or bars. Consider an example of how ψ measurement relates to plant functions.

Photosynthesis is sharply reduced in most plants at 1.9 MPa. Even in a relatively drought tolerant crop like sorghum, potentials of -1.5 to -1.9 MPa can cause partial stomatal closure. However, in real life, as stomata close leaf temperature usually rises, which increases vapor pressure and tends to maintain or even increase transpiration. More examples will appear later in the chapter.

Plant water is often visualized in terms of movement or flow in plants. Our leaf with a ψ of -1.94 MPa may have a lower capacity to transpire water than a leaf at -0.5 MPa (a higher ψ). Transpirational cooling may be impaired at -1.94 MPa because of both reduced leaf water vapor pressure and partially closed stomata.

TEMPERATURE STRESSES

Freezing Temperatures

Freezing temperature problems differ with general climatic patterns. A plant persistence problem exists during the fall and winter where winters are severe, particularly for winter annuals such as wheat, and rye and for perennial forages such as alfalfa and many grasses. Late spring frosts and freezes in these same climates sometimes destroy flowers in early blooming plants and cause great damage to economic crops such as apples. The latter is more a plant protection than a plant persistence problem. Plant protection problems are common in more southerly USA locations such as Florida and south Texas where frost has been known to destroy large portions of the citrus crops. Similar freeze damage problems occur in northern and central USA with apple and peach. Economic losses from freezing damage are staggering. Volumes of freezing damage literature have been reviewed by Levitt (1980a, b, 1972), Mazur (1968), and Olein (1967).

Supercooling of the cell and its external (extracellular) solution often precedes freezing. Ice usually forms first in the extracellular solution. As the amount of ice increases, solute concentration increases. Consequently, the mole fraction of water decreases (Appendix eq. 1) and vapor pressure of the extracellular solution is lowered (Appendix eq. 2). When vapor pressure of the extracellular solution declines below the vapor pressure of the solution inside (intracellular solution), the cell water will flow out of the cell. In other words, ψ (Appendix eq. 4) of the extracellular solution is lower (more negative) than ψ of the intracellular solution, and water flows down that gradient to the extracellular solution (Fig. 8.1). This flow concept should be studied carefully since it is pertinent to many water relations studies. As water continues to move out of the cell, the intracellular solute concentration increases and further reduces the intracellular solution freezing point. The plasma membrane excludes seed ice crystals from the cell, and no ice forms within the cell in many plants until a temperature of about $-10°C$ is reached.

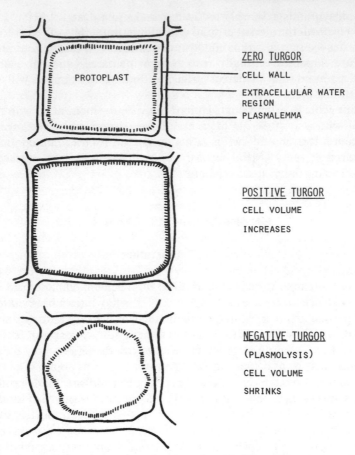

Fig. 8.1. Cell condition at zero turgor (incipient plasmolysis), positive turgor, and negative turgor.

   Dehydration from extracellular freezing eventually causes severe contraction of the cell and sometimes collapse of the protoplast (Fig. 8.1). Tissues contract simultaneously since cells are interconnected by plasmodesmata. If the cell is dead, the collapsed protoplast will not reabsorb water when thawing occurs, even though the cell becomes turgid. Rapid thawing and low permeability of the plasma membrane to water may force separation of the protoplast from the wall of a living cell. Intracellular freezing, contrasted to extracellular freezing, is usually lethal. Cause of injury is presumed to be protoplast disruption. Intracellular freezing is more likely to occur at very rapid cooling rates.
   Injuries resulting from extracellular freezing are usually indirect injuries resulting from dehydration. Possible injuries induced by dehydration include decreased cell volume below some critical level, decreased separation of functional macromolecules, increased intracellular and extracellular solute concentrations, solute precipitation, adverse pH changes, and gas exchange interference (smothering) from the ice itself. Obviously a number of

these events are interrelated, which contributes to the uncertainty of which one or more of the events inflict injury. Undoubtedly, different injury mechanisms operate in different plants depending on what the freezing conditions are. The relative importance of some of the above can be deduced from examples of resistance to extracellular freezing damage.

Resistant plants either avoid or tolerate freezing temperatures. Cells more permeable to water permit a more rapid flow of water out when extracellular freezing occurs. This lowers the freezing point by increasing cell solute concentration and avoids freezing up to a point. Damage may still result in some cells when a critical minimum size limit is reached even though freezing does not occur. Hardy cells often exist with a greater fraction of their water frozen than nonhardy cells. The important quantitative interrelationships between avoidance and tolerance mechanisms are illustrated in Fig. 8.2. Note that a hardy plant with a freezing point lowering of 3°C (freezes at −3°C) will kill at −6°C if its dehydration tolerance is 50%. However, if hardening increased the plant's dehydration tolerance to 80%, the killing point goes from −6 to −15°C. The lowest freezing point lowerings known in hardy plants are about −4°C. If plants are to survive freeze-killing temperatures in the −40°C range, they must tolerate a freeze dehydration of 90% of the plant's water. Freeze dehydration tolerance in the order of 85–89% is known in wheat (Johansson and Krull, 1970, and Johansson, 1970). Tolerance of dehydration strain appears to account for a sizeable portion of freezing resistance.

What are the factors contributing to cold hardening? A generality to keep in mind is that hardening usually involves an expenditure of metabolic

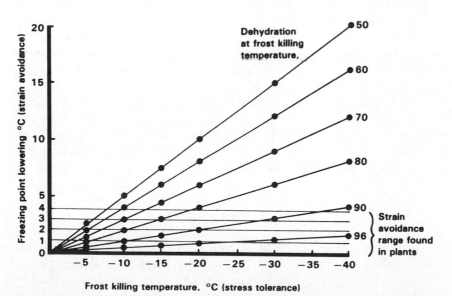

Fig. 8.2. Dependence of freezing stress tolerance (abscissa) on dehydration strain avoidance (ordinate) and dehydration strain tolerance (percentage dehydration at frost-killing temperature). Reproduced by permission, Levitt (1972).

energy and parallels a decrease in growth. Metabolic energy expenditure comes either from photosynthesis or some energy store when temperatures are much below those accommodating normal growth. Winter cereals, for example, often harden just above freezing. Various kinds of solutes are likely to accumulate under these low–temperature–low–growth conditions. Frequently, freezing tolerance increases as total solute accumulation rises naturally. Artificially increasing solutes by infiltration of sugars also increases frost hardiness in plants already potentially frost tolerant.

Fall sugar accumulation is important in some cereals but not in others. Where it is important, apparent sugar losses from respiration in the spring may cause loss of hardening. The same plant hardened in the fall may be incapable of hardening in the summer even when artificially exposed to hardening temperatures. Growth stage obviously influences hardening in plants.

Levitt summarizes the influence of sugars on freezing tolerance by concluding 1) that sugars accumulating in the vacuole could decrease the potential amount of ice formed and, therefore, contribute to the avoidance component and 2) that sugars, as a result of being metabolized, exert some unknown protective influence on cells that increases their tolerance of freeze-induced dehydration.

Accumulation of soluble proteins in cortical cells of trees sometime correlates with freezing hardiness, but many exceptions exist. Similar results can be cited in crops like winter wheat and alfalfa.

Fatty acid content of alfalfa roots (Gerloff et al., 1966) and locust trees (Siminovitch et al., 1967, 1968) and the degree of unsaturation increased during hardening. More specifically, the phospholipid content increased with hardiness in some of the trees. St. John and Christiansen (1976) showed that chemically blocking syntheses of linolenic acid markedly increased the chilling sensitivity of cotton seedlings. The function of all these accumulated solutes remains difficult to visualize. Generally, irreversible precipitation of protein is suspected.

Freezing death is not attributed so much to enzyme protein denaturation as it is to membrane damage (protein aggregation) resulting in loss of selective permeability. The relative amount of damage can be estimated from solute efflux from the cells as measured by electrical conductivity changes in a solution bathing the cells.

Cool or Chilling Temperatures

Lyons (1973) and Levitt (1972) both have excellent recent reviews on chilling injury. Lyons reiterated the fact that physiological dysfunction occurs in many tropical and subtropical plants when they are subjected to temperatures of 10–12°C or lower but above freezing. This chilling injury is quite important to agronomists due to the tropical origins of major crops like cotton, corn, rice, sorghum, and millet. These crops are sensitive to chilling temperatures. Such problems likewise relate to the production of citrus species, tomato, avocado, banana, pineapple, and mango. Chilling temperatures in horticultural crops produce problems in growth and development and storage (fruit discoloration, pitting, abnormal ripening, etc.). Problems in agronomic crops are mostly in metabolic systems associated with growth and development. Development problems in grain crops usually relate to yield loss due to the reduced seed number component of yield.

Chilling injury is generally induced by temperatures of 1 to 5°C persisting for a few hours to several days. Definite chilling effects, though not necessarily immediately injurious, occur in chill-sensitive plants at about 10 to 12°C. Many chill-induced effects relate to changes in cellular permeability. Figure 8.3 shows a schematic pathway summarizing various aspects of chilling injury. The sequence of events illustrated may not appropriately represent the order in which related events occur in all cells. Neither do all the events illustrated necessarily occur in all chill-sensitive cells. Nonetheless, the processes illustrated occur in a variety of species, and many have been illustrated within a single species.

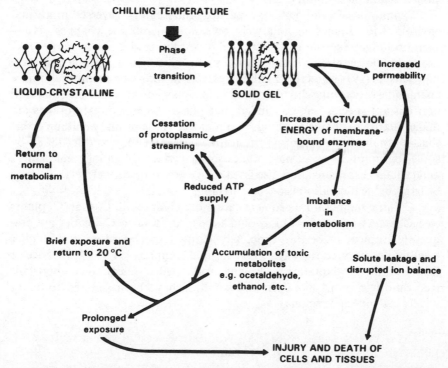

Fig. 8.3. Schematic pathway of the events leading to chilling injury in sensitive plant tissues. The membrane model is an adaptation by Lyons of the model proposed by Singer; by permission, Lyons, 1973.

As in many types of injury, membrane problems have been implicated. Note in Figure 8.3 that a membrane phase transition from liquid-crystalline to solid gel is postulated and that there are several effects. More than a century ago researchers noted that protoplasmic streaming ceases at about 10 to 12°C in susceptible plants. More recently it has been noted that protoplasmic streaming continues in resistant plants to near 0°C. Protoplasmic streaming changes, being energy dependent, probably relate to the reduced ATP supply common to chilling injury in some susceptible plants. Increased activation energy of membrane-bound enzymes correlates closely with the membrane phase changes (liquid-crystalline to solid gel, Fig. 8.3), which coincides with an abnormal accumulation of pyruvate, ethanol, and acetaldehyde. This is as anticipated if glycolysis remains functional (occurs in soluble enzyme systems) while membrane-dependent mitochrondrial respiration is disrupted. Reduced ATP supply follows with subsequent cessation of protoplasmic streaming. Most of the metabolic chilling effects mentioned have been demonstrated in mitochrondria. The phase change in susceptible plants coincides with a sharply increased energy of activation requirement at temperatures below 10°C, which suppresses mitochondrial oxidation. This sharp increase in energy of activation does not occur in chill-resistant plants until near 1°C.

Symptoms of brief chilling-exposure injury can be reversed in normal environments, suggesting relatively reversible membrane changes. However, prolonged exposure and increased permeability give rise to solute leakage and toxic metabolite accumulation sufficient to cause permanent injury.

Raison (1974) described chilling effects in terms of membrane phase changes in roots, mitochondria, and chloroplasts for several resistant and nonresistant species. He concluded that phase changes in membranes are due primarily to changes in lipids and that this is the primary chilling effect that is usually reversible. Importance of a phase change is questioned by some, but implication of membranes seems certain. Metabolic imbalances and cellular disorganization discussed above are secondary effects that may be irreversible if conditions persist too long.

Chilling temperature problems are more common in C_4 than C_3 plants (see Chapters 5 and 6). This would be expected since C_4 plants are frequently tropical in origin. Tropically adapted species often have higher levels of unsaturated fatty acids in their lipid complexes. The significance of this in terms of adaptation and plant productivity is clear since important tropical-origin crops like corn and sorghum have been selected to adapt widely in temperate regions.

High Temperatures

Levitt evaluated the work of many scientists on heat stress. The molecular bases for heat tolerance and susceptibility are reviewed by Alexandrov (1977). A 45–65°C temperature range for inducing heat stress in higher land plants (classified as moderate thermophiles) is suggested. Mesophiles, considered to be "lovers of moderate temperatures," grow and develop in the 10–30°C range and are heat stressed at 35–45°C. Most crop plants presumably fall in the mesophile or lower thermophile range. While it may be useful to classify plants by temperature ranges, that alone is not adequate since the time of exposure to a given temperature is critical. Alexandrov (1964) demonstrated that the heat-killing effect of a stress temperature varies inversely with time in an exponential manner. Consequently, upper temperature as a killing limit for a plant cannot be as precisely defined as a lethal low temperature limit, which is not particularly time dependent. In fact, experimental methods developed to test for heat damage are based on the effects of different time-temperature exposure combinations. Importance of recognizing the time factor in assessing temperature damage in the field becomes obvious since exposure time varies when sun angle changes at different times and seasons in relation to leaf angle and fruit shape or position.

Plant organs may be either above or below the surrounding air temperature. Their temperature depends largely on color, position relative to the sun, surface area in relation to weight and transpiration capacity, and water status as it relates to transpiration rate. Fleshy fruits can suffer burns (Huber, 1935) from excessive temperatures. Such fruits do lose water but do not control temperature via transpirational cooling at levels realized in leaves, partly because of comparatively low surface area to weight ratios.

The most obvious direct heat damage usually occurs when germinating seeds or seedling plants in bare, dark soils are exposed to high radiation levels. Wang (1972) considered results of various workers and noted black sand temperatures at 41°C compared to 35°C in white sand under similar radiation conditions. The 41°C temperature is lethal for most germinating vegetables and is borderline for corn.

Levitt (1972) effectively presented data from many sources illustrating deleterious, high plant organ temperatures relative to the surrounding air (Table 8.1). While injury occurs in nature, it is probably uncommon, oc-

Table 8.1. Plant temperatures in relationship to temperatures of surrounding air. Adapted by permission from Levitt (1972).

Plant	Plant part	Plant temperature	Air temperature
		°C	
Sempervivum spp.	Succulent leaves	48–51	31
Tomato	Ripe fruit	38–42	27–29
Various	Leaves	44.25	36.5
Various	Leaves	37.6	24.7
Conifer	Needles, twigs, stems (herbaceous and woody)	9–11.8 above air	
Various	Thin leaves	6–10 above air	20–30
Herbaceous plants	Thick leaves	20 above air	20–30

curring only in exceptionally hot summers. This is presumably true because plants growing in a given location are there because they have become adapted over the centuries to the type of environment peculiar to that location or a similar location.

Alexandrov (1977) reviewed the effect of growth on heat resistance in plant cells and concluded that, generally, fully differentiated cells are more resistant than growing cells. He concluded that thermostability in leaves was associated with growth but not with developmental stage in *Gagea lutea*. Thermostability in cereal grain inflorescences may well vary with developmental stage, judging from differing sensitivities in seed numbers produced when plants are exposed to high temperatures at several inflorescence stages.

The nature of heat injury is complicated. Initially, as temperatures rise, a point is reached where trapping of photosynthetic energy no longer exceeds respiratory expenditure and growth ceases. Usually, at some higher temperature, metabolic lesions or disruptions of normal metabolic pathways occur. Levitt (1972) collected literature implicating protein denaturation and perhaps lipid denaturation as the direct consequences of heat damage. The primary effect is likely membrane damage, which means that a electrolyte leakage occurs and that normal functioning of membrane-bound enzymes is impaired. Presumably some essential metabolite(s) is not formed or is blocked, and essential biochemical reaction sequences cannot occur. Nitrogen metabolism may be disrupted permitting accumulationg of toxic NH_3 levels. Anaerobic respiration may result, producing toxic quantities of ethanol, acetaldehyde, and other chemicals.

Before considering temperature effects on physiological processes, you should recognize that high temperatures and water stress often occur together in the field. Since high temperatures contribute to desiccation in nature, the separate effects of each are hard to differentiate. Sullivan et al. (1977) reviewed selected literature on this subject. Earlier investigators sometimes concluded that heat and drought resistances were essentially a single stress resistance and that, in selecting for thermal stability, they likely would be selecting for drought resistance also. While there is apparent correlation between heat and drought resistance, obvious examples of direct heat injury apart from desiccation injury are known.

Effects of temperature on many physiological processes are fairly well documented. A major process of universal interest is photosynthesis, even though it may not necessarily be the most yield-limiting developmental or physiological process in a given environment. Cotton and sorghum data in Fig. 8.4 illustrate a temperature effect on C_3 and C_4 photosynthesis. Table 8.2 illustrates minimum, optimum, and maximum temperature values for photosynthesis and growth in C_4 grasses and C_3 plants. These data provide reasonable insight into the influence of temperatures on photosynthesis and growth in a range of economic plants. An additional point requiring emphasis is the kind of variability expected within a species. This knowledge is essential before seriously considering the possibility of genetic manipulation

Table 8.2. Maximum, minimum, and optimum temperatures for photosynthesis and growth in C_4 grasses and C_3 plants. Adapted by permission from Ludlow (1976).

Process and type of plant	Minimum	Optimum	Maximum
		°C	
Photosynthesis			
C_4 grasses	5–10	35–45	45–60
C_3 plants (herbaceous)	−10–0	15–30	35–45
Growth			
C_4 grasses	10–15	30–40	40–50
C_3 plants (herbaceous)	0–10	10–30	30–40

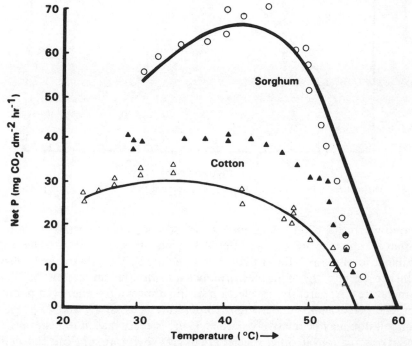

Fig. 8.4. Effect of temperature on photosynthetic rates of sorghum and cotton. The solid line is the average of 3 experiments. Solid triangles represent 1 experiment with cotton of higher Net P at 2.0 ly⁻¹ min⁻¹. (1 ly = μmol m⁻² s⁻¹). From El-Sharkawy and Hesketh (1964).
Approximate ly conversion to μmol m⁻² s⁻¹: 1 ly/min = 697 WM⁻² (solar radiation)
697 WM⁻² × 4.6 (fraction of radiation between 0.4 to 0.7 μm = 320 WM⁻²
320 WM⁻² × 4.6 (μɛi/m² s)/WM⁻² = 1475 μmoles/m² s⁻¹

to increase field productivity by enhancing photosynthetic rate in harsh environments. Sullivan et al. (1977) reported maximum photosynthesis for RS 626 and N 9040 sorghums at 37 and 41°C, respectively (Fig. 8.5). A comparison was then made (Fig. 8.6) at temperatures above optimum among RS 626, RS 691, N 9040, and Redlan. Nebraska (NE) 691 is a hybrid utilizing the Redlan and N 9040 parents. Two points are of interest. NE 691 maintains photosynthetic productivity from 37 to 43 + °C, which is well

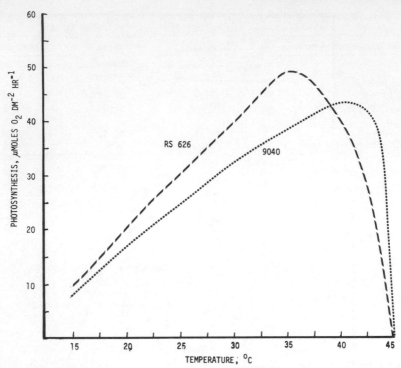

Fig. 8.5. Differences in temperature optimum for photosynthesis of two sorghum genotypes. From Norcio (1976).

beyond where RS 626 declines and stops. Second, note that the curve for Redlan is similar to RS 626 while N 9040 holds up well at 43°C. The performance level in the NE 691 hybrid was held up apparently by the N 9040 influence despite the poor performance of the Redlan parent at high temperatures. Apparently, 1) selection within a species for improved photosynthetic performance under high temperature is possible; and 2) genetic manipulation may be relatively simple. There is every reason to believe that similar observations in other species are likely. Therefore, one can expect to breed for higher photosynthetic rates under high temperature conditions.

WATER STRESS

Plant response to water stress has been the subject of many books and reviews. Levitt (1972) pointed out that the primary drought-induced strain in crop plants is simply cell dehydration, from which many effects arise. Resulting damage to essential plant processes can be reversible or irreversible depending on the severity of dehydration. As in chilling or high temperature situations, consequences of membrane damage can be of several types. Damage to membrane-bound enzymes usually means accumulation of metabolic intermediates or waste products (alcohol, acetaldehyde, pryu-

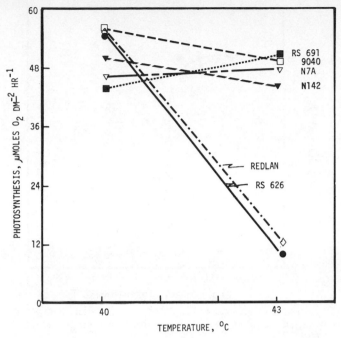

Fig. 8.6. High temperature effects on photosynthesis of sorghum RS626, RS691, N9040, and Redlan and corn N7A and N142. From Norcio (1976).

vate, NH_3, etc.) to the extent that some are toxic. At the same time, photosynthesis can be slowed enough so that the plant literally starves or else the respiratory chain and its associated energy production are sufficiently disrupted to block normal cellular maintenance. If this stress persists long enough, effects are irreversible, and the plant dies. More often, mild to intermediate stress levels are important. In this regard, three excellent reviews by Hsiao (1973) and Hsiao and Acevedo (1974) and Begg and Turner (1976) are available.

CO₂ Exchange

Photosynthetic rate under many conditions relates closely to CO_2 concentration at the carboxylation site. Total resistance to CO_2 influx to the carboxylation site is usually broken into several components, as treated by Boyer (1976) and others. Resistances associated with CO_2 diffusion in the gaseous portion of the diffusion pathway are cuticle, boundary layer, stomatal, and mesophyll intercellular space. Cuticular resistance is very large compared to the other resistances. Stomatal resistance is large compared to boundary layer and intercellular space resistance. Resistance to CO_2 transfer in the intracellular space is called mesophyll resistance, which has a physical (liquid) component and sometimes a biochemical (CO_2 fixation) component.

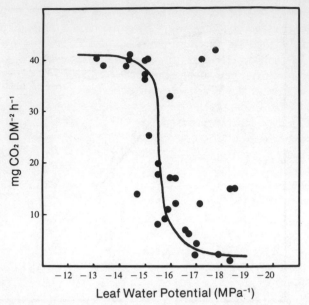

Fig. 8.7. Photosynthesis vs. leaf water potential at floret initiation. Samples included fully watered controls and plants 6 days without water. Developed from Hultquist (1973).

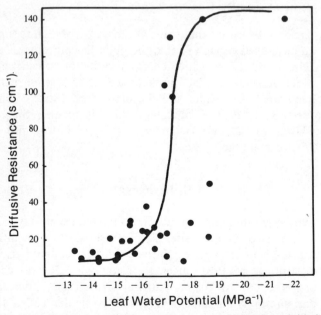

Fig. 8.8. Water vapor diffusive resistance vs. leaf water potential at floret initiation. Samples included fully watered controls and plants 6 days without water. Developed from Hultquist (1972).

Hultquist's plots of photosynthesis vs. leaf water potential (Fig. 8.7) and diffusive resistance vs. leaf water potential (Fig. 8.8) in sorghum at the floret initiation stage are typical of many species. The curves are essentially the inverse of each other, suggesting that diffusive resistance is a primary controller of photosynthesis over a range of stress levels. The importance of stomatal function is obvious at this growth stage in sorghum. However, after bloom, Hultquist found no significant correlation between CO_2 uptake and leaf water potential, suggesting appreciable loss of stomatal control. Therefore, stage of development influences stomatal control of photosynthesis.

The graphs of photosynthesis, viscous flow, resistance, and stomatal conductance in sunflower shown in Fig. 8.9 are representative of results from a number of investigations for several species. The plot of stomatal conductance (inverse of resistance) correlates well with photosynthetic rate.

Turner et al. (1978) evaluated photosynthesis, leaf conductance, and water potential in both irrigated soybean (*Glycine max*) and soybean from which water was withheld 27 days. Soil water depletion in the dryland crop was 27 mm in the upper 120 cm of soil.

Measurement	Irrigated	Dryland	% Reduction
Photosynthesis (mg $^{14}CO_2$ cm^{-2} sec^{-1})	154	70	55
Leaf conductance (cm sec^{-1})	1.03	0.52	50
Water potential (MPa)	−1.5	−1.9	31

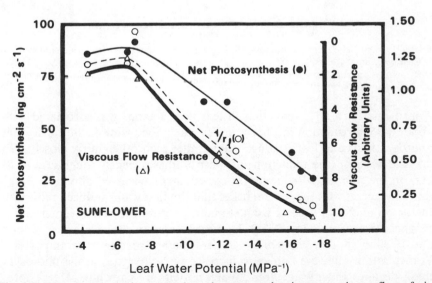

Fig. 8.9. Net photosynthesis, stomatal conductance, and resistance to viscous flow of air through leaves in sunflower at the indicated water potentials. From Boyer (1976), by permission.

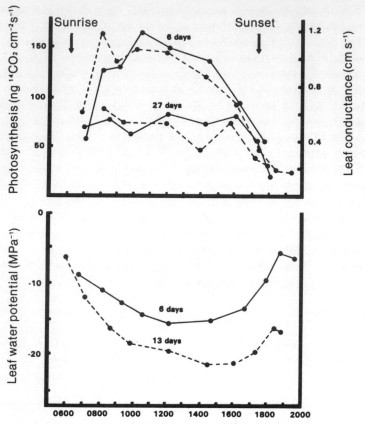

Fig. 8.10. Diurnal changes in photosynthesis, leaf conductance and leaf water potential in Ruse soybean at grain filling after water had been withheld in the field for 6, 13, or 27 days. By permission from Turner et al. (1978).

Approximately a 50% reduction in leaf conductance was associated with about a 55% reduction in photosynthesis at noon. Note in Fig. 8.10 that photosynthesis and leaf conductance parallel each other fairly closely. On dryland soybean the reduction in photosyntheses began roughly at −15 bars in the morning. Reduction in conductance occurred about the same time and agrees with the general belief that the first water-induced reduction factor on photosynthesis is from stomatal closure rather than internal resistances (Slatyer, 1973a, 1973b). However, experimental evidence supports the likelihood that resistances other than stomatal are important. Transpiration rates are known to increase or hold steady while photosynthesis declines under water stress. Water deficits are known to affect photosynthetic rate in isolated chloroplasts where stomates are not a factor.

Biochemically, water deficits express their negative effect within the photosynthetic apparatus in photosystem II at the same time that total leaf photosynthesis is affected (Boyer, 1976). Photosystem I was affected adversely to a lesser degree at the same time. According to Keck and Boyer

(1974), limitation orders in sunflower shift from electron transport at mild stress to photophosphorylation at severe desiccation (-1.7 MPa or below water potential). Generally the carboxylating enzymes are fairly resistant to heat and water stress conditions. Activity reductions of rhibulose-1,5 diphosphate carboxylase and phosphoenolpyruvate carboxylase decline with water stress but not sufficiently to account for the reduction in photosynthesis (Huffaker et al., 1970; Shearman et al., 1972). Phosphoenolpyruvate carboxylase is less affected than ribulose-1,5 diphosphate carboxylase.

Additional indirect evidence supporting the existence of photosynthesis control other than by stomatal resistance comes from results of Hultquist (1973). Sorghum stomata became functionally relatively insensitive after the flowering stage, but photosynthesis was still regulated at about the same rate whether plants were fully watered or drought stressed.

Sullivan and Eastin (1975) reviewed literature stressing that preconditioning or previous history effects must be considered in analyzing plant response. Generally, when plants grown at a high water level or with no stress history were subjected to water stress over a short period, reduction in photosyntehsis often occurred at about -0.5 MPa leaf water potential. However, when water was withheld from field plants or where drying out was controlled at a slow rate, photosynthesis was not likely to be reduced until the -1.5 to -2.0 MPa range was reached. Similar trends occur in O_2 uptake by chloroplasts isolated from plants exposed to rapid vs. gradual drying conditions. The slower, preconditioning desiccation pattern apparently permits plant adaptation or stress hardening to the reduced water supply.

Stomatal response has been mentioned with respect to its relative place in controlling photosynthesis. The authors found that corn and millet generally close their stomata at higher water potentials than sorghum. Stomate closure shuts off evaporative cooling and results in higher leaf temperatures. Most likely corn and millets in general have, because of their stomatal response characteristics, become better adapted than sorghum to higher temperatures. Heat tolerance testing suggests that corn and millet are indeed more heat tolerant than sorghum (Sullivan, 1972).

These results can be considered meaningfully in terms of Levitt's (1972) stress avoidance and tolerance terminology (Fig. 8.11). Corn and millet appear more heat tolerant than sorghum probably because their stomates close sooner, and they have developed mechanisms to withstand higher leaf

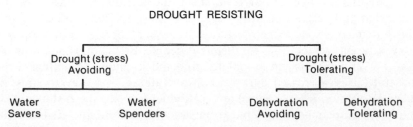

Fig. 8.11. The nature of xerophytism. By permission from Levitt (1972).

temperatures. Their heat tolerance mechanisms seem to contribute to drought-tolerating characteristics. By contrast, sorghum acts more as a drought avoider. Having a lower heat tolerance level, sorghum successfully avoids high temperatures by keeping its stomates partially open so that some transpirational cooling persists even at low soil water levels. This works so long as sorghum's extensive root system is able to keep exploring and extracting whatever limited soil water is available. Keeping stomates partially open and extensively ramifying root systems can both be considered examples of drought resistance mechanisms of the avoiding type.

Considering again differences within species, Hultquist (1973) tested a normal sorghum hybrid vs. a sorghum hybrid stress tolerant to the more severe field drought cycles of the U.S. Great Plains. Plants were water stressed in greenhouse pots and fed $^{14}CO_2$ photosynthetically to compare photoassimilate distribution between the developing panicle and roots. The stress resistant hybrid allocated a higher proportion of available assimilate to the developing panicle compared to roots. When stress reached an extreme level, sufficient to damage the panicle, the whole plant died. These were greenhouse pot experiments and stress levels imposed exceeded those generally found in Great Plains field plantings. Under field conditions, this hybrid nearly always gives higher yields under stress than the second hybrid, due to emphasis on panicle development.

The second hybrid, considered less stress resistant than the first one in the Great Plains, suffered partial panicle damage due to the water stress but exerted a partial panicle upon rewatering. The assimilate-proportioning characteristics of the second hybrid permitted it to persist under the extreme water stress conditions, but it yielded very little grain. Persistence is of little value to the farmer. The drought resistance of the second hybrid may have been primarily through an avoidance mechanism because of greater proportionate assimilate transfer to the root and lower aerial plant parts compared to the panicle. Stress resistance in the first hybrid was centered on making grain, while stress resistance in the second hybrid was primarily plant persistence. The most desirable type of stress resistance depends on the severity of the stress environment.

Other Plant Processes

Hsiao's (1973) review of respiration and nitrate reductase literature shows that dark respiration often declines when stress is moderate to severe. This statement is somewhat dangerous because moderate and severe stress carry different connotations for different people. Brix (1962) and Shearman et al. (1972) demonstrated in loblolly pine and sorghum, respectively, that moderate stress increased dark respiration rate and very severe stress reduced respiration below control levels. Reductions have been shown in O_2 uptake by mitochondria isolated from tissue at more than -0.4 to -0.5 MPa.

Stress-induced nitrogen metabolism imbalance is usually characterized by formation and accumulation of amides, which constitute a detoxification system for excess NH_3 produced due to metabolic lesions. One of the first enzymes to suffer reduced activity under stress is nitrate reductase. Protein synthesis is also very sensitive. Another system sensitive to water stress is symbiotic nitrogen fixation (Huang et al., 1975).

Hsiao (1973) and Hsiao and Acevedo (1974) conveniently reviewed some of the general effects of water stress on key plant responses discussed (Table 8.3). Note again that nitrogen metabolism and cell wall synthesis, both growth-related processes, are very sensitive.

Growth

The most notable point in Table 8.3 is that growth is the overall process most sensitive to water stress. The treatment of growth by Hsiao and Acevedo (1974) is hard to improve on beginning with their definition.

Cell growth can be defined as irreversible enlargement or expansion of cells. Although seemingly purely physical, this definition incorporates implicitly all metabolic aspects of growth. Cell enlargement cannot be sustained without concomitant synthesis of membranes, organelles, proteins and cell-wall material, and is almost always accompanied by differentiation at the subcellular level. Even a minute irreversible increment in cell size presumably involves enzyme-catalyzed reactions, at least in the cell wall.

Given that growth depends on so many integrated metabolic processes, it is not surprising that growth is so sensitive to small reductions in water potential.

Green (1968) and Green et al. (1971) demonstrated that growth is proportional to a function of cell extensibility (a cell wall property undoubtedly conditioned by metabolic events). The level of cell wall flexibility is obviously important. Salisbury and Ross (1969) produced a Hofler diagram

Table 8.3. Generalized sensitivity to water stress of plant processes[†]
(by permission from Hsiao and Acevedo, 1974).

Process affected	Reduction in tissue ψ required to affect process[‡]			Remarks
	0 MPa	10 MPa	20 MPa	
Cell growth	———— — — —			
Wall synthesis	————			growing tissue
Protein synthesis	————			tissue growing
NO_3^- reductase level	—————			
ABA accumulation	— — — — ———			
Stomatal opening	— — — — ——————————— — — —			species differ
CO_2 assimilation	— — — — ——————————— — — —			species differ
Respiration	— — — ————			
Proline accumulation	— — — —————			
Sugar accumulation	————————			

† Length of lines represents range of stress when a process becomes first affected. Dashed lines signify more tenuous data.

‡ With ψ of well-watered plants under mild evaporative demand as reference point.

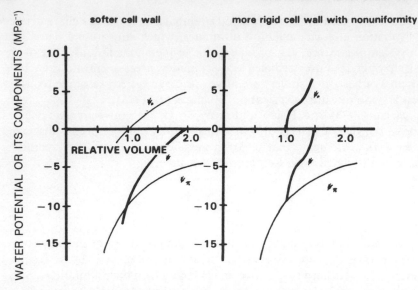

Fig. 8.12. The Hofler diagram using the concept of water potential as discussed. The components of water potential are shown as they change with changing volume: for a cell with a soft cell wall (left) and for one with a more rigid cell wall and with nonuniform expansion properties (right). By permission from Salisbury and Ross (1969).

(Fig. 8.12) that helps visualize how cell volume (growing may or may not be involved) might be influenced by cell wall rigidity. When osmotic potential ($\psi\pi$ or ψ_s used earlier) is held constant, the softer cell obviously has a greater volume increase (2.0 vs. 1.5 at zero ψ) and ψ_p is relatively lower.

Boyer (1968) demonstrated some leaf enlargement in sunflower between -0.3 and -0.4 MPa water potential. However, the enlargement rate was only 15% or less of the enlargement rate at water potentials above -0.3 MPa.

Analysis of growth characteristics requires considering two components of water potential in a cell.

$$\psi = \psi_p + \psi_s$$

where ψ_p is pressure or potential and ψ_s is osmotic potential (arising from solute particles). Turgor pressure inside the cell is necessary to permit cell expansion (expansive growth). Green (1968) demonstrated that any change in turgor pressure effects an immediate growth rate change or cessation as turgor pressure goes from positive to negative ($\psi_p = \psi - \psi_s$).

Plants regulate ψ and ψ_p to maintain growth by adjusting the levels of osmotically active solutes (after sucrose) in cells. This adjusting mechanism called osmoregulation is a means of minimizing differences between the plant and its environment.

Water Use Efficiency

Improvement of water use efficiency (WUE) is a concern in most production or research operations involving water stress and usually high temperature stress. Many water- and temperature-mediated plant responses influence WUE. Arnon's (1975) chapter on WUE is recommended to students.

The most common definition of WUE is yield (Y) of economic crops per unit of water used (evapotranspiration or ET).

$$WUE = \frac{Y}{ET}$$

Obviously any factor that increases yield (such as increasing production period or increasing the efficiency of production by improved fertility or metabolic efficiency) will improve WUE. Likewise, any factor reducing evapotranspiration that has no seriously deleterious effect on yield will increase WUE. A second WUE expression is net CO_2 uptake per unit of water transpired on an individual leaf or a smaller leaf segment basis. Generally, C_4 species (corn, sorghum, etc.) have higher WUE's than C_3 crop species (wheat, cotton, etc.), partly because C_4 species have lower mesophyll resistances and higher photosynthetic rates than C_3 species (Bierhuizen, 1976).

COMBINED STRESS EFFECTS

Temperature and water effects are not independent in real production situations. The following crop examples illustrate how temperature, water, and/or both influence economic production and include limited reference to plant factors influencing resulting yields. Salter and Goode (1967) summarized extensive data on plant sensitivity to water stress at different stages of development. Grain crops, for example, are particularly sensitive to yield reductions if stressed during the last half of the inflorescence development period.

The generality to bear in mind when assessing environmental effects on yields is that seed number in grains nearly always correlates more positively with yield than seed size. Therefore, all environmental factors influencing the seed number component of yield must be considered. However, water stress during grain fill is also very costly in terms of seed size. Fischer's (1973) water stress investigations in wheat (Fig. 8.13) illustrate generally what happens. Obviously the stress influence on seed number is much more severe than on photosynthetic area. Note that about a -1.5 MPa stress for a few days just prior to ear emergence (EE) caused a 25–30% reduction in grains per spikelet. No reduction in photosynthetic area had occurred but possibly photosynthesis would have been reduced some by then.

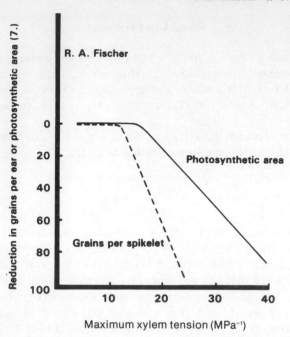

Fig. 8.13. Major yield parameters and their responses to plant water stress. The response of grains per spikelet comes from data restricted to stresses commencing in the period 15 to 5 days before ear emergence. The photosynthetic area response curve is derived from results at all stages of development tested. By permission from Fischer, 1973.

Warrington et al., 1977 conducted a growth room experiment on temperature effects during the vegetative stage (GS1), inflorescence development (GS2), and grain filling (GS3) (Table 8.4). Temperatures during vegetative and grain-filling periods had little influence on seed number except at 25°C during GS3. As expected, seed number was mostly influenced during the spike development period (GS2). The high of 70.5 main-stem kernels developed at 15°C over 62 days compared to 19.5 kernels developed at 25°C over 29 days. This represents about a 72% reduction in kernels developed in 53% less time. By contrast, vegetative production was reduced only 28%. Invariably, as environmental limitations arise, vegetative development flourishes (though diminished some) at the expense of reproductive development.

This growth root work is a good complement to earlier wheat field research of Cackett and Wall (1971). High and low elevation sites were chosen at similar latitudes where temperatures averaged 3.3–3.5°C lower in the high elevation area. The 3.5°C lower temperature resulted in over 35% more fertile spikelets having 57% more grains per spikelet. A 19% higher 1000 grain weight gave 88% more kg ha⁻¹. Judging from both field and growth room yield results, temperature effects deviating from a presumed fairly narrow optimum during inflorescence development are very significant in terms of yield reduction.

Table 8.4. Total grain number per main-stem wheat head as influenced by temperature at each growth stage.† Adapted from Warrington et al., 1977.

Growth stage	Number of grains per ear		
	Day temperature (°C)‡		
	15	20	25
GS1 (vegetative)	57.8	59.8	56.8
GS2 (influorescence development)	70.5	59.8	19.5
GS3 (grain fill)	57.2	59.8	48.2
LSD (P < 0.05)	10.1		

† Other growth states held at 20/15°C.
‡ Night temperature was 5°C cooler than the day temperature indicated.

Table 8.5. Grain yield of corn when water was withheld throughout most of the grain fill period. A low water vapor (LVP) pressure vegetative stage pretreatment (−26 mbar) was compared to a high vapor pressure pretreatment of −5 mbars (Boyer and McPherson, 1975).

Plants†	Low VP pretreatment	High VP pretreatment
	kg ha⁻¹	kg ha⁻¹
Control	11 700	10 500
Desiccated	7 970	4 930

† Leaf water potentials were −0.3 to −0.4 MPa and −1.8 to −2.0 MPa in control and desiccated plants, respectively, throughout most of the desiccation period.

Boyer and McPherson (1975) reviewed the effects of water deficits on physiology and yield of cereals. Of special interest was corn grown through the vegetative stage in growth rooms under low and high humidities (Table 8.5). The low-humidity pretreatment yield was only modestly different from the high-humidity pretreatment yield when plants were given full water after tasseling (controls in Table 8.5). However, when water after tasseling was cut to one-seventh of the controls, yield reduction in the high-humidity pretreatment lot was reduced 54% compared to only 29% reduction for plants preconditioned by low water vapor pressure during vegetative growth. When water was reduced after tasseling, the low-vapor-pressure plant leaves maintained higher water potentials and higher photosynthetic rates for a longer time compared to high-vapor-pressure plants. The authors assume that adaptation of the low humidity plants to desiccation was through drought avoidance rather than tolerance characteristics. They conclude that avoidance characteristics are often developed at the expense of photosynthetic capacity.

For example, leaf area shed in response to drought tends to reduce transpiration but also reduces photosynthesis. Likewise, increased root size and activity to explore for limited water may be done at the expense of leaf area production. They consider selection for drought tolerance to be potentially more desirable than drought avoidance. Perhaps the merits of the two drought resistance characteristics depends on the frequency and severity of drought in the production area. Probably some combination of both avoidance and tolerance mechanisms will prove to permit the highest, reasonably stable economic production level.

We referred to the time and metabolic efficiency components of yield earlier in this chapter. Yield can be expressed as a function of production rate and production time: Yield = Production rate × time in production. Work by Eastin et al. (1976) and Eastin (1976) illustrate these relationships. We grew a range of sorghum hybrids from temperate to cool-tolerant in different day/night temperature combinations, and found the following results. First, the temperature optimum for the temperate hybrid was about 5°C higher than for the cool-tolerant hybrid. Second, both hybrids suffered a 25% yield reduction when their approximate optimum temperature was exceeded by 5°C. Temperatures 10°C above optimum reduced yield 47% in the cool-tolerant hybrid. Third, the 25% yield reduction in the temperature hybrid corresponded to a 32% reduction in seed number. Seed size loss rather than seed number loss accounted for the yield reduction in the cool-tolerant hybrid at 5°C above optimum, but at the 10°C elevation seed number reduction was 35% (47% yield reduction).

Reduction of the panicle expansion period (GS2—Eastin, 1972) generally related to seed number reduction. Relatively small time reductions related to sizeable seed number reductions. However, time alone likely does not account for the above observations. Temperature-induced shifts in assimilate partitioning are also likely involved.

Reduction in length of the grain filling period from a 5°C night increase was much larger (18–20%) than the reduction during panicle development. In addition, the rate of grain fill was reduced on the order of 25–30% at 5–10°C above optimum night temperatures. Therefore, the crop suffered from both a reduced grain fill period and lower grain fill rates. As expected, genotypes were not affected the same so selection for desirable traits can be accomplished to minimize stress effects.

We have recently shown that night temperature controlled 5°C above ambient in the field during panicle development reduced grain yield 28%. Therefore, our growth chamber results appear to apply in real production situations.

Temperature has been discussed partly in terms of how it influences metabolic efficiency or growth efficiency as described by Tanaka (1977). He feels growth efficiency in rice, corn, and soybean is not much influenced within normal temperature ranges. Reason exists to further evaluate the possible influences of temperature and water stress on metabolic efficiency in production of economic products within a genotype.

Studies of growth or metabolic efficiencies necessarily involve considerations raised by Penning de Vries (1972, 1974) on respiration and maintenance costs. He defines maintenance as including

. . .the processes which maintain cellular structures and gradients of ions and metabolites, and also the processes of physiological adaptation that maintain cells as active units in a changing environment. Formation of new enzymes at the expense of others and salt accumulation in some stress conditions are examples of such adaptations. Hence maintenance is not as conservative a process as its name suggests.

Maintenance respiration rates have been estimated by a number of means, all of which have definite limitations. Literature estimates cover a range of 8 to 60 mg glucose per g dry weight per day (a 30 mg estimate is common) at 25°C, a modest temperature when thinking of stress conditions. A reasonable cost estimate on a dry weight basis might be 1 to 3% daily of the total biomass, which is an appreciable energy expenditure.

MINERAL STRESS IN PLANTS

The soil environment may impart stresses to plants apart from temperature and water effects. Problems arise with either excesses or deficiencies of various salts. Bernstein (1977) reviewed this subject, pointing out that growth problems arise primarily through osmotic effects or specific ion effects. Some problem soils are alkaline, saline, (excessive soluble salts) or sodic (exchangeable sodium occupies 15% or more of the available cation sites). Other problem soils are acid. Acidity or alkalinity is largely a function of precipitation in relation to evapotranspiration (ET). When ET normally exceeds precipitation, soils tend toward alkalinity. If the reverse is true, soils are leached and usually acidic.

In saline soils a mixture of salts contributes to osmotic effects usually larger than single salt effects. As osmotic potential in soils decreases due to salinization, plant osmotic potential must decrease to maintain an osmotic gradient for water entry. By contrast, specific ion effects occur in sodic soils since exchangeable Na decreases exchangeable magnesium (Mg) and calcium (Ca), leading to deficiency symptoms. Acid soils can also cause deficiency and toxicity problems. Aluminum toxicity problems are well defined in some acid southeast USA soils and in tropical acrisols and ferralsols.

Clark (1980) and Clark and Brown (1980) recently reviewed plant response to mineral element toxicities and deficiencies. Mineral stresses occur in about one-fourth of the world's soils. Clark's discussion centers more on mineral use efficiency in economic production as contrasted to Bernstein's emphasis on osmotic and specific ion effects. The most extensive soil deficiencies in the USA are nitrogen (N), phosphorous (P), and potassium (K). The next most common deficiencies are Mg, copper (Cu), iron (Fe), zinc (Zn), boron (B), manganese (Mn), and molybdenum (Mo). Correcting deficiencies can usually be accomplished by addition of fertilizer or soil amendments. Sometimes only foliar applications are successful. Adding individual elements results in interactions with other elements and can cause problems. For example, added P can interact and cause deficiencies of Mn, Cu, Fe, Zn, and Mo. Phosphorus can also bind elements such as Al, Fe, and Mn under circumstances where they tend to be toxic. Clearly, the balance of elements is critical (see Chapter 7).

The costs of fertilizer elements needed to correct soil deficiencies or toxicities have risen dramatically in the past decade. This has given rise to interest in selecting plants to fit existing harsh soil environments rather than modifying the soil so extensively. To this end, Clark and Brown (1980) considered variation in plant genotype abilities to adapt to harsh soil environments as it relates to breeding and genetics. Considerable variability exists in plants to exclude elements found in excess and extract elements from the soil for which deficiency symptoms normally would occur in harsh soil environments. For example, scientists in Brazil have selected both wheat and sorghum for Al tolerance and adaptation to low-pH soils, conditions where most cultivars display classical phosphorus deficiencies. Also Clark's students have demonstrated differences in Al toxicity tolerances in sorghum using nutrient solutions. Both seminal and adventitious roots of the tolerant sorghums were about twice the size of intolerant plants. Proper liming of acid soils and phosphorus additions can overcome the toxicity problems, but the costs are very high. Phosphorus nutrition is part of the problem. Clark (1980) and Clark and Brown (1980) have demonstrated differences in the abilities of sorghum to take up P and utilize it efficiently, including its remobilization within the plant over time. Plant properties and characteristics relating to efficient uptake and utilization of mineral elements are heritable and, will, no doubt, be exploited more completely as fertilizer costs continue to rise.

Many soil-related problems in the past have been overcome by correcting soil deficiencies or toxicity factors. Abruptly higher energy costs and scarcity of quality water are forcing agriculturists to use these approaches more sparingly. An alternative, which is more plausible currently, is to fit the plant to the environment through breeding and genetics.

AIR POLLUTANTS

Plant stress from phytotoxic compounds in the air has been estimated to cause over $500 million worth of damage and loss of crop production each year. A general awareness of the way stresses are imposed by air pollutants will be illustrated. Several review papers and books on the subject are available for those interested in more detail (Mudd and Kozlowski, 1975; Ting and Heath, 1975; Mudd, 1975; Dugger, 1974; Dugger and Ting, 1970).

A great deal of the air pollution causing injury to plants arises from the combustion of high-sulfur fossil fuels and from oxidants arising from incomplete combustion in the internal combustion engine. The latter is the greatest single source of air pollution. Damage to vegetation due to petrochemical pollutants alone was about $113 million in 1970 (U.S. Environmental Protection Agency, 1978) from 110 million t of pollutant released by automobiles. Industrial pollution was second with 21 million t released. Plants themselves, when burned or decomposing, may contribute to pollution by releasing chemicals, including sulfur dioxide.

The incomplete combustion of the internal combustion engine forms hydrocarbon fragments, which catalyze the formation of oxidizing compounds and nitrogen oxides such as NO_2. The NO_2 may photochemically cleave to form NO and O. The O then reacts with molecular oxygen (O_2) to form ozone (O_3) which may cause severe plant injury even at concentrations less than 0.5 ppm. The nitrogen oxides may also react with hydrocarbon fragments and ozone in a complex way to form peroxyacetyl nitrate, a compound highly toxic to plants (Ting and Heath, 1975; Dugger, 1974).

Plants grown at warm temperatures (27–32°C) prior to ozone exposure are generally more sensitive to ozone injury than those maintained at cooler temperatures (10–16°C) before exposure. However, during exposure more injury occurs at lower temperatures. Plants grown at high relative humidities are more sensitive to ozone injury than plants grown at low relative humidities. Plants grown at low light intensities (10–20% of full sunlight) are generally more sensitive to ozone than plants grown at high intensities, although the injurious effects are not light dependent. Ozone injury characteristically occurs on the upper surface of plant leaves and may cause a water logged appearance, slight wilting, spotting, and chlorosis of cells near the stomates. Injury from sulfur dioxide (SO_2) derived from high-sulfur fuels causes interveinal burns on the leaves. Alfalfa is considered one of the most sensitive plants to SO_2. The symptoms are bleaching or burns of both interveinal leaf tissue and along the margins. Sulfur dioxide is derived primarily from fossil fuels and organic materials, from the ore-refining industry, and from the manufacture of sulfuric acid. It is usually accepted that SO_2 enters plants through open stomates. Once inside the plant, the SO_2 is oxidized to the highly toxic sulfite (SO_3) and then converted to the less toxic sulfate (SO_4).

Moisture-stressed plants are usually more tolerant to air pollutant injury than those grown under ample moisture conditions. And, in general, plants grown with optimum soil nutrients are more sensitive to pollutant injury than those grown with nutrient deficiencies or excess. The exact influence of mineral nutrition on susceptibility to air pollutants is not clear.

Injury from peroxyacetyl nitrate is more characteristically on the lower surface of the leaves where a bronzing appearance occurs, again, near the stomates. In some cases the upper surface is also affected. Peroxyacetyl nitrate is both phytotoxic and eye irritating. Plant injury caused by this highly toxic compound is light dependent, whereas that caused by ozone is not. Among plants very sensitive to peroxyacetyl nitrate are pinto bean, lettuce, and oats. Intermediate resistant plants include wheat and tobacco. Resistant plants include cucumber, cotton, and corn. Photosynthesis is usually inhibited by peroxyacetyl nitrate. It has been suggested that peroxyacetyl nitrate reacts with sulfhydryl groups (SH), and that the plants are more susceptible when more SH groups are available, as during the reductive process of photosynthesis.

Air pollutant injury is most evident to leaf tissue, however, it is known that root growth may also be reduced by air pollutants (Tingey et al., 1971). Also, it is expected that the pollutants may not always act alone but in combinations, producing different effects than when the individual pollutants are studied alone. The effects may be additive, greater than additive, or less than additive (Mudd and Kozlowski, 1975). Much research is needed on both the control of air pollutants and on different plant responses to their presence.

SUMMARY

The extremely broad subject matter covered in such brief forms does not give justice to all the authors in the respective areas. Of primary importance is membrane alteration and sometimes damage due to dehydration and temperature problems. From that point, a host of metabolic imbalances or lesions can arise. One key to overcoming or neutralizing these environmentally induced stresses is finding individual genotypes that possess the desirable exceptions to the metabolic imbalance or lesion. Success will depend largely on whether selection pressure techniques can be developed to expose desirable traits and on the numbers of genotypes screened.

A second key to overcoming environmentally induced stresses is to modify the environment. A success story of that type is the combination minimum–tillage-chemical–fallow system developed in western Nebraska, western Kansas, and eastern Colorado. That system has successfully minimized erosion, maximized water intake and storage and often increased yield 50–100%. Cooler soils induced by this system demand different genotype temperature responses. Cultural manipulations designed to improve the environment are not likely to achieve maximum gain without genotype alteration to fit the new environment. Approaches to both the genotype and environment alteration need to be pursued simultaneously, usually jointly, to achieve adequate progress in ameliorating influences of environmental stresses on plant persistence and production.

APPENDIX: WATER POTENTIAL DERIVATION

Water potential is an energy term based on thermodynamic concepts. Klotz (1967) provides reasonable thermodynamic treatment in simplified form regarding energy changes in biochemical reactions for the serious student. Many suitable physical chemistry texts are also available. Salisbury and Ross (1969, 1978) present abbreviated discussions of water potential for quick references. Here, consider briefly how water potential describes the potential work capacity of water in a system. The free energy of a component in a system relates to its capacity to do work (a measure of component activity). How is the free energy of water in a system determined? Free energy of a component depends on the amount of the component (mole fraction) in a system and the mean free energy per component molecule.

The mole fraction of water (n_w) can be calculated from vapor pressures (using Raoult's law for ideal solutions of nonelectrolytes) since

$$p = p° \frac{N_w}{N_w + N_s} = p° n_w \qquad [1]$$

where p is the vapor pressure of water in the solution, p° is the vapor pressure of pure water at a specified temperature, N_s is the moles of solute, and N_w is the moles of water. Therefore n_w is the mole fraction of water in the solution. Vapor pressure of pure water (p°) can be obtained from a chemistry handbook and p can be determined experimentally by measuring the vapor pressure of a solution in a closed container. Rearranging equation 2 shows how the mole fraction of water (n_w) can be determined from relative vapor pressure (or relative humidity if multiplied by 100).

$$\frac{p}{p°} = n_w \qquad [2]$$

The chemical potential of water in a solution ($\mu_w - \mu°_w$) is calculated by substituting the mole fraction estimate, p/p°, into an equation for chemical potential:

$$\mu_w - \mu°_w = RT \ln (p/p°) \qquad [3]$$

Chemical potential expressed in terms of energy per unit volume is defined as water potential (ψ).

$$\psi = \frac{\mu_w - \mu°_w}{\overline{V}_w} = \frac{RT \ln (p/p°)}{\overline{V}_w} \qquad [4]$$

where ψ is water potential expressed as energy per unit volume (ergs or joules cm^{-3}) and $\overline{V}_w$ is the partial molal volume of water (18 cm^3 $mole^{-1}$). The universal gas constant, R, given as 8.314 joules $mole^{-1}$ $°K^{-1}$ is basically an energy expression. Water potential can also be expressed in pressure units (bars) as will be shown in the problem.

The chemical potential of pure water is zero; that is, when μ_w is for pure water, p = p° and p/p° is 1. The ln of 1 is zero as is $\mu_w - \mu°_w$. When p/p° is less than one, the ln is negative indicating the work potential of water in the diluted system (negative value) is less than in pure water (zero).

While ψ is a difficult concept, a good intuitive feeling for it is possible using the water mole fraction estimate from solution vapor pressure measurements (p/p°) in a chemical activity equation. Thus water potential describes the potential work capacity of water in a system. Working through the following example will add some appreciation of the concept.

Consider a leaf section placed in a closed container with a small air volume at 25°C. Assume that when the leaf water is fully equilibrated with the air in the chamber the vapor pressure (p) of water is experimentally determined to be −31.22 mbars (−3.122 MPa). The following values are then available from which ψ can be calculated.

p = 31.22 mbars
p° = 31.67 mbars (*Chemistry Handbook* source)
p/p° = 0.986
  R = 8.314 joule mole^{-1} °K^{-1}
  T = 298 °K
  $\overline{V}_w$ = 18.02 cm^3 mole^{-1}

$$\psi = \frac{(8.314 \text{ joule mole}^{-1} \text{ °K}^{-1}) (298 \text{ K}) (\ln 0.986)}{18.02^3 \text{ cm mole}^{-1}}$$

The partial molar volume of water is 18.02 cm^3 mole^{-1} because water weighs 1 g cm^{-3}, and its molecular weight at 298 °K is 18.02 g (or cm^3) mole^{-1}. In the new international units partial molal volume is expressed in m^3 mole^{-1} or 18.02 × 10^{-6} m^3 mole^{-1}. Therefore,

$$\psi = \frac{(8.314 \text{ joule mole}^{-1} \text{ °K}^{-1}) (298 \text{ K}) \ln .986}{18.02 \times 10^{-6} \text{ m}^3 \text{ mole}^{-1}}$$

$$= \frac{(8.314 \text{ joule}) (298) (-0.0141)}{18.02 \times 10^{-6} \text{ m}^3}$$

$$= \frac{-34.93 \text{ joule}}{18.02 \times 10^{-6} \text{ m}^3}$$

$$= -1.939 \times 10^6 \text{ joule m}^{-3}$$

This energy term is converted to the pressure term, bars, as follows:

$$-1.939 \times 10^6 \text{ joule m}^{-3} = -1.939 \times 10^6 \text{ Newton m}^{-2}$$

$$= -1.939 \times 10^6 \text{ Pascal (Pa)}$$

$$= -1.939 \text{ MPa}$$

$$= -19.4 \text{ bars}$$

With the ψ of pure water arbitrarily set at zero, water in a plant leaf at -1.939 × 10^6 joule m^{-3} has a lower work capacity by comparison. Use of this energy expression is logical when thinking in terms of a water effect on a physiological process such as photosynthesis. However, most literature to date expresses ψ in the pressure term bars and more recently Pa or MPa.

The preceding leaf problem considered only the influence of the osmotic component (ψ_s − potential due to solute effects) in a solution. Another major component of cell ψ is pressure potential (ψ_p) arising from turgor pressure in the cell when cell solutes cause water to osmose into the cell and exert pressure against the cell wall. A third component, matric potential (ψ_m), relates to adsorption of water to hydrophilic surfaces of cell walls and contents.

$$\psi = \psi_s + \psi_p + \psi_m \qquad [5]$$

Water potential in plant cells contains all the components in the above expression. A vapor pressure determination measures their collective effect.

REFERENCES

Alexandrov, V. Y. 1964. Cytophysiological and cytoecological investigations of heat resistance of plant cells toward the action of high and low temperature. Quart. Rev. Biol. 39:35–77.

Alexandrov, V. Y. 1977. Cells, molecules and temperature. Springer-Verlag, New York.

Arnon, Isaac. 1975. Physiological principles of dryland production. p. 3–145. *In* U.S. Gupta (ed.) Physiological aspects of dryland farming. Oxford and IBH Publishing Co., New Dehli, India.

Begg, J. E., and N. C. Turner. 1976. Crop water deficits. *In* N. C. Brady (ed.) Adv. Agron. 28: 161–217.

Bernstein, Leon. 1977. Physiological basis of salt tolerance in plants. p. 283–290. *In* Amir Muhammed, Rustem Aksel, and R. C. von Borstel (ed.) Genetic diversity in plants. Plenum Press,

Bierhuizen, J. B. 1976. Irrigation and water use efficiency. p. 421–430. *In* W. B. Billings, F. Golley, O. L. Lange, and J. S. Olson (ed.) Water and plant life, Ecological Studies 19, Springer-Verlag, New York.

Boyer, J. S. 1968. Relationship of water potential to growth of leaves. Plant Physiol. 43:1056–1062.

Boyer, J. S. 1976. Water deficits and photosynthesis. *In* T. T. Kozlowski (ed.) Water deficits and plant growth. 4:154–190. Academic Press, New York.

Boyer, J. S., and H. G. McPherson. 1975. Physiology of water deficits in cereal crops. *In* N.C. Brady (ed.) Adv. in Agron. 28:1–23. Academic Press, New York.

Brix, H. 1962. The effect of water stress on the rates of photosynthesis and respiration in tomato plants and loblolly pine seedlings. Physiol. Plant. 15:10–20.

Cackett, K. E., and P. C. Wall. 1971. The effect of altitude and season length on the growth and yield of wheat (*Triticum aestivum* L.) in Rhodesia.

Clark, R. B. 1980. Plant response to mineral element toxicity and deficiency. *In* C. F. Lewis and M. N. Christiansen (ed.) Breeding plants for marginal environments. John Wiley & Sons, Inc., New York. In press.

Clark, R. B., and J. C. Brown. 1980. Role of the plant in mineral nutrition as related to breeding and genetics. *In* L. S. Murphy, E. C. Doll, and L. M. Walsh (ed.) Moving up the yield curves; advances and obstacles. Am. Soc. of Agron., Madison, Wis. In press.

Dugger, W. M. (ed.). 1974. Effect of Air Pollutants on Plants, Am. Chem. Soc. Symp. Series 3. ACS, Washington, D.C.

Dugger, W. M., and I. P. Ting. 1968. The effect of peroxyacetyl nitrate on plants: photoreductive reactions and susceptibility of bean plants to PAN. Phytopathology 58:1102–1107.

Dugger, W. M., and I. P. Ting. 1970. Air pollution oxidants—Their effects on metabolic processes in plants. Annu. Rev. Plant Physiol. 21:215–234.

Eastin, J. D. 1972. Photosynthesis and translocation in relation to plant development. p. 214–246. *In* N. G. P. Rao and L. R. House (ed.) Sorghum in the seventies. Oxford and IBH Publishing Co., New Delhi, India.

Eastin, J. D. 1976. Temperature influence on sorghum yield. *In* J. I. Sutherland and R. J. Falasca (ed.) Proc. 31st Annual Corn and Res. Conf., Am. Seed Trade Assn., Washington, D.C.

Eastin, Jerry, Ian Brooking, and A. O. Taylor. 1976. Influence of temperature on sorghum respiration and yield. Agron. Abstracts. p. 71.

El-Sharkawy, M. A., and J. D. Hesketh. 1964. Effects of temperature and water deficit on leaf photosynthetic rates of different species. Crop Sci. 4:514–518.

Fischer, R. H. 1973. The effect of water stress at various stages of development on yield processes in wheat. p. 233–241. *In* R. O. Slatyer (ed.) Plant response to climatic factors. UNESCO, Paris.

Gerloff, E. D., T. Richardson, and M. A. Stahman. 1966. Changes in fatty acids of alfalfa roots during cold hardening. Plant Physiol. 41:1280–1284.

Green, P. B. 1968. Growth physics in *Nitella*: a method for continuous in vivo analyses in extensibility based on a micro-manometer technique for turgor pressure. Plant Physiol. 43: 1169–1184.

Green, P. B., R. O. Erickson, and J. Buggy. 1971. Metabolic and physical control of cell elongation rate. Plant Physiol. 47:432–430.

Hsiao, T. C. 1973. Plant responses to water stress. Annu. Rev. Plant Physiol. 24:519–570.

Hsiao, T. C., and Edmundo Acevedo. 1974. Plant responses to water deficits, water-use efficiency, and drought resistance. Agric. Meteorology 14:59–84.

Huang, Chi-Ying, J. S. Boyer, and L. N. Vanderhoef. 1975. Limitation of acetylene reduction (nitrogen fixation) by photosynthesis in soybean having low water potentials. Plant Physiol. 56:228–232.

Huber, H. 1935. Der Warmehaushalt der Pflanzen. (In German). Naturwiss, Landwirtsch 17: 148.

Huffaker, R. C., T. Radin, G. E. Kleinkoph, and E. L. Coy. 1970. Effects of mild water stress on enzymes of nitrate assimilation and of the carboxylative phase of photosynthesis in barley. Crop Sci. 10:471–474.

Hultquist, J. H. 1973. Physiologic and morphologic investigations of grain sorghum (*Sorghum bicolor* (L.) Moench). I. Vascularegation; II. Response to internal drought stress. Ph.D. Thesis. Univ. of Nebraska, Lincoln, Nebr.

Johansson, N.-O. 1970. Ice formation and frost hardiness in some agricultural plants. Natl. Swed. Inst. Protection Contrib. 14(132):364–382.

Johansson, N.-O., and E. Krull. 1970. Ice formation, cell contraction and frost killing of wheat plants. Natl. Swed. Inst. Plant Protection Contrib. 14(131):343–362.

Keck, R. W., and J. S. Boyer. 1974. Chloroplast response to low leaf water potentials. III. Differing inhibition of electron transport and photophosphorylation. Plant Physiol. 53: 474–479.

Klotz, I. M. 1967. Energy changes in biochemical reactions. Academic Press, New York.

Kramer, P. J., and H. Brix. 1965. Measurement of water stress in plants. UNESCO Arid Zone Research 25:343–351.

Levitt, J. 1980. Responses of plants to environmental stresses. Vol. 1, Chilling, freezing and high temperature stresses. Academic Press, New York. 497 p.

Levitt, J. 1980b. Responses of plants to environmental stresses. Vol. 2, Water, radiation, salt and other stresses. Academic Press, New York, 607 p.

Levitt, J. 1956. The hardiness of plants. Academic Press, New York.

Levitt, J. 1972. Responses of plants to environmental stresses. Academic Press, New York.

Ludlow, M. M. 1976. Ecophysiology of C_4 grasses. p. 364–380. *In* O. L. Lange et al. (ed.) Water and plant life. Springer-Verlag, New York.

Lyons, J. M. 1973. Chilling injury in plants. Annu. Rev. Plant Physiol. 24:445–466.

Mazur, Peter. 1969. Freezing injury in plants. Annu. Rev. Plant Physiol. 20:419–445.

Mudd, J. B. (ed.). 1975. Effect of air pollution on plants. Academic Press, New York.

Mudd, J. B., and T. T. Kozlowski (ed.). 1975. Responses of plants to air pollution. Academic Press, New York.

Norcio, N. 1976. Effects of high temperature and water stress on photosynthesis and respiration rates in grain sorghum. Ph.D. Thesis. The University of Nebraska, Lincoln, Nebr.

Olein, C. R. 1967. Freezing stresses and survival. Annu. Rev. Plant Physiol. 18:387–408.

Penning, De Vries, F. W. T. 1972. Respiration and growth. p. 327–347. *In* A. R. Rees (ed.) Crop processes in controlled environments. Academic Press, London.

Penning, De Vries, F. W. T. 1974. The cost of maintenance processes in plant cells. Annu. Bot. 39:77–92.

Raison, J. K. 1974. A biochemical explanation of low-temperature stress in tropical and sub-tropical plants. p. 487–497. *In* R. L. Bieliski, A. R. Ferguson, and M. M. Cresswell (ed.) Mechanisms of regulation of plant growth. Bull. 12. The Royal Society of New Zealand.

Salter, P. J., and J. E. Goode. 1967. Crop responses to water at different stages of growth. Resources Res. Series 2. East Malling, Maidstone, Kent Commonwealth Bur. of Hort. and Plantation Crops.

Salisbury, F. B., and Cleon Ross. 1969. Plant physiology. Wadsworth Publishing Co., Inc., Belmont, Calif.

Salisbury, F. B., and Cleon Ross. 1978. 2nd ed. Plant physiology. Wadsworth Publishing Co., Inc., Belmont. Calif.

Shearman, L. L., J. D. Eastin, C. Y. Sullivan, and E. J. Kinbacher. 1972. Carbon dioxide exchange in water stressed sorghum. Crop Sci. 12:406–409.

Siminovitch, D., B. Rheaume, and R. Sachar. 1967. Seasonal increases in protoplasm and metabolic capacity in the cells during adaptation to freezing. p. 3–40. *In* C. L. Prosser (ed.) Molecular mechanisms of temperature adaptation. Pub. No. 84, Am. Assoc. Adv. Sci., Washington, D.C.

Siminovitch, D., B. Rheaume, K. Ponvoy, and M. Lepogo. 1968. Phospholipid, protein, and nucleic acid increases in protoplasm and membrane structures associated with development of extreme freezing resistance in black locust tree cells. Cryobiol. 5:202–225.

Slatyer, R. O. 1973a. The effect of internal water status on plant growth, development and yield. p. 177–191. *In* R. O. Slatyer (ed.) Plant response to climatic factors. Proc. Uppsala Symp. 1970. UNESCO, Paris.

Slatyer, R. O. 1973b. Effects of short periods of water stress on leaf photosynthesis. p. 271–276. *In* R. O. Slatyer (ed.) Plant response to climatic factors. Proc. Uppsala Symp., 1970. UNESCO, Paris.

St. John, J. B., and M. N. Christiansen. 1976. Inhibition of linolenic acid synthesis and modification of chilling resistance in cotton seedlings. Plant Physiol. 57:257–259.

Sullivan, C. Y. 1971. Techniques for measuring plant drought stress. p. 1–18. *In* K. L. Larson and J. D. Eastin (ed.) Drought injury and resistance in crops. CSSA Special Publication No. 2, Crop Sci. Soc. of Am., Madison, Wis.

Sullivan, C. Y. 1972. Mechanisms of heat and drought resistance in grain sorghum and methods of measurement. p. 247–264. *In* N. G. P. Rao and L. R. House (ed.) Sorghum in the seventies. Oxford and IBH Publishing Co., New Delhi, India.

Sullivan, C. Y., and J. D. Eastin. 1975. Plant physiological responses to water stress. *In* J. F. Stone (ed.) Plant modification for more efficient water use. Agric. Meteorology 14:113–128.

Sullivan, C. Y., N. V. Norcio, and J. D. Eastin. 1977. Plant responses to high temperature. p. 301–318. *In* Amir Muhammed, Rustem Aksel, and R. C. von Borstel (ed.) Genetic diversity in plants.

Tanaka, Akira. 1977. Photosynthesis and respiration in relation to productivity in crops. p. 213–229. *In* A. Mitsu, S. Miyachi, A. San Pietro, and Saburo Tamura (ed.) Biological solar energy conversion. Academic Press, New York.

Ting, I. P., and R. L. Heath. 1975. Responses of plants to air pollutant oxidants. Adv. Agron. 27:89–121.

Tingey, D. T., W. W. Heck, and R. A. Reinert. 1971. Effect of low concentrations of ozone and sulfur dioxide on foliage, growth and yield of radish. J. Am. Soc. Hort. Sci. 96:369–371.

Turner, N. C., J. E. Begg, H. M. Rawson, S. D. English, and A. B. Hearn. 1978. Agronomic and physiological responses of soybean and sorghum crops to water deficits. III. Components of leaf water potential, leaf conductance, $^{14}CO_2$ photosynthesis, and adaptation to water deficits. Aust. J. Plant Physiol. 5:169–177.

Wang, J. Y. 1972. Agricultural meteorology. Milieu Information Service. San Jose, Calif.

Warrington, I. J., R. L. Dunstone, and L. M. Green. 1977. Temperature effects at three developmental stages on the yield of the wheat ear. Aust. J. Agric. Res. 28:11–27.

Weatherley, P. E. 1950. Studies in the water relations of the cotton plant. I. The field measurement of water deficits in leaves. New Phytol. 49:81–97.

9 Flowering

F. G. DENNIS, JR.
Department of Horticulture
Michigan State University
East Lansing, Michigan

The cryptogams (Gk. *kryptos,* hidden, plus *gamos*, marriage), including the ferns and other spore-producing plants, were preeminent in the plant
kingdom during the Paleozoic era, between 230 and 600 million years ago.
During the subsequent Mesozoic and Cenozoic eras, the phanerogams
(*phaneros,* visible), including both gymnosperms (*gymnos,* naked, plus
sperma, seed) with sporophylls and angiosperms (*angios,* case) with true
flowers, gained ascendance and today comprise the bulk of our economic
plants. The appearance of the flower gave a distinct advantage to these
plants, for out-crossing and hybridization became much more common,
with a marked acceleration of evolution.

As far as the plant is concerned, flowers serve one major function—reproduction. Without flowers, and the seeds produced in the resultant fruits,
few of our economic crops could survive. Only those that can be propagated
vegetatively could be successfully grown, and improvement of these would
be slow and tedious for lack of genetic variation in the characters desired.
Flowering plants, therefore, provide man with an almost unlimited opportunity for genetic improvement.

Flowers serve man in many ways, some aesthetic, such as production of
perfumes and beautification of the environment, some mundane, such as
the production of cotton fiber (actually extensions of the seed coat).[1]
Flowers are also used as food, either directly, as in broccoli and cauliflower
(both *Brassica oleracea* L. var. *botrytis*) and artichoke (*Cynara scolymus*
L.), or indirectly through the production of a wide variety of fruits, including apples (*Pyrus malus* L.), bean, rice, wheat, and maize. The purpose of
this chapter is to describe some of the factors that control the flowering process and to discuss the use of cultural practices in controlling flowering.

[1] Scientific names of important crop plants are given in Table 1.1, Chapter 1.

Published in *Physiological Basis of Crop Growth and Development,* © American Society of
Agronomy—Crop Science Society of America, 677 South Segoe Road, Madison, WI 53711,
USA.

COMPONENTS OF YIELD

Seed or fruit yield, as volume or weight per unit of land area, is a product of the number of plants per unit area, the number of inflorescences per plant, the number of fruits or seeds (pods, kernels, etc.) per inflorescence, and the weight per fruit or seed (e.g., Lelley, 1976; Adams, 1967). These factors are recognized as components of yield; they are not independent, for an increase in one component often leads to a decrease in another. In general, the number of inflorescences per plant declines as the number of plants per unit area (planting density) increases. Similarly, the weight per fruit declines as the number of fruits developing per inflorescence increases. All four factors, therefore, must be considered when evaluating potential yield. The second factor (number of inflorescences per plant) involves flowering and will be discussed in this chapter; the third and fourth relate to fruit development and will be covered in Chapter 10.

Although tiller formation is not a part of flowering per se, it cannot be ignored when discussing the flowering of most cereal crops, for it largely determines the number of inflorescences per plant and therefore the yield potential. Production of additional tillers can compensate for winter injury and therefore is more important in winter than in spring wheat. In spring grains, yield is a function of seeding rate, particularly under irrigation. In fact, Donald (1968) suggested that a spring wheat without tillers ("uniculm" wheat) would be ideal for maximum production, for this would reduce "plasticity," or variation in yield as a function of compensation between tiller number, spikelet number, etc.

As the number of tillers per plant increases beyond a certain point, spike size and number of kernels per spike decline. Tiller number varies with cultivar and with spacing and other cultural practices. For example, the number of tillers per plant in IR8 rice grown in pots increased from 3.7 with no nitrogen to 57 with 100 mg L^{-1} nitrogen in the culture solution. In a spacing trial under field conditions, tiller number per plant in the same cultivar increased from 3.4 to 15.7 as spacing between plants increased from 10 × 10 to 50 × 50 cm (Chandler, 1969).

DEVELOPMENTAL STAGES

Plants are not capable of flowering upon germination but must pass through a juvenile period lasting from a few days to many years depending upon species. Treatments given to induce flowering in this period will fail. Annual plants—those that survive only one year—exhibit relatively short juvenile periods. Some weed species, such as lamb's quarters (*Chenopodium album* L.) can be induced to flower 4 to 5 days after germination. On the other hand, forest trees often exhibit very long juvenile periods and may not flower for 25 or more years after seed germination.

An appreciation of the phenomenon of juvenility is important if problems are to be avoided in commercial production of certain crops. In areas with mild winters, cole crops such as cauliflower and cabbage (*Brassica oleracea* L. var. *capitata*) are sown in the fall for winter production. If sown at the proper time, cabbage will be in the juvenile stage during exposure to temperatures low enough to induce flower stalk elongation ("bolting") in adult plants (see section below on effects of temperature), and therefore produce heads, rather than seedstalks. If sown too early, bolting will occur. Similarly, cultivars of sugar beet used for early planting must be "bolting resistant" to prevent flower induction in response to cool spring temperatures. Biennial varieties of cauliflower, on the other hand, should be sown early enough in the spring to assure exposure to temperatures low enough to induce flowering. Kohlrabi has a very short juvenile phase and will bolt whenever exposed to temperatures of less than 10°C for 1 week or longer.

Juvenility is also important to the tree breeder, for 25, or even 10 years is a long time to wait to make a cross. For this reason, progress in breeding tree species has been slow compared with that obtained with herbaceous plants.

A distinction should be made between true juvenility, in which flowering is prevented by internal conditions, and a failure to flower because of unfavorable environmental conditions. For example, plants of some cultivars of tobacco will never flower if they are continuously exposed to long photoperiods; once the photoperiod is shortened below a critical point, flowering ensues (see below). These plants are not juvenile but are fully capable of flowering given appropriate conditions.

Following the juvenile period, the adult phase begins and may last for as short a time as a summer in annuals or as long as several thousand years in sequoias (*Sequoiadendron giganteum* Buchholtz) and bristle-cone pines (*Pinus aristata* Engl.). Senescence then begins, the plant dies, and the cycle is repeated. Senescence may occur as a "programmed" change, as in monocarpic (Gk. *monos,* one, plus *karpos,* fruit) plants such as wheat, which flower and fruit only once, then die, or as a gradual weakening in polycarpic plants, such as forest trees, which flower repeatedly. Monocarpic plants are often annuals, but may be biennials which flower in the second year after germination (cabbage), or may live for even longer periods before flowering, fruiting, and dying, as bamboo [*Bambusa arundinacea* (Retz.)] and agave (*Agave americana* L.). In polycarpic plants, senescence probably results from the ravages of insects and diseases and from the difficulty of supporting the bulk of the plant, rather than from specific internal changes in metabolism. Scions may be taken from an old tree, for example, and either rooted or grafted on a young seedling, and the resultant tree is capable of repeating the cycle. This is, in fact, the manner in which fruit trees are propagated; e.g., all 'Delicious' apple trees have descended from one original seedling tree. The apical meristem of such plants may, therefore, be considered "immortal."

CONTROL OF FLOWERING

As noted above, plants are incapable of flowering during the juvenile phase, regardless of environmental conditions. Thereafter, environmental factors such as light, temperature, and to some extent moisture and nutrition, can exert profound effects. In addition, treatments with growth regulators and certain other chemicals can either promote or retard flowering in some plants.

Light

One of the most important factors affecting flowering is light. Light effects may be divided into three different components—responses to light duration (photoperiod), to light quality (wave-length), and to irradiance (radiant energy). These three components often interact to affect flowering.

The discovery of photoperiodism is commonly attributed to W. W. Garner and H. A. Allard of the U.S. Department of Agriculture, although previous workers were aware of this phenomenon. Although Garner and Allard are best remembered for their work with Maryland Mammoth tobacco, a plant that failed to flower except when grown in the greenhouse during the winter, they made numerous observations on other plants, notably soybean. They divided plants into three groups, based upon their responses to photoperiod, viz., short-day plants (SDP), long-day plants (LDP), and day-neutral plants (DNP) (Fig. 9.1). Subsequent experiments have shown that the length of the dark period, as well as length of the light period, is crucial for flowering under most conditions. Therefore, one might call SDP's long-night plants and LDP's short-night plants, but the original terminology has been retained for convenience. The light and dark periods actually act together. A practical definition of a SDP is one that will not flower or is retarded in flowering unless the dark period exceeds a critical length; conversely, a LDP is one that will not flower or is retarded in flowering unless the dark period is less than a critical length. Thus certain strains of both *Perilla ocymoides* L. and darnell (*Lolium temulentum* L.) will flower if the dark period is 8 to 10 hours long, yet the former is a SDP, the latter a LDP (Fig. 9.2). Shortening the dark period to 2 hours prevents flowering in *Perilla,* while lengthening it to 14 hours prevents flowering in darnel.

A distinction should be made between qualitative, or absolute, vs. quantitative responses to photoperiod. Plants such as *Perilla* and darnel exhibit qualitative responses; they do not flower if the dark period is too long or too short. More commonly, the response is quantitative; if photoperiod is not optimum, the plants require longer to flower, or do not flower as profusely, but eventually they do flower. This concept is illustrated in Fig. 9.3. Several other response types are now recognized, including plants that require, for example, long days prior to short days (LSDP) or low tempera-

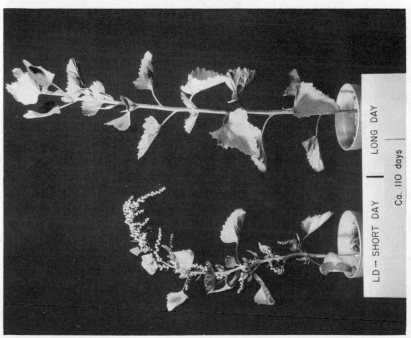

Fig. 9.1. Effect of photoperiod on flowering response of the short-day plant lambs-quarters (left) and the long-day plant spinach (right). Short day = about 8 hr; long day = about 20 hr. From Salisbury (1963), by permission.

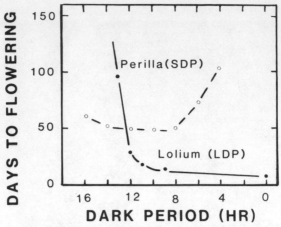

Fig. 9.2.   Time to initiation (*Lolium*) or appearance (*Perilla*) of flowers in a LDP (*Lolium temulentum* L., cv. Ceres) and in a SDP (*Perilla ocymoides* L., strain 324) as affected by length of the dark period in a 24-hour cycle. Data from Evans (1958) and Gaillochet et al. (1962), by permission.

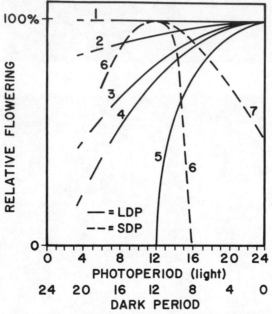

Fig. 9.3.   Diagram illustrating flowering responses to various day-lengths, each line representing a different type of response. 1. A truly day-neutral plant, flowering about the same at all daylengths. (With many species, there is little or no flowering when days are unusually short; e.g., from 6 to 8 h long.) 2. Plant slightly but probably insignificantly promoted in its flowering by long days. 3 and 4. Both plants quantitatively promoted by long days (to different degrees), although they flower on any daylength. 5. Qualitative or absolute long-day plant such as henbane (flowers only when days are longer than about 12 h). 6. Qualitative short-day plant such as cocklebur (flowers only when days are shorter than about 15.6 h; nights longer than 8.3 h). Note that cocklebur also fails to flower if days are shorter than 3 to 5 h. 7. Quantitative short-day plant, flowers on any daylength but better under short days. Note that different species have different critical day and night lengths, not just the 12 and 15.6 h shown for henbane and cocklebur. From Salisbury (1982), by permission.

Table 9.1. Photoperiod and temperature requirements for flowering of some economically important plants.† (Adapted from Salisbury, 1963. Used by permission.)

Temperature requirement	Photoperiod requirement		
	Day neutral	Short day	Long day
None	Bean (*Phaseolus vulgaris* L.)	Coffee (*Coffea arabica* L.)	Carnation cvs. (*Dianthus caryophyllus* L.)
	Buckwheat (*Fagopyron esculentum* Moench.)	Pigweed (*Amaranthus retroflexus* L.)	Clover (*Trifolium* spp.)
	Corn (*Zea mays* L.)	Sweet potato (*Ipomoea batatas* (L.) Lam.)	Foxtail (*Alopecurus pratensis* L.)
	Cucumber (*Cucumis sativus* L.)		Oat cvs. (*Avena sativa* L.)
	Kentucky bluegrass (*Poa pratensis* L.)		Ryegrass cvs. (*Lolium perenne* L.)
			Spinach cvs. (*Spinacia oleracea* L.)
High			
Absolute	--‡	--	--
Quantitative	Fuchsia (*Fuchsia* spp.)	Cosmos (*Cosmos bipinnatus* Gav. Ann.)	Phlox (*Phlox drummondii* Hook)
	Summer rice (*Oryza sativa* L.)	Chrysanthemum cvs. [*Chrysanthemum morifolium* (Ramat.) Hemsl.]	
		Winter rice (*Oryza sativa* L.)	
Low			
Absolute	Cabbage (*Brassica oleracea* var. *capitata*)	Chrysanthemum cvs. [*Chrysanthemum morifolium* (Ramat.) Hemsl.]	Carnation cvs. (*Dianthus caryophyllus* L.)
	Celery (*Apium graveolens* var. *dulce*)		Ryegrass cvs. (*Lolium perenne* L.)
	Hydrangea [*Hydrangea macrophylla* (Thunb.) Ser.]		Spinach cvs. (*Spinacia oleracea* L.)
			Sugar beet (*Beta vulgaris* L.)
Quantitative	Broad bean (*Vicia faba* L.)	Chrysanthemum cvs. [*Chrysanthemum morifolium* (Ramat.) Hemsl.]	Carnation cvs. (*Dianthus caryophyllus* L.)
	Sweet pea (*Lathyrus odoratus* L.)		Oat cvs. (*Avena sativa* L.)
	Vetch (*Vicia angustifolia* L.)		Spinach cvs. (*Spinacia oleracea* L.)
			Winter barley (*Hordeum vulgare* L.)
			Winter wheat (*Triticum aestivum* L.)

† Several species occur in more than one category because of cultivar differences in response. A somewhat different classification system is used by Vince-Prue (1975).
‡ No species are known which have these requirements.

ture prior to long days (Table 9.1). Salisbury (1963) provides an extensive coverage of this topic.

Some plants flower following exposure to one long night or one long day but most require more than one cycle. Strawberry [*Fragaria* × *ananassa* (Duch.)], a SDP, requires six to 20 cycles depending upon the cultivar, while early flowering chrysanthemum [*Chrysanthemum morifoli-*

Table 9.2. Number of inductive cycles required to saturate the flowering response in several species. Response varies with strain and cultivar.†

Response type	Species	No. of cycles
Short day plants	Japanese morning glory (*Pharbitis nil* Chois.)	1
	Cocklebur (*Xanthium strumarium* L.)	1
	Soybean (*Glycine max* L.)	3–7‡
	Strawberry [*Fragaria* × *ananassa* (Duch.)]	6 to 20
	Perilla ocymoides L.	8.5
	Early flowering chrysanthemum [*Chrysanthemum morifolium* (Ramat.) Hemsl.]	12§
Long day plants	Spinach (*Spinacea oleracea* L.)	1
	Darnel (*Lolium temulentum* L.)	1 to 4
	Henbane (*Hyoscyamus niger* L.)	4 to 5

† Data from various studies cited in Evans (1969).
‡ Three cycles are sufficient to induce flowering, but the percentage of nodes flowering increases with further cycles up to 7 or more.
§ Twelve cycles are sufficient for induction, but longer exposure to short photoperiods is necessary to prevent axillary buds from resuming vegetative growth on transfer to long photoperiods.

um (Ramat) Hemsl.] requires 12 cycles. If chrysanthemum plants are transferred to long days too soon, the inflorescence stops developing and the axillary buds grow vegetatively. In at least one soybean cultivar ('Biloxi'), all plants flower after three cycles, but additional cycles up to seven increase the number of nodes bearing flowers (Table 9.2).

A list of some economically important plants in which flowering is markedly affected by photoperiod is given in Table 9.1. A more extensive listing of the photoperiodic behavior of agronomic crops may be found in Martin et al. (1976). The commercial importance of photoperiodism is readily evident in the florist industry, in which such species as poinsettia (*Euphorbia pulcherimma* Wild.) and chrysanthemum (SDP) are covered with black cloth during part of the day to exclude light and thereby induce flowering in mid-summer. Normal daylength is either extended with artificial lighting, or the dark period is interrupted with a light break, to induce flowering of LDP's such as carnation (*Dianthus caryophyllus* L.) and coneflower (*Rudbeckia bicolor* Nutt.) or to prevent flowering of SDP's such as chrysanthemum until the plants reach a desirable size. Commercial growers of field crops and vegetables must take photoperiodism into account when choosing cultivars. For example, early cultivars of lettuce, an LDP, form flower stalks ("bolt") and flower under the long photoperiods in mid-summer, and soybean cultivars differ markedly in time of flowering in response to photoperiod. Because daylength changes as one moves toward or away from the equator (Fig. 9.4), soybean cultivars suitable for Michigan are unsuitable for Mississippi because they flower before the

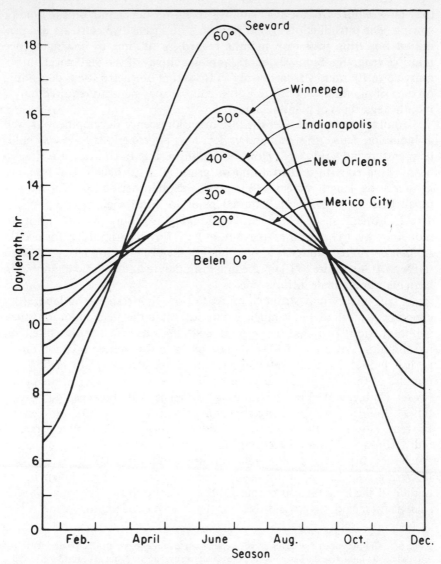

Fig. 9.4. Effect of latitude on photoperiod (sunrise to sunset). Figures indicate degrees from the equator. From Leopold (1964), by permission.

plants are of sufficient size, resulting in low yields. Conversely, when grown in Michigan, cultivars suitable for Mississippi will flower late and be killed by frost before the pods mature. Such cultivar differences account for the appearance of some species, e.g., spinach (*Spinacia oleracea* L.) and rye-grass in more than one location in Table 9.1. At lower latitudes, soybean is normally not planted until 1 month before the longest day of the year to avoid premature flower induction.

Rice is a major crop in the tropics. Traditional cultivars are short day plants and are sensitive to the small changes in photoperiods which occur;

they may require from 4 to 11 months to flower depending on the time of sowing. The introduction of new photoperiod-insensitive cultivars has permitted flowering in 3 to 5 months regardless of time of sowing, hence multiple cropping. Similarly, an Ethiopian cultivar of the SDP sorghum required 3 to 10 months from sowing to flowering in Puerto Rico, depending on date of planting, whereas one U.S. cultivar (day neutral) required only 2 months regardless of planting date.

Photoperiod may affect the rate of inflorescence development as well as induction per se. In wheat and barley, long photoperiods following cold-temperature induction (see below) hasten flowering. In contrast, the time to heading in smooth-stalked meadow grass (*Poa pratensis* L.) becomes greater as daylength increases. Similarly, the photoperiod to which quantitative LDP's or SDP's are exposed prior to inflorescence initiation can affect subsequent development. Prolonging exposure to SD prior to induction by LD increases the number of flower primordia formed in *Lolium perenne* L. and wheat, both quantitative LDP's. In maize, a quantitative SDP, exposure to LD preceding initiation increases flower number in both male and female inflorescences.

Light quality may strongly influence flowering response in laboratory experiments, although it is much less important under field conditions than is photoperiod. Red light (wavelength 660 nm) is as effective as white light in inhibiting flowering of SDP's when given in the middle of a long dark period, but far-red light (730 nm) has little or no effect. However, if a few minutes of far-red light are given immediately after the red light, the plants flower (Table 9.3). The red/far red treatments can be repeated several times, flowering/no flowering depending upon the last treatment (red vs. far-red). However, the ability of far-red to reverse the promotive effect of red declines with time, becoming nil after 1 to 90 min, depending upon species (Table 9.4). Similar relationships hold for LDP's, except that flowering response is the opposite. For example, red light given in the middle of the long night promotes, rather than inhibits, flowering. Studies such as these laid the groundwork for the discovery of phytochrome, a pig-

Table 9.3. Inhibition of flowering in cocklebur by a 2 min exposure to red light (R) in the middle of an inductive dark period, and the effect of subsequent 2 min exposures to far-red (FR), or alternating R and FR, light in counteracting the inhibition. Note that the effect of far-red declines with time. (From Downs, 1956. Used by permission.)

Exposure	Floral stage†
Control—no night-break	6.0
R	0.0
R-FR	5.6
R-FR-R	0.0
R-FR-R-FR	4.2
R-FR-R-FR-R	0.0
R-FR-R-FR-R-FR	2.4
R-FR-R-FR-R-FR-R	0.0
R-FR-R-FR-R-FR-R-FR	0.6

† 0 = Vegetative; 8 = maximum flowering response.

Table 9.4. Effect of length of interval between exposure to red and far-red light during inductive dark period on flowering of cocklebur. (From Downs, 1956. Used by permission.)

Treatment	Floral stage†
Control—no interruption	7.0
Control—red without far-red	0.0
Red, followed by far-red after (min):	
0	6.5
20	3.8
30	1.8
40	0.5
60	0.0

† 0 = Vegetative; 8 = maximum flowering response.

ment that apparently plays a crucial role in photoperiodism (see below). Many species are not as responsive as soybean to short interruptions of the dark period. In strawberry, for example, the duration of the light interruption must be greater than 1 hour to inhibit flowering completely. However, responses generally parallel those of soybean and other sensitive species such as cocklebur (*Xanthium strumarium* L.) and Japanese morning glory (*Pharbitis nil* Chois.), in which flowering is inhibited by as little as a few seconds to a few minutes of exposure to light during the long dark period.

Some SDP's can flower when grown in total darkness, provided they are supplied with an energy source, such as sucrose. Under normal circumstances, however, light is essential for growth and development and, therefore, for flowering. Soybean, for example, will flower on photoperiods between four and 14 to 15 hours. Within limits, vigor is proportional to irradiance.

Photoperiodic responses are often induced by very low levels of light. For example, flowering of cocklebur may be inhibited by as little as 0.02 to 0.20 mW m^{-2} nm^{-1} of red light during the inductive dark period. (For comparison, a full moon provides about 0.002 mW m^{-2} nm^{-1} at 660 nm.) But much higher irradiances—on the order of 10 mW cm^{-2}—are necessary during the light period for photosynthesis and optimum flowering.

Flower production can be drastically reduced by artificial shading. With young apple trees in England, when light level in the outer crowns of the trees was reduced to 37, 25, and 11% of full sunlight, flowering was reduced 40, 56, and 79%, respectively, in comparison with unshaded trees. Similar reductions have been observed in wheat, carnation, apricot (*Prunus armeniaca* L.), blueberry (*Vaccinium corymbosum* L.), peach (*Prunus persica* Batsch.), pear (*Pyrus communis* L.), grape (*Vitis vinifera* L.), and pine (*Pinus* spp.). In contrast, although short-term shading can reduce flowering, mature plants of cacao (*Theobroma cacao* L.), vanilla (*Vanilla planifolia* Andr), and coffee (*Coffea arabica* L.), are intolerant of high light levels. Under full sun, leaf abscission and death of twigs result in reduced yields. Banana (*Musa acuminata* Colla) or plantain (*Musa paradisiaca* L.) trees are often used for shading coffee plantations in the tropics.

Temperature

In many species several days to several weeks of low temperatures (0 to 10°C) promote flowering, and, as in photoperiodism, there are qualitative and quantitative responses depending upon species and strain (Table 9.1). The promotive effect of cold treatment (0 to 10°C) on subsequent flowering is called vernalization (literally, "springization"). A large group of herbaceous plants, termed biennials and including carrots (*Daucus carota* L.), have an absolute requirement for vernalization. They normally do not flower until the second year of growth (Fig. 9.5), being induced to flower by the low temperatures of winter between first and second growing seasons.

Fig. 9.5. Effects of cold- or gibberellin-treatment on flowering of an early carrot variety. Left: control receiving neither cold- nor gibberellin-treatment; center: no cold, but treated with 10 μg gibberellin daily for 4 wk; right: 6 wk of cold, no gibberellin. From Lang (1957), by permission.

As mentioned above in discussing juvenility, the response of biennial plants to low temperatures often creates problems for the farmer. If celery, grown for its petioles, or beets, grown for their roots, flower prematurely, the crop is often worthless. Entire celery fields have been known to flower when temperatures of 5 to 10°C have prevailed for 10 or more days in the early part of the season. The importance of local conditions is illustrated by the fact that sugar beets are sown in September in both the Salt River Valley of Arizona and the Imperial Valley of California, some 403 km to the west, yet seed is produced in Arizona, sugar in California. The difference in response is primarily a consequence of lower temperatures in Arizona, which promote bolting.

A related problem, sometimes observed in sugar beet, is the ability of seeds developing on the mother plant to be vernalized by low temperatures late in the growing season. Such seeds may flower prematurely when planted. To prevent this, seed production should be restricted to areas where temperatures remain above 12°C during seed maturation.

Certain groups of cultivars within a species often differ in their requirements for vernalization. Winter cultivars of wheat and barley tiller freely without vernalization, but will not form heads unless imbibed seeds or young plants are subjected to temperatures of 3 to 10°C for 2 to 10 weeks. Plants older than 3 months do not respond. These cultivars are sown in the fall in northern latitudes; a mild winter may delay flowering the following spring. Spring cultivars do not require vernalization, although low temperature may hasten flowering. Certain cultivars classified as intermediate require only a short period at cold temperatures; they head if planted early in the spring, but not if planted late.

Some bulb crops require chilling, not for flower induction, but for stimulating subsequent elongation of the inflorescence. Hyacinths (*Hyacinthus orientalis* L.) and tulips (*Tulipa gesneriana* L.) form flower primordia readily in the fall, but subsequent emergence of the flowers is prevented or greatly retarded in the absence of chilling. Bulb forcers have developed well-defined schedules for producing plants for sale at Easter and other holidays. Flowering of many temperate zone trees is likewise dependent on adequate chilling to break bud dormancy.

The second effect of temperature is its effect upon the plant's response to daylength. Flowering will not occur in Petkus winter rye under SD, regardless of germination temperature. When grown under LD, plants developing from seeds germinated at 18°C require 20 weeks to flower, but those from seeds germinated at 1°C require only 9 weeks. The requirement for cold treatment is quantitative in this case, rather than qualitative, in that lack of cold only delays rather than prevents flowering under LD. Similar interactions between temperature and photoperiod are common (Table 9.1). For example, long days hasten flowering in vernalized sugar beet and lettuce plants. Short photoperiods can replace vernalization in some wheat cultivars, provided plants are subsequently transferred to long days. Winter wheat is a quantitative long-day plant (Table 9.1) because heading of many cultivars following vernalization is hastened by long days. This characteristic delays heading during the short photoperiods of spring, thus reducing the danger of frost injury. Cultivars grown at low latitudes are generally less sensitive to photoperiod.

Another interesting interaction between temperature and photoperiod occurs in one species of morning glory [*Ipomoea purpurea* (L.) Roth] which is a LDP at low temperature (13°C), a SDP at high temperature (21 to 24°C) and day-neutral at intermediate temperature (17 to 18.5°C).

High, rather than low, temperatures are required for flower induction in some species, although the requirement is generally a quantitative one. Again, temperature interacts with photoperiod. Summer rice and fuchsia (*Fuchsia* spp.) are day neutral, hence the promotive effect of high temperature is not daylength-dependent. However, high temperatures also hasten flowering in the short-day species winter rice and cosmos (*Cosmos bipinnatus* Cav. Ann.), and in phlox (*Phlox drummondii* Hook)—a LDP. On the other hand, camellias (*Camellia japonica* L.) exhibit an absolute requirement for high temperature and a quantitative requirement for long days. Additional response types are discussed by Salisbury (1963) and Vince-Prue (1975).

Aside from its effects upon flower induction, temperature is important in determining the time of flowering. Data are available on the number of "heat units" required from sowing to heading in wheat and rye (Nuttonson, 1955, 1958). Heat units are calculated by determining the mean temperature for each day, subtracting a base temperature (usually 40°F or 5°C), and accumulating the remainders ("degree-days") for each day from sowing (spring wheat) or 1 March (winter wheat) to heading. In the USA as a whole, spring wheat requires an average of 1142 units from sowing to heading, whereas winter wheat requires 978 from 1 March to heading. Note that only spring and summer temperatures are used in these calculations, and that winter wheat, already well established by 1 March, heads more rapidly than does spring wheat, which may be sown in March, April, or May, depending upon zone of cultivation. The number of heat units required decreases as latitude increases, apparently because of the longer summer photoperiods at high latitudes; spring wheat values for the Soviet Union, Canada, and Finland are 1123, 983, and 866, respectively. These values were obtained using data from cultivars adapted to each climatic zone and are therefore not strictly comparable. Data for a single cultivar grown at several latitudes (Table 9.5) clearly indicate the importance of long

Table 9.5. Effects of photoperiod and accumulated growing-degree days on time from emergence to full heading in 'Marquis' wheat. (From Nuttonson, 1955.)

Location	Latitude (°N)	Days to heading	(a) Degree-days base 32°F	(b) Avg. day-length (hour)	(a) × (b)
Fairbanks, Alaska	64	33	955	20.3	19 386
Winnipeg, Manitoba	50	46	1 402	16.2	23 712
Lincoln, Nebraska	41	60	1 743	14.2	24 751
Tlalnepantla, Mexico	19	106	2 121	11.6	24 604

photoperiods in hastening heading at the higher latitudes. Data for both day-degrees and daylength are included, together with their product. The product increases 27% while degree-days alone increases 122% from one extreme to the other, indicating less variation when both temperature and photoperiod are included in the calculation.

Water Supply

Water availability markedly affects yields of grains, although effects on flowering per se are not as well documented. Moisture deficits delay flower primordia formation in barley and grain sorghum. If the deficit is prolonged, spikelet numbers can be reduced. Sorghum is relatively tolerant of moisture stress, and development resumes on rewatering, although anthesis may be delayed for 10 days or more. In rice, withholding moisture reduces both number of spikelets per panicle and spikelet fertility; greatest effect on fertility is observed when treatment is applied during the 2 week period prior to heading. Sugarcane is grown for its stems, rather than its fruit. Flowering inhibits growth and reduces sugar content, but can be prevented by imposing a mild moisture stress in August-September in Hawaii.

Nutrition

Many early plant physiologists considered nutrition to be the determining factor in controlling flowering. G. Klebs, a German botanist, proposed in 1918 that a high ratio of carbohydrate to inorganic nutrients, particularly nitrogen (C/N ratio), favored flowering. E. J. Kraus and H. R. Kraybill of Oregon State University are often cited as having provided evidence for this proposal in tomato (*Lycopersicon esculentum* Mill.), but their work actually dealt with fruiting, rather than flowering (see Chapter 10). Others have found some correlations between nutrition and flowering, however. For example, heavy applications of nitrogen (= low C/N ratio in the plant) will delay flowering in young fruit trees, while old, weak trees (having a high C/N ratio) often flower profusely. In sour cherry (*Prunus cerasus* L.) the percentage of lateral buds flowering declines with increasing shoot vigor. Despite these correlations, there appears to be no critical C/N ratio, similar to a critical photoperiod, for flowering. The effect of nitrogen status varies with species in photoperiodically induced plants, nitrogen deficiency often promoting flowering in LDP's, while delaying it in SDP's.

Nitrogen may affect flower development following induction. In certain species of pine, application of nitrogen stimulates production of female "flowers" (strobili), but has little or no effect on production of male strobili. Similar results have been obtained with tung (*Aleurites fordii*

Hemsl.). In Douglas-fir [*Pseudotsuga menziesii* (Mirb.) Franco] nitrogen application increases the production of both male and female strobili. In wheat and rice, nitrogen deficiency reduces the numbers of spikelets and fertile florets. Nitrogen application is most effective in increasing the numbers of spikelets per panicle in rice when applied 3 to 8 weeks prior to heading.

Growth Regulators

Chemicals capable of inducing flowering under environmental conditions not conducive to flowering would be of considerable interest in agriculture. For example, many hours of labor could be saved if poinsettias and other SDP's could be sprayed once or twice with a chemical, rather than having to cover them with black cloth each evening and to remove the cloth every morning. Unfortunately, few such chemicals exist. The pineapple (*Ananas comosus* Merr.) industry in Hawaii has used naphthaleneacetic acid (NAA) successfully to induce more concentrated flowering and fruiting, thereby facilitating harvesting. Their success was once considered to be evidence for the role of natural auxins in flowering, but we now know that the effect is indirect, through the stimulation of ethylene production (see below), and ethylene-generating compounds such as ethephon are currently used. Many other species in the same family (Bromeliads) can be induced to flower by treating with ethylene, but the response does not occur in most plants. Other chemicals that exert a marked stimulative effect on flowering are the gibberellins (GA's), which promote flowering both in many LDP's and in plants that require vernalization (Fig. 9.5), and certain synthetic growth retardants, which promote flowering in woody plants. Thus gibberellin A$_3$ (GA$_3$) has been used experimentally to induce flowering of the LDP's lettuce and spinach under short days, and of the cold-requiring plants beet, carrot, cabbage, and parsley (*Petroselinum crispum* Nym.) in the absence of chilling temperatures. Commercial application of GA's has been limited, however, by variability of response among and within cultivars, and the response of cold-requiring cereals such as wheat and oat has been poor.

Growth retardants, some of which may inhibit the synthesis of endogenous gibberellins, promote flowering in some herbaceous species such as tomato, and in many woody plants, including azalea (*Rhododendron* spp.), apple, and stone fruits. Daminozide [butanedioic acid mono-(2,2-dimethylhydrazide)] hastens flowering in young fruit trees when applied early in the growing season. However, the compound slows fruit growth and cannot be used at this time on bearing trees if fruit size is a problem.

Other compounds promote flowering of particular species under some circumstances. For example, 2,3,5 triiodobenzoic acid (TIBA), a chemical that interferes with auxin transport, promotes flowering in apple, tomato, and several other species.

While not affecting flower induction per se, certain chemicals affect the subsequent development of the flower primordia. Cucurbits are particularly responsive to treatment with chemicals that alter the ratio of male to female flowers. The flower primordia contain both pistil and stamen initials, but one or the other is suppressed in the formation of pistillate or staminate flowers. Treatment with auxin favors the production of the former; treatment with gibberellin the latter (Table 9.6). Gibberellins A_4 and A_7 are particularly effective and are used by breeders to induce gynoecious (female) mutants to produce pollen. Responses to both auxins and gibberellins are probably mediated through their effects on ethylene biosynthesis.

Treatment of cucurbits with ethylene gas or ethephon markedly increases the proportion of female flowers produced. Furthermore, preventing the accumulation of endogenously produced ethylene in the tissues by growing the plants under low atmospheric pressure can reduce or completely prevent the production of female flowers in some species. Inhibitors of ethylene synthesis or action, such as $AgNO_3$, are also effective. Most commercial cucumber (*Cucumis sativus* L.) cultivars produce only male flowers for the first 10 to 15 nodes. Ethephon may have potential for hastening commercial production by shortening this period of sterility (Table 9.6). It may also be used by breeders to cause androecious plants (which produce only male flowers) to form some female flowers, thereby allowing selfing and maintenance of genetically pure lines.

Certain other chemicals also affect sex expression. For example, cytokinin application causes the development of perfect flowers in some "male" (androecious) grape clones. The effects of nutrition on sex expression have already been discussed.

THE PHYSIOLOGY OF FLOWERING

Phytochrome

The isolation of phytochrome (Gk. *phytos*, plant, and *chroma,* a color) in 1959 was preceded by extensive investigations into the effects of different wavelengths of light on physiological processes that were light-dependent, including flowering of both SDP's and LDP's and germination of lettuce

Table 9.6. Effects of ethephon, an ethylene-releasing chemical, and gibberellin on sex expression in cucumber flowers. (From Robinson, 1969. Used by permission.)

Treatment	Node of appearance of first male or female flower	
	Male	Female
None	2.0	8.0
Ethephon, 250 mg L^{-1}	14.3	2.0
GA_3, 2000 mg L^{-1}	1.3	>17.0

and other seed. Under the leadership of Sterling B. Hendricks and Harry Borthwick, a group was organized at the USDA Plant Research Station at Beltsville, Maryland, in 1944. A large spectrograph for separating light into its various components by means of a prism was constructed, and partially defoliated plants were exposed to the various wavelengths. Flowering response was plotted as a function of incident energy at each wavelength. A crucial finding was that the wavelengths most active in promoting the germination of lettuce seeds at 660 nm (red light) were also maximally effective in inhibiting the flowering of SDP's, and promoting the flowering of LDP's, when given in the middle of a long night (Fig. 9.6). In addition, far-red light (730 nm) given immediately after the red nullified its effects; that is, lettuce seeds did not germinate, SDP's flowered, and LDP's did not flower.

These results were taken as evidence that one pigment was responsible for all of these phenomena. The pigment was assumed to exist in two forms, one (P_r) predominately absorbing red light, the other (P_{fr}) predominately absorbing far-red light. The P_r was converted to P_{fr} on absorption of red light; P_{fr} was converted back to P_r on absorption of far-red light (Fig. 9.7). In many plants P_{fr} converts to P_r also during long periods of darkness. This conversion of P_{fr} to P_r was considered to be one of the time-measuring reactions in photoperiodism, with LDP's requiring certain relatively high levels of P_{fr} for flowering, while SDP's required relatively low levels. This concept was an oversimplification, however, for we now know that (a) P_{fr} may be destroyed, as well as being converted to P_r, (b) the P_r/P_{fr} ratio may be more important than are absolute levels of either, and (c) other time-measuring reactions are far more important than is conversion of P_{fr} to P_r.

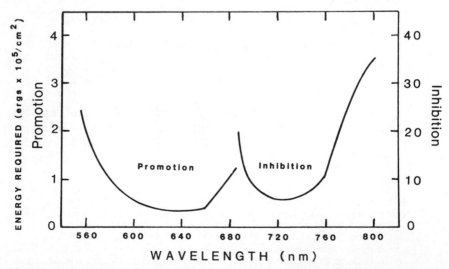

Fig. 9.6. Action spectrum for promotion of flowering in cocklebur by light applied during an inductive long night ("promotion") and for inhibition of the promotive effect of red light ("inhibition") by a subsequent exposure to far-red. The effectiveness of the light is inversely proportional to the incident energy required. Note that the scales differ for the two curves. Adapted from Salisbury (1963), by permission.

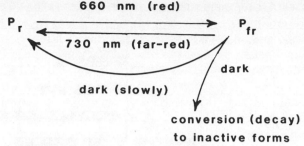

Fig. 9.7. Interconversion of the forms of phytochrome (P_r and P_{fr}) in response to light of different wavelengths and in darkness.

Sequence of Events in the Photoperiod Induction of Flowering

The photoperiodic induction of flowering has been studied intensively in an attempt to determine why plants respond as they do. Table 9.7 summarizes the most important facts known concerning the effects of photoperiod on flowering of SDP's, as exemplified by cocklebur. The information gained from studies with other species is included in some instances.

Table 9.7. Major facts known about photoperiodic induction of flowering in SDP's as exemplified by cocklebur.

A. Light period preceding critical dark period.
1. Minimum of 4 hours of light is required preceding a 12-hour period to permit flowering; response is saturated at 12 hours (Fig. 9.8, cases 4 and 5). As the light period is shortened, the critical dark period is lengthened (Fig. 9.8, cases 14 to 17).
2. Minimum light intensity required is 10^4 mW m^{-2}.
3. White or red light is effective; far-red light is not effective (Fig. 9.8, cases 6 to 9).
4. Response is independent of temperature between 5 and 44°C (Fig. 9.8, cases 10 to 13).
5. In some species, e.g., Japanese morning glory, flowering can occur in total darkness.

B. Critical dark period for promoting flowering.
1. Minimum duration is slightly more than 8 hours in a 24-hour cycle; rate of response is saturated at 12 hours (Fig. 9.8, cases 1, 2, 6).
2. The critical dark period becomes shorter as the light period preceding it is extended (Fig. 9.8, cases 14 to 17).
3. In some species other than cocklebur (soybean, lamb's quarter), as the period is extended from 16 to 64 hours, oscillations occur in response, with a periodicity of approximately 24 hours (Fig. 9.9, bottom).
4. A brief interruption (minimum of 2 to 3 sec) of white or red light reduces or eliminates the response, depending on timing. The maximal effect is obtained at 8 hours.
5. Flowering will not occur if light is applied for up to 1 hour after 8 hours of darkness, even though the plant is subsequently held in the dark for 12 hours (Fig. 9.8, case 4). Extending the subsequent dark period to 15 hours permits flowering (Fig. 9.8, case 14).
6. If the light interruption is greater than 4 hours, flowering will occur following a subsequent 12-hour dark period (Fig. 9.8, case 5).
7. A brief exposure to far-red light during the dark period does not inhibit flowering. However, far red will reduce or eliminate the inhibitory effect of a preceding red light flash if applied within 30 min, the effect decreasing as the time interval increases (Table 9.4).
8. In some species other than cocklebur (soybean, lamb's quarter), the effect of light interruption during an extended dark period (16 to 64 hours) oscillates between inhibitory and promotive no effect depending upon timing (Fig. 9.10).

C. Light period following dark period.
1. A light period is generally required following the dark period before flowering can occur. This period can be replaced by feeding sucrose.

HOURS BEFORE OR AFTER BEGINNING OF INDUCTIVE DARK PERIOD

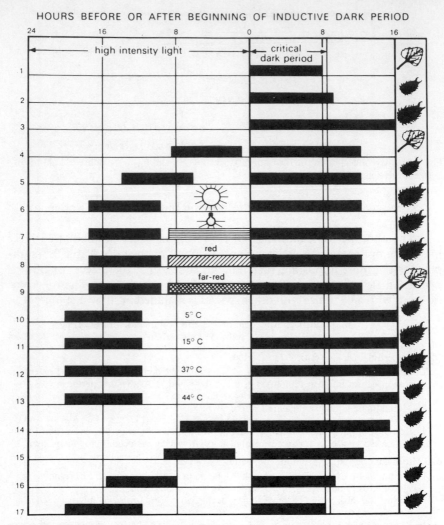

Fig. 9.8. Effects of various photoperiodic treatments on flowering of cocklebur (see Table 9.7). From Salisbury (1966), by permission.

The process as a whole in cocklebur may be separated into 3 major sequences—the light period preceding the critical dark period, the critical dark period itself, and the subsequent light period (Fig. 9.10). Under normal circumstances, plants will have been exposed to long photoperiods prior to the inductive long night. Note (Table 9.7) that (a) the duration of the previous light period has a profound effect upon response to the critical dark period, and (b) a light flash during the critical dark period eliminates the plant's response to a subsequent inductive dark period.

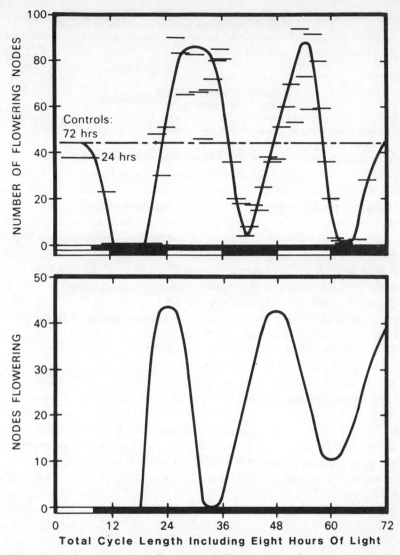

Fig. 9.9. Rhythmical responses in flowering of 'Biloxi' soybean in response to (top) light breaks (short horizontal bars) during a 64-hour dark period or (bottom) dark periods of varying length following an 8-hour day. All plants received 7 cycles. Bars at the bottom of upper figure indicate total light-dark cycle (top bar) or postulated photophile ("light-loving", white) and skotophile ("dark-loving", black) phases of the cycle. Data for both figures from various publications of K. C. Hamner. Figures from Salisbury (1963), by permission.

These observations have led to several hypotheses as to the mechanisms whereby photoperiod affects flowering in SDP's. These hypotheses deal with the means by which (a) light initiates the photoperiodic reaction, (b) the length of time in darkness is measured, and (c) the message is relayed

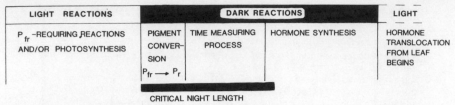

CRITICAL NIGHT LENGTH

Fig. 9.10. Possible partial processes occurring during photoperiodic induction of flowering in a short-day plant. From Vince-Prue (1975), by permission.

from the leaf to the meristem where flowering actually occurs. The explanation most often advanced for the first mechanism is based on the observation that regular oscillations or "endogenous rhythms" occur in plants and other organisms. For example, movement of leaves of many plants follows an approximate 24-hour (*circadian,* from the Latin *circa* = approximate, and *dies* = day) cycle, and time of illumination can alter this pattern. Students of the flowering process therefore have proposed that the effect of prior illumination on the plant's response to a critical dark period varies, depending on the timing of the exposure relative to an endogenous rhythm. Furthermore, both the "light on" or "dawn" signal and the "light off" or "dusk" signal affect response, and the two must be properly timed with respect to one another. Although this explanation may seem fanciful, further analysis of the process may prove it to be correct.

One of the original hypotheses to explain time measurement, or how the plant senses when the dark period has exceeded a critical length, was that this period represented the time required for conversion of P_{fr} to P_r (the "hour-glass" hypothesis). Thus flower induction (or inhibition of induction in the case of LDP's) could proceed only after the level of P_{fr} fell below a critical level. But indirect evidence indicates that this conversion is complete a short time after the dark period begins. If this is true, some other mechanism must measure time. For example, the ratio of P_{fr} to P_r could be the determining factor in "starting the timer" in an oscillating system dependent on endogenous rhythms. Provided at least 6 hours have elapsed in darkness, irradiation with red light will set the clock back to zero, while subsequent irradiation with far red restores the clock to its position prior to irradiation.

The processes in LDP's, while similar to those occurring in SDP's, are not as well-defined (Table 9.8). The most notable difference, aside from the reversal in response to light/dark cycles, is the failure of far red light to overcome the inhibitory effect of a red light break when the dark period is prolonged. This failure is probably related to the long period of exposure required, in contrast with the short period needed for cocklebur.

Once sufficient time has elapsed in either light (LDP) or darkness (SDP), a message must be "written" for transmission to the meristem where flowering occurs. In SDP's translocation of this message requires exposure to light following the critical dark period. This presumably allows the leaf to export sucrose to the meristem as a carrier for the message, for light can be replaced by feeding the leaf with sucrose.

Table 9.8. Major facts known about photoperiodic induction of flowering in LDP's
as exemplified by darnel.

A. Light period preceding critical dark period.
  1. Plant will flower in continuous light. Some species, e.g., radish (*Raphanus sativus* L.)
      will flower when grown in total darkness.
  2. Flowering of plants exposed to 8 hours of daylight is little affected by a subsequent ex-
      posure to 8 hours of far red light in a 24-hour cycle. Red light is somewhat effective in
      stimulating flowering, but a mixture of red and far red is markedly promotive (Fig.
      9.11).

B. Critical dark period for inhibiting flowering.
  1. Minimum duration is 8 to 14 hours in a 24-hour cycle depending upon strain of darnel.
      Response is nearly saturated at 14 hours (Fig. 9.2).
  2. A 5-min red light break is only effective in promoting flowering if the dark period is
      slightly longer than the critical length. However, with longer dark periods (16 hours)
      light breaks must be longer than 2 hours to be effective, and effectiveness increases with
      time of exposure. In strain Ba 3081, exposure to 1 hour of red light during 10 long nights
      induces flowering, and response increases with time of exposure.
  3. Red–far red reversibility cannot be demonstrated because of the long period of exposure
      necessary. In some other LDP's (e.g., barley and henbane), exposure to far red light fol-
      lowing exposure to red light prevents the promotive effect of the former, provided the
      dark period is near the critical length.
  4. If plants are exposed to an 8-hour photoperiod, followed by a dark period exceeding 16
      hours, the effect of a 4-hour interruption with red light oscillates from inhibitory to pro-
      motive depending on timing.
  5. The effects of red and far-red light in promoting and inhibiting flowering differ
      markedly depending on timing during the light/dark cycle.

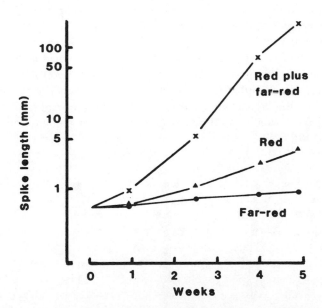

Fig. 9.11. Effects of red, far-red, and mixed red plus far-red light on flowering in *Lolium
temulentum* L. strain Ba 3081. Plants received 8 hours of daylight plus supplementary light
of indicated wavelengths from 1600 to 2400 hours daily. From Vince-Prue (1975), by permission.

Florigen

A hormone, florigen, was first postulated for the control of flowering in 1937 by M. K. Chailakhyan, a Russian plant physiologist, on the basis of grafting experiments with chrysanthemum and other plants. He and several other workers had demonstrated that if a non-induced plant were grafted onto an induced plant, the former would flower, even if it were never exposed to inductive daylengths. In some plants, such as cocklebur, the effect can be transmitted from plant to plant through several graft "generations". Thus, grafting a lateral shoot from an induced plant on a non-induced plant induces flowering of the latter (Fig. 9.12). If another shoot is then removed from the second plant and grafted on a third, it too flowers, and so on. In other plants, such as *Perilla,* a member of the mint family, only tissues from the original photoinduced plant are effective. Those from the graft-induced plant are ineffective when grafted on a third plant. In several genera, notably *Nicotiana,* LD species or cultivars will flower under SD when grafted to induced plants of SD species, and vice versa. These observations indicate that a hormone (florigen) is produced during induction and can move through a graft union to induce flowering in the graft partner. This hormone appears to be at least physiologically identical in LDP's, SDP's, and DNP's and is non-specific with regard to genus and species.

Further evidence for florigen arises from studies of the effect of removing the photoinduced leaf at various intervals after induction. If one leaf of a cocklebur plant is darkened to provide a "short day" while the remainder of the plant is illuminated, flowering will result provided all leaves between the darkened leaf and the apex are removed. If the induced leaf is removed soon after the beginning of the inductive dark period, no flowering occurs. Removal between 20 and 36 hours results in progressively greater responses. Thus, the hormone appears to be translocated from the induced leaf to the apical meristem, which responds by changing from the vegetative to the reproductive state.

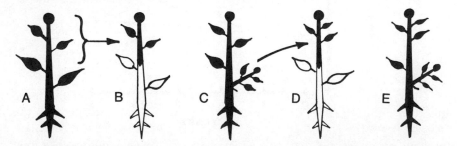

Fig. 9.12. Transmission of the flowering stimulus by grafting in cocklebur. (A) Plant induced to flower by exposure to short days. (B) Scion from (A) grafted on noninduced plant. (C) Same as (B) with lateral shoot induced to flower by grafting. (D) Scion from lateral shoot of (C) grafted on second noninduced plant. (E) Same as (D) with lateral shoot induced to flower by grafting. Adapted from Lang (1965), by permission.

Although many attempts have been made to isolate florigen, none has provided unequivocal evidence for a specific chemical. This failure has given rise to alternative hypotheses, one being that flowering is controlled by a balance between promoters produced by photoinduced leaves and inhibitors produced by leaves exposed to non-inductive conditions. This concept is supported by the fact that the presence of non-induced leaves reduces the flowering response when only one portion of the plant is exposed to inductive photoperiods. Their removal early in the inductive cycle permits maximum response. This evidence for inhibitors has been discounted by some investigators on the grounds that leaves between the induced leaf and the apex act as competing "sinks" for assimilates, thus reducing the quantity of florigen reaching the apex. However, direct evidence for such inhibitors was provided by Anton Lang and Soviet co-workers in 1977. They demonstrated that grafting scions of *Nicotiana sylvestris,* a LDP, on a day-neutral cultivar of tobacco inhibited flowering under SD. Inhibitors produced under SD therefore must have been translocated from the LDP.

Another multiple-hormone system has been postulated by Chailakhyan (1968) who first proposed the name "florigen" for the flowering hormone. He suggests that high levels of two hormones—gibberellins and "anthesins" (Gk. *anthein,* to bloom)—must be present in a plant before flowering can occur. Short-day plants kept on long photoperiods contain sufficient gibberellins, but little if any anthesins. Transfer to short days stimulates the production of anthesins and the plants flower. Conversely, long day plants kept on SD produce anthesins in abundance, but are deficient in gibberellins. On long photoperiods, gibberellins are synthesized, and flowering ensues. While this thesis is an attractive one, it predicts that if a LDP were to be grafted on a SDP and both partners kept on their respective non-inductive photoperiods, gibberellins would be translocated from the SDP to the LDP and anthesins from the LDP to the SDP, and both graft partners should flower. In fact, they do not, making this hypothesis untenable.

Despite the difficulties in isolating florigen or other flowering promoters or inhibitors, the concept of such hormones is a challenging one. Perhaps future work with more sophisticated techniques will provide us with a chemical basis for flowering.

Other Hormones

Research in the 1940s and early 1950s stressed the role of auxins in the photoperiodic induction of flowering, and many studies were made on the relative levels of these compounds under long vs. short days. Although auxin levels increase temporarily when plants are moved from short to long days, no conclusive evidence has been obtained for their role in flowering. The discovery by Lang (1956) in 1956 that gibberellic acid could induce flowering in certain long-day plants gave great impetus to research on the role of the native gibberellins in flowering. As with auxins, comparisons of the gibberellin content of plants kept under long vs. short photoperiods had not until recently revealed differences sufficient to account for flowering responses. However, Metzger and Zeevaart (1980), using more sophisticated techniques, have now demonstrated that concentrations of gibberellins increase dramatically in spinach plants when they are transferred from SD to LD. Spinach grows as a rosette under SD and bolts under LD, thus changes in GA content may be associated with the elongation of the stem, rather than with flowering, per se. Although high levels of abscisic acid, a potent growth inhibitor, have been reported to occur in plants grown under short days, subsequent investigations with more refined techniques have not confirmed this observation.

SUMMARY

In the course of evolution, plants have developed mechanisms to permit survival in harsh environments; one of these is the ability to flower at a time when environmental conditions will allow fruit and seed maturation. Environmental signals control or modify flowering in many species, temperature and light intensity and duration being the most important. Farmers must be aware of such signals and plant responses to them in selecting species and cultivars for their particular location and considerable use is made of alterations in daylength and temperature in regulating flowering of ornamental plants. Several chemicals can promote or inhibit flowering or modify development of flower parts to affect sex expression. Some of these are useful commercially.

Although the mechanisms by which environment controls flowering are only partially understood, a number of facts have been established. Phytochrome is the pigment that receives the photoperiodic signal and, therefore, plays a key role in flowering, as well as many other processes. Its structure and properties change with the wavelength of incident radiation. Both promoters and inhibitors of flowering occur in plants and can be transmitted from plant to plant by grafting. These chemical messengers are non-specific in terms of both species and response types, for they can be translocated from SDP's to LDP's and from one species to another. No chemicals are presently known which can induce flowering consistently when applied to SDP's under LD conditions. Although gibberellins can induce flowering in LDP's under SD, they are ineffective on SDP held under LD and, therefore, cannot be the hypothetical florigen.

Numerous questions remain unanswered. Some of the most important are the following: How does the plant measure the length of sequential light and dark periods? How is the signal that is received by phytochrome translated into chemical messenger? What is florigen? What is the chemical nature of inhibitors of flowering? How can systems which have the same receptor (phytochrome) and messengers (hormones which promote/inhibit flowering) have opposite effects in different plants (i.e., LDP's vs. SDP's)? Answering these questions remains the goal of further research in this important field.

REFERENCES

Adams, M. W. 1967. Basis of yield component compensation in crop plants with special reference to the field bean, *Phaseolus vulgaris*. Crop Sci. 7:505–510.

Chailakhyan, M. Kh. 1937a. Hormone theory of plant development. [In Russian] Moscow, Leningrad. Acad. Science, URSS. (Cited by L. T. Evans, 1969.)

----. 1937b. Concerning the hormonal nature of plant development processes. [In Russian] C. R. (Dokl.) Acad. Sci. URSS. 16:227–230. (Cited by L. T. Evans, 1969.)

----. 1968. Internal factors of plant flowering. Annu. Rev. Plant Physiol. 19:1–36.

Chandler, R. F., Jr. 1969. Plant morphology and stand geometry in relation to nitrogen. p. 263–285. *In* Physiological Aspects of Crop Yield. Am. Soc. of Agronomy-Crop Sci. Soc. of Am., Madison, Wis.

Donald, C. M. 1968. The design of a wheat genotype. p. 377–378. *In* Proc. Third Int. Wheat Genetics Symposium, Canberra, Australia.

Downs, R. J. 1956. Photoreversibility of flower initiation. Plant Physiol. 31:279–284.

Evans, L. T. 1958. *Lolium temulentum* L., a long day plant requiring only one inductive photocycle. Nature 182:197.

----. (ed.) 1969. The induction of flowering. Some case histories. Cornell Univ. Press, Ithaca, N.Y.

Gaillochet, J., C. C. Mathon, and M. Stroun. 1962. Nouveau type de reaction et changement du type de reaction au photoperiodisme chez *Perilla ocimoides* L. C. R. Acad. Sci. (Paris) 255:2501.

Garner, W. W., and H. A. Allard. 1920. Effect of the relative length of night and day and other factors of the environment on growth and reproduction in plants. J. Agric. Res. 18: 553–606.

Klebs, G. 1918. Uber der Blutenbildung von Sempervivum. Flora (Jena) 111–112, 128–151.

Kraus, E. J., and H. R. Kraybill. 1918. Vegetation and reproduction with special reference to the tomato. Oregon Agric. Exp. Stn. Bull. 149.

Lang, A. 1956. Induction of flower formation in biennial *Hyoscyamus* by treatment with gibberellin. Naturwissenschaften 43:284–285.

----. 1957. The effect of gibberellin upon flower formation. Proc. Natl. Acad. Sci. USA 43:709–717.

----, M. Kh. Chailakhyan, and I. A. Frolova. 1977. Promotion and inhibition of flower formation in a dayneutral plant in grafts with a short-day plant and a long-day plant. Proc. Natl. Acad. Sci. USA 74:2412–2416.

Lelley, J. 1976. Wheat breeding. Theory and practice. Akademiai Kiado, Budapest.

Leopold, A. C. 1964. Plant growth and development. McGraw-Hill Book Co., New York.

Martin, J. H., W. H. Leonard, and D. L. Stamp. 1976. Principles of field crop production. 3rd ed. MacMillan Co., New York.

Metzger, J. D., and J. A. D. Zeevaart. 1980. Comparison of the levels of 6 endogenous gibberellins in roots and shoots of spinach in relation to photoperiod. Plant Physiol. 66: 679–683.

Nuttonson, M. Y. 1955. Wheat-climate relationships and the use of phenology in the thermal and photo-thermal requirements of wheat. American Institute of Crop Ecology, Washington, D.C.

----. 1958. Rye-climate relationships and the use of phenology in ascertaining the thermal and photo-thermal requirements of rye. American Institute of Crop Ecology, Washington, D.C.

Robinson, R. W., S. Shannon, and M. D. de la Guardia. 1969. Regulation of sex expression in the cucumber. BioScience 19:141–142.

Salisbury, F. B. 1963. The flowering process. 1st ed. Pergamon Press, New York.

————. 1966. Die Bluhenbildung. (Flowering.) Naturwiss. Med. 12:48.

————. 1982. Photoperiodism. Hort. Rev. 4:66–105.

————, and C. Ross. 1978. Plant physiology. 2nd ed. Wadsworth Publishing Co., Belmont, Calif.

Vince-Prue, D. 1975. Photoperiodism in plants. McGraw-Hill Book Co., New York.

SUGGESTED READING

Bernier, G., J. M. Kineb, and R. M. Sachs. 1981. The physiology of flowering. Vol. 1. The initiation of flowering. Vol. II. Transition to reproductive growth. CRC Press, Boca Raton, Fla.

Chailakhyan, M. K., and J. Heslop-Harrison. 1979. Genetic and hormonal regulation of growth, flowering, and sex expression in plants. Am. J. Bot. 66:717–736.

Evans, L. T. 1975. Daylength and the flowering of plants. Benjamin, Menlo Park, Calif.

Hillman, W. S. 1962. The physiology of flowering. Holt, Rinehart, and Winston, Inc., New York.

Lang, A. 1965. Physiology of flower initiation. Encyclop. Plant Physiol. 15(1):1380–1536.

Leopold, A. C., and P. E. Kriedemann. 1974. Plant growth and development. 2nd ed. McGraw-Hill Book Co., New York.

Wareing, P. F., and I. D. J. Phillips. 1978. The control of growth and differentiation in plants. Pergamon Press, New York.

Schwabe, W. W. 1971. Physiology of vegetative reproduction and flowering. In F. C. Steward (ed.) Plant physiology: A treatise. Vol. 6A:233–411, Academic Press, New York.

Zeevaart, J. A. D. 1976. Physiology of flower formation. Annu. Rev. Plant Physiol. 27:321–348.

Zeevaart, J. A. D. 1978. Phytohormones and flower formation. p. 291–324. In D. S. Letham, P. B. Goodwin, and T. V. J. Higgins (ed.) Phytohormones and related compounds—A comprehensive treatise. Vol. II. Elsevier, Amsterdam.

10 *Fruit Development*

F. G. DENNIS, JR.
Department of Horticulture
Michigan State University
East Lansing, Michigan

The value of flowers to a plant derives solely from their reproductive function, i.e., seed production. Seeds are borne in fruits, some relatively simple in structure, some complex, which arise from flowers. Fruits can be defined morphologically as "mature ovaries, together with closely associated parts," or physiologically as "those tissues which support the ovules." Since some fruits, such as banana (*Musa acuminata* Colla) and navel orange (*Citrus sinensis* L.), are seedless, having been selected for cultivation, the second definition speaks of "ovules," which are present even in seedless fruits, rather than seeds. In grasses, "fruits" and "seeds" are practically synonymous, for the seed occupies all the space within the fruit (caryopsis) and is inseparable from the ovary wall.

Seeds are useful in a number of ways: for propagation; for food for human or animal consumption; for conversion into other useful products, such as whiskey prepared from rye and beer made from barley.[1] Vegetable oils from soybeans, maize, oil palm (*Elaeis guineensis* Jacq.), and olive (*Olea europaea* L.) can be used for margarine and shortening, or in paints and varnishes. Fibers of cotton seeds provide the raw material for textiles. However, many fruits have greater commercial value than the seeds which they contain. One of the obvious uses of fruits is food, as peaches (*Prunus persica* Batsch.), grapes (*Vitis* spp.), and tomatoes (*Lycopersicon esculentum* Mill.), and many others. Other fruits are used for flavorings [hop, vanilla bean (*Vanilla planifolia* Andr.), red pepper (*Capsicum annuum* L.)], for dyes [indigo (*Indigefera tinctoria* L.)], for waxes [wax myrtle (*Myrica cerefera* L.)], and for ornamental purposes [holly (*Ilex* spp.), bittersweet (*Celastrus scandeus* L.)].

---

[1] Scientific names of important crop plants are given in Table 1.1, Chapter 1.

Published in *Physiological Basis of Crop Growth and Development,* © American Society of Agronomy—Crop Science Society of America, 677 South Segoe Road, Madison, WI 53711, USA.

POLLINATION

The development of most fruits is dependent upon pollination and fertilization, together with subsequent seed development. However, exceptions occur in the case of parthenocarpic (Gk. *parthenos,* virgin, plus *karpos,* fruit) fruits, which are seedless. Parthenocarpy may be vegetative, in which case no external stimulus is required, or stimulative as a result of pollination alone or chemical treatment. Seedless bananas, navel oranges, and certain varieties of figs (*Ficus carica* L.) are products of vegetative parthenocarpy, while numerous species can be induced to produce seedless fruits by treatment with growth regulators. Many of the "seedless" grapes actually require fertilization, but the embryos abort soon after the fruits begin to develop.

Pollen may be transferred from the anther to the stigma in a variety of ways. In most plants, pollen is shed at or after anthesis and is transferred to the stigma by insects or by mechanical means, primarily wind. In violets (*Viola odorata* L.), sorrels (*Rumex* spp.), and lespedeza anthers dehisce to release pollen prior to anthesis, and thus pollination occurs within the bud (cleistogamy, from Gk. *kleistos,* closed, and *gamos,* marriage). Most fruit crops are dependent upon insect pollination. Many legumes are self-pollinated, but in some, including alfalfa, the structure of the flower is such that insect visitation is necessary to release ("trip") the style and stigma, thus allowing pollination.

Although wind pollination can occur under a range of environmental conditions, pollination of crops dependent upon insects is markedly affected by weather conditions. Bees are the most efficient agents of pollen transfer, and their activity is sharply curtailed by low temperatures or rain. A short period of good weather in an otherwise poor blossoming period can result in a commercial crop provided bees are available at a critical time. Thus, honeybees are often rented by orchardists and seed producers to aid in pollination.

Cross pollination is the transfer of pollen from one cultivar or strain to a second; pollination within a cultivar, even though more than one plant may be involved, is termed self-pollination. Cross pollination is necessary for the production of significant yields in many crops, while others produce abundant yields when selfed. Examples of both types of crops are given in Table 10.1. Cross pollination may be required for a number of reasons. In dichogamous plants, (Gr. *dicha,* in two) the pollen is shed either before or after the pistils are receptive. Flowers may be either protandrous (Gk. *proto,* first and *andros,* male) or protogynous (Gk. *gyne,* female). In the former case, pollen is shed before the stigma is receptive, in the latter case, after the stigma has ceased to be receptive. Dichogamy occurs in onion (*Allium cepa* L.) and carrot (*Daucus carota* L.) (protandrous), in several of the nut crops (generally protandrous), in pearl millet [*Pennisetum americanum* (L.) K. Schum.] (protogynous), and in corn (somewhat pro-

Table 10.1. Classification of some crops by pollination requirements. Because cultivars may differ in requirements, some crops are listed in more than one category.

A. Self-pollinated

Monocotyledons: Barley (*Hordeum vulgare* L.), foxtail millet [*Setaria italica* (L.) Beauvois], oats (*Avena sativa* L.), rice (*Oryza sativa* L.), sorghum [*Sorghum bicolor* (L.) Muench], wheat (*Triticum aestiva* L.).

Dicotyledons: Alfalfa (*Medicago sativum* L.), yellow sweetclover [*Melilotus officinalis* (L.)], vetch (*Vicia angustifolia* L.), field pea (*Pisum arvense* L.), cowpea [*Vigna unguiculata* (L.) Walp], soybean [*Glycine max* (L.) Merrill], most beans (*Phaseolus* spp.), peanut (*Arachis hypogaea* L.), flax (*Linum usitatissimum* L.), cotton (*Gossypium hirsutum* L.), tobacco (*Nicotiana tabacum* L.), tomato (*Lycopersicon esculentum* Mill.), lettuce (*Lactuca sativa* L.), sweetpea (*Lathyrus odoratus* L.), sour cherry (*Prunus cerasus* L.), most peach (*Prunus persica* Batsch.) cultivars, plum (*Prunus domestica* L.), walnut (*Juglans regia* L.), pecan [*Carya illinoinensis* (Wagenh.) C. Koch], most apricot (*Prunus armeniaca* L.) cultivars, grape (*Vitis* spp.), strawberry (*Fragaria* × *ananasa* Duch.), orange (*Citrus sinensis* L.).

B. Cross-pollinated

Monocotyledons: corn (*Zea mays* L.), rye (*Secale cereale* L.), timothy (*Phleum pratense* L.), pearl millet [*Pennisetum americanum* (L.) K. Schum], many perennial grasses.

Dicotyledons: alfalfa (*Medicago sativum* L.), white sweetclover (*Melilotus alba* Desr.), red clover (*Trifolium pratense* L.), ladino clover (*Trifolium repens forma dodigense* Hort. ex Gams.), birdsfoot trefoil (*Lotus corniculatus* L.), buckwheat (*Fagopyrum esculentum* Moench), cole crops including cabbage and cauliflower (both *Brassica oleracea* L.), apple (*Pyrus malus* L.), pear (*Pyrus communis* L.), sweet cherry (*Prunus avium* L.), plum (*Prunus domestica* L.), almond (*Prunus amygdalus* L.), filbert (*Corylus avellana* L.), mandarin (*Citrus* × *nobilis* Lour.), tangelo (*Citrus* × *Tangelo* J. Ingram and H. E. Moore).

C. Apomictic (may or may not require pollination)

Monocotyledons: Kentucky bluegrass (*Poa pratensis* L.), dallisgrass (*Paspalum dilatatum* Poir.).

Dicotyledons: *Citrus* spp., avocado (*Persea americana* L.).

D. Parthenocarpic

Monocotyledons: Banana (*Musa acuminata* Colla).

Dicotyledons: certain cultivars of cucumber (*Cucumis sativus* L.), orange (*Citrus sinensis* L.), grape (*Vitis* spp.) and fig (*Ficus carica* L.).

---

tandrous). In other plants, pollen fails to germinate on the stigma of a flower of the same genotype, or the pollen tube fails to grow to the base of the style or grows so slowly that the ovules degenerate before fertilization can occur. Thus, inbreeding is impossible, and the plants are termed self sterile. Buckwheat, white clover cultivars, and many cultivars of fruit trees are self-sterile. The term self infertile indicates that some fertilization occurs, but not to an extent sufficient for a commercial crop. Incompatibility refers to the failure of plants of different genotypes to produce fruit or seeds when crossed. In apple (*Pyrus malus* L.), for example, the cultivar 'Cortland' cannot be used as a pollen source for 'Early McIntosh' and vice versa. Most plants are incompatible with members of other species, although exceptions occur.

Certain cultivars are capable of producing commercial crops of fruit when cross pollinated, but they produce nonviable pollen. Pollen from triploid plants germinates very poorly, and pollen tube growth is slow and irregular. Triploids have three times the haploid number (n) of chromosomes, and irregularities occur during meiosis when the unequal numbers of chromosomes are divided between two daughter cells.

In crops which require cross-pollination, pollen of the male parent must mature when the stigma of the female parent is receptive; otherwise, fruit set will not occur.

After deposition on the stigma, the pollen grain germinates to produce a pollen tube which elongates through the style, enters the micropyle of the ovule, and penetrates the embryo sac. One of the generative nuclei in the pollen tube unites with the nucleus of the egg cell in the embryo sac to produce the zygote, a single diploid cell which subsequently divides to become the embryo. The second generative nucleus generally unites with the two polar nuclei in the embryo sac, giving rise to a triploid endosperm. The endosperm is a source of nutrients for the growing embryo.

In certain plants, embryos may develop from the maternal nucellar tissue, a phenomenon termed apomixis (Gr. *apo,* from, and *mixis,* a mixing). Pollination and fertilization are required in some species, e.g., *Citrus,* but not in others, e.g., Kentucky bluegrass and dallisgrass. Several apomictic embryos may develop in a single seed. These are genetically identical with the female parent.

ENVIRONMENTAL AND CULTURAL CONTROL OF FRUIT SET

Grain yield per plant is determined by the number of inflorescences, the number of florets per inflorescence, the proportion of florets which set seed, and the weight per seed. Of these, seed set is generally the most important. Seed development is essential for fruit development except in parthenocarpic fruits, and final fruit size of multi-seeded fruits is often proportional to seed number. The precise mechanism by which seeds exert their effect on fruit development is unknown. Immature seeds contain relatively high concentrations of hormones, however, and these are thought to control the growth of the fruit.

Definition and Extent of Set

Fruit set refers to the proportion of flowers which give rise to fruits. To prevent misinterpretation, the terms "initial set" (initial swelling of the ovary) and "final set" (production of mature fruits) are often used. In many species, flowers which do not develop into fruits abscise as a result of the formation of an abscission layer at the base of the pedicel (flower stalk) or peduncle (stem of inflorescence). In some species, e.g., cucumber (*Cucumis sativus* L.), strawberry (*Fragaria* spp.), and grasses, no abscission layer forms in non-fertilized flowers, yet the flower fails to develop further.

Thus, merely preventing formation of the abscission layer does not assure fruit development. On the other hand, abscission can occur even after fruit growth is initiated, the dropping of young fruits being particularly noticeable in fruit trees, cotton, and bean.

The proportion of flowers which develops into mature fruits varies greatly among species and cultivars. In cereal grains, up to 70% of the florets may set fruits; however, the percentage is usually lower, partially because of sterility of some of the florets. In soybean, 20 to 60% of the pods reach maturity. Deciduous fruit trees may set from 5 to 50% of a full complement of flowers. In some species, however, the production of flowers is so great that less than 1% set is adequate for a good yield. For example, an avocado (*Persea americana* L.) tree may bear a million individual blossoms, while a single mango (*Mangifera indica* L.) inflorescence may contain 5000 flowers. Percentage fruit set is often inversely correlated with flower density, as the flowers compete with one another and with other organs for nutrients (and hormones?) supplied by other plant parts. In both wheat and apple, partial flower removal increases the percentage set of the remaining flowers. Similar effects can be demonstrated by apex removal (tipping) in both grape and apple, and potato breeders often remove developing tubers to improve fruit set. In some species, competition between flowers is less important; their partial removal does not improve set in cherry, almond, or cotton.

Nutrient Supply

The effect of vigor upon fruit set in tomato was documented in a classic paper by E. J. Kraus and H. R. Kraybill in 1918, in which they noted the need for a balance between carbohydrates and nitrogen for maximum fruit production. When excessive amounts of nitrogen were supplied, plants were over-vigorous and set few fruits. If such plants were transferred to a medium containing a moderate amount of nitrogen, fruit set was abundant; if transferred to a medium with a very low level of nitrogen, they were again unfruitful. Excessive applications of nitrogen reduce fruit set of rice indirectly by interfering with anther dehiscence and shedding of pollen. In summary, fruit set is greatest when vigor is moderate. Vigor can be affected by numerous environmental factors, chief among them being mineral nutrition, light, and water supply.

Environmental Control

Although shading sometimes increases fruit set, probably by reducing the temperature, shaded plants or portions of plants generally yield less fruit than those exposed to full sun. In apple trees, shaded limbs in the center of the tree normally produce relatively few fruits. Similar effects of shading have been noted in cotton, wheat, tomato, and coffee (*Coffea arabica* L.). Young cotton bolls abscise within 2 to 3 days after the plants are artificially shaded. Photoperiodic effects on fruit set are not as well documented as are the effects of irradiance, but long days reportedly stimulate pod abscission in certain soybean cultivars, and failure of some Mexican and South American cotton accessions to set fruit at College Station, Texas has been attributed to the relatively long daylength (15 hours) there.

Persistent rain during bloom reduces cross-pollination and therefore fruit set; on the other hand, drought during early phases of fruit growth may reduce the number of fruits developing in many grain and other agronomic crops, including cotton and soybean. Corn is particularly sensitive to drought stress, probably because the long style ("silk") wilts, slowing pollen tube growth. An extended period of flowering provides some protection from isolated periods of drought stress.

Humidity can affect fruit set of certain species. The pistil in the bean (*Phaseolus* spp.) flower is particularly sensitive to low humidity. Artificial misting of the plants during the fruit setting period thus increases yields. Lima beans are grown on the California coast to take advantage of the cool, moist climate there. Fruit drop in the 'Washington Navel' orange in California has been ascribed to water stress resulting from low humidity, indicating that humidity may be important after, as well as during, fertilization.

In general, fruit set is favored by moderate temperatures; low temperatures may reduce pollen tube growth and pollen viability and/or may induce flower and/or fruitlet abscission in some species, e.g., tomato. High temperatures may also be detrimental, particularly in such cool-season crops as peas. Navy beans are concentrated in an area adjacent to Lake Huron in eastern Michigan because of cool temperatures during the flowering period. In one experiment with pepper (*Capsicum annuum* L.), no fruit set occurred in plants grown continuously at 20 to 27°C, while plants grown at 10 to 16°C after flower buds developed set practically all of their flowers. (in this case, all fruits were seedless.)

Climate represents the sum of the environmental parameters in a given region, and establishes the geographical limits of growth of all crops. For example, the combined effects of higher light intensity, few cloudy days, and cooler nights in the Yakima Valley of Washington, as compared with the Great Lakes region, allow Yakima apple growers to produce heavier crops of fruit without size reductions. A classic example of the effect of climate on fruit set is the 'Bartlett' pear (*Pyrus communis* L.), which is parthenocarpic in certain areas of California, yet must be pollinated in the Eastern states. Parthenocarpic development is associated with much greater tree vigor in California.

Chemical Treatment

Much research has been devoted to the use of growth regulating chemicals to increase fruit set, but few are being used commercially at present. Auxins such as 2-naphthoxyacetic acid (NOA) induce fruit set (parthenocarpy) in tomato in the absence of pollination. This chemical is used to a limited extent by growers of greenhouse tomatoes for increasing yields in the absence of pollinating insects. Auxins also induce parthenocarpy in holly, cucumber, figs, and seedless grapes. Many other species are not responsive to applied auxins, however, and commercial use of auxins is limited to certain varieties of grapes and pears. Gibberellins are effective in some species, including pome and stone fruits. Limited use is made of gibberellic acid for increasing fruit set of pear in Holland and Belgium and of navel oranges in Florida. Large acreages of 'Thompson Seedless' grapes are sprayed with gibberellic acid, although it actually reduces fruit set, but stimulates growth of the berries. Growth retarding chemicals, such as daminozide [butanedioic acid mono-(2,2-dimethylhydrazide)] are used to increase fruit set on seeded grapes in the Great Lakes states. The increased set is apparently due to inhibition of vegetative growth which competes with the flower cluster for nutrients. Triiodobenzoic acid (TIBA), an anti-auxin which appears to interfere with the movement of natural growth regulators in the plant, has increased yields of soybeans, but commercial application has been limited by inconsistent response. Although fruit set may be increased, a major portion of the effect is due to increased branching, permitting better penetration of sunlight.

MECHANISMS CONTROLLING FRUIT SET

The effects of pollination and fertilization on the growth of the ovary and surrounding tissues may be mediated by hormones arising from pollen tubes, ovary tissue, and fertilized ovules. A Swiss botanist, H. Fitting, in 1909, reported the effect of pollination on the swelling of orchid (several genera, including *Phalaenopsis*) ovaries, and showed that dead pollinia (structures containing pollen) and water extracts of pollinia had similar effects, even though no mature fruits developed. S. Yasuda, a Japanese worker, made similar observations on cucurbits in 1934. F. G. Gustafson of the University of Michigan showed in 1936 that pure chemicals (auxins) could induce fruit set. Furthermore, he noted that swelling ovaries contained relatively large amount of similar chemicals in comparison with non-swelling ovaries. Subsequent experiments with tobacco and other species suggested that the amount of auxin supplied by pollen was too small to account for fruit set, but that the pollen tube stimulated the production of large quantities of auxin in the style and ovary, and thus initiated fruit growth. Some of these experiments were later criticized because they were not performed under sterile conditions. We now know that bacteria can convert the amino acid tryptophan into the auxin indole-3-acetic acid (IAA).

The auxin content of the ovaries of some species increases rapidly as growth begins. This was demonstrated by S. H. Wittwer in corn and by J. P. Nitsch in grape. Furthermore, unpollinated ovaries of parthenocarpic cultivars of grapes and oranges contain more auxin than do those of cultivars dependent upon pollination. Although this correlation suggests a causal relationship, we have no proof at present that increase in auxin content is responsible for fruit growth, rather than the reverse.

Another well-documented source of auxin is the fertilized ovule. However, most data indicate that the auxin content of the ovary and/or ovules does not rise appreciably until several weeks after fertilization. Furthermore, the ovaries of many species do not respond to auxin treatment. In addition to auxins, both cytokinins and gibberellins occur in ovaries and seeds but their levels are no better correlated with fruit set.

Fertilization converts the flower, a slow-growing organ at anthesis, into a rapidly growing and metabolizing one which has the capacity to mobilize nutrients to it at the expense of vegetative growth. In okra [*Abelmoschus esculentus* (L.) Moench] for example, fruit set results in virtual cessation of vegetative development, and the photosynthates are thereafter diverted to fruit production. The fruit is referred to as a "metabolic sink," the leaf as a "source" of the carbohydrates, amino acids, and other compounds required for continued fruit growth. Many synthetic growth regulators have the capacity to mobilize nutrients to the point of application on a leaf or stem, and fruits probably exert their effects via similar compounds produced in the seed and/or ovary tissues. However, the crucial experiments necessary for establishing a cause and effect relationship remain to be performed.

FRUIT GROWTH

Prevention of Abscission

Initiation of fruit growth does not insure fruit maturity. In dicotyledons, several waves of abscission occur early in fruit development. The first consists mainly of flowers with defective or unfertilized ovules; subsequent waves of abscission are associated with abortion of the developing embryos, although this may be an effect rather than a cause of abscission. Conditions which affect fruit set—vigor, light, etc.—affect abscission in the opposite way, i.e., good vigor tends to promote fruit set and reduce abscission. Abscission does not readily occur in most grasses, an exception being reed canarygrass, but embryo abortion may occur, resulting in empty glumes.

As noted above, grain yield is more dependent upon fruit set than upon fruit size. The same is true for soybean yield. In fact, yield is often negatively correlated with fruit size, for yield generally increases, while fruit size decreases with increasing numbers of fruit per plant. Fruit size is important primarily for those crops in which a premium is paid for the larger sizes.

Growth Rate

Fruits differ in type and rate of growth. Tomato fruits expand almost entirely as a result of cell enlargement; pome fruits such as apple and pear undergo extensive cell division as well as cell enlargement during the 4 to 6 weeks immediately following fertilization, after which almost all growth is by cell enlargement. Cell division continues in the avocado fruit until it is harvested. Both tomato and pome fruits exhibit a sigmoid cumulative growth curve (Fig. 10.1), the initial growth rate being slow in absolute terms, then increasing to a steady state and finally decreasing toward maturity. Stone fruits such as peach exhibit a double sigmoid growth curve (Fig. 10.1), with two periods of growth (Stages I and III) each resembling the single curve for tomato. The period of slow growth (Stage II) is associated with rapid growth of the embryo and lignification of the endocarp (pit). However, seedless peach fruits, as well as fruits lacking a stony endocarp such as grape and strawberry, also exhibit a double sigmoid growth curve, suggesting that something more fundamental than embryo or endocarp development is involved. Rates of growth, as contrasted with cumulative growth in diameter, are shown in the lower part of Fig. 10.1.

Environmental Control

Many factors interact to determine the final size of fruits. The need for adequate moisture and light are obvious. While light is often the limiting factor in yield, CO_2 enrichment of the air in commercial greenhouses improves yields in some cultivars of tomato, indicating that CO_2 can be limiting under certain conditions (Wittwer and Robb, 1964).

Low night temperatures increase yields of some crops such as potato and corn, apparently by reducing the rate of respiration and thus maintaining carbohydrate reserves. This temperature effect is partially responsible for the heavier crops of apples grown in Washington in comparison with the eastern USA. Water stress hastens leaf senescence in wheat, thereby reducing yields. Plant physiologists are currently attempting to find chemicals which will either reduce drought stress or delay senescence; either response could increase yields.

Correlative Effects

In addition to external limitations, internal limitations to fruit growth such as competition for photosynthates within the plant are extremely important. Once a threshold number is reached, further increases in fruit number per plant reduce fruit size, a fact which dictates removal of some fruits (see above). This competitive effect holds for grains as well. Although total yield per wheat plant increases with tiller number, both spike size and number of kernels per spike decrease. On the other hand, as the number of kernels per spike increases, their size decreases. Even within spikelets, cor-

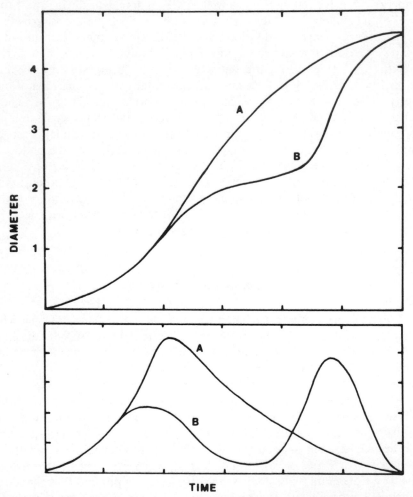

Fig. 10.1. Growth curve of tomato (A) vs. peach (B), representing berry and drupe fruits, respectively. Above: cumulative diameter; below: rate of change in diameter over time. Units are arbitrary (schematic).

relation effects are noticeable, for basal florets yield larger kernels than apical ones. Competition between grains appears to be less important in barley, for removal of some kernels does not increase the weight of those remaining. In rice, final kernel size is limited by the size of the hull, which is determined shortly before heading.

In dry bean, pod number per plant increases dramatically with planting distance. The unpublished data of Jose Montoya (cited by Erdman and Adams, 1978) indicate an increase in number of pods per plant from 11 to 48 as within-row spacing increased from 2 to 12 inches. Whether increased flowering or heavier fruit set was responsible for this response is not evident from the data. However, neither the number of seeds per pod nor the weight per seed was influenced appreciably, indicating that beans differ greatly from cereals in their response to changing plant density. Within limits, the more vigorous the plant, the more fruits it can support. In an overly vigorous plant, on the other hand, shoots may compete with fruits for metabolites.

Position on the plant or inflorescence has a marked effect on ultimate fruit size. The terminal flower on a strawberry or apple inflorescence (cyme) is generally the largest, while seeds (fruits) from the basal florets of spikelets of grasses such as wheat and oat are heavier than those from the terminal florets. In part, this difference may be explained by the structure of the first-formed flowers. More achenes are present on the primary flowers (terminal on the cyme) of strawberry than on secondary or tertiary flowers, providing the potential for larger receptacles. In grasses, basal florets are formed earlier on the meristem and emerge earlier than more apical florets. Early flower receptivity may provide a distinct advantage to the developing fruit. Seeded cucumber fruits developing at basal nodes on the plant inhibit the development of fruits at subsequent nodes; their removal permits the later formed fruits to develop normally. Seedless fruits generally do not have a comparable effect.

Seeds markedly affect fruit development. In many species, fruit size is correlated with the number of seeds developing in the ovary. Seedless grapes are in most cases quite small. Berry size increases progressively as seed number increases from one to three or more. The shape of fruits can also be affected by placement of seeds. Cucumbers with all or most of their seeds at one end of the fruit are altered in shape, the seeded end enlarging to a greater extent than the seedless end. If an apple fruit contains seeds on only one side, that side tends to be larger. In stone fruits, two ovules are present in the ovary at anthesis, but one normally aborts. Such fruits are asymmetric. H. B. Tukey pointed out in 1936 that the side to which the seed is attached is larger (Fig. 10.2). A classic example of the effect of seeds on fruit development is the work of J. P. Nitsch (1950) then a young French scientist working on his doctorate at the California Institute of Technology. He removed the achenes (ovaries, rather than true seeds) from strawberry fruits at various intervals after fertilization, and demonstrated that re-

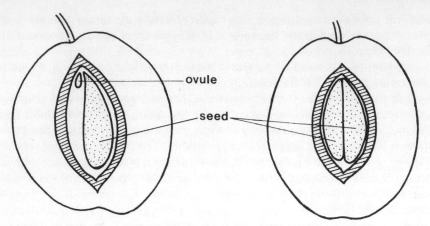

Fig. 10.2. Effect of seed attachment on growth of plum fruits. Fruit on left contains one mature seed, one aborted ovule. Note that side of fruit to which seed is attached is larger. Fruit on the right is symmetrical, for both ovules have developed into seeds (schematic).

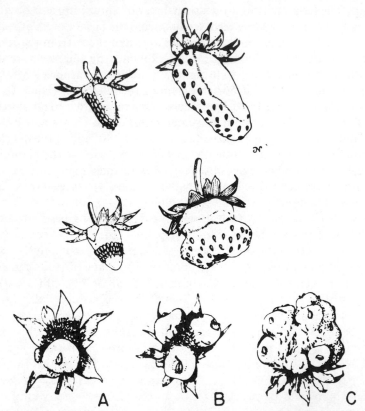

Fig. 10.3. Effect of removing achenes from receptacle (top) or of pollinating only one or several achenes (bottom) on the growth of strawberry receptacles. At the top, flower is shown after removal of part of achenes, together with resulting fruit at right. At the bottom: (A) 1 achene pollinated, (B) 3 achenes pollinated, (C) several achenes pollinated. From Nitsch (1950, 1963).

ceptacle enlargement stopped immediately. Fruits of various shapes and sizes were produced at will by removal of achenes from different portions of the fruit (Fig. 10.3).

With the exception of parthenocarpic fruits, fruit growth is strongly dependent on seed development only during the early phases of ovary enlargement. In several species, seeds can be removed or destroyed without affecting subsequent fruit development. In apple, this can be done after the "June" drop, or approximately 6 to 7 weeks after flowering; in peach toward the middle of Stage II of enlargement. Thus, the fruit appears to become independent of the seed in these species when approximately half-grown. Nevertheless, such fruits do not compete well with seeded fruits, indicating that seeds may still be important as long as other fruits are present (see below).

Chemical Treatment

Some of the same growth regulating chemicals which induce fruit set also stimulate fruit growth. Gibberellic acid is very effective in increasing the size of seedless cultivars of grapes, such as 'Thompson Seedless' (Fig. 10.4), and is presently used commercially in California for this purpose. However, seeded fruits are generally not responsive to treatment. An exception is the fig, a fruit with the double sigmoid type of growth, in which the period of retarded growth can be dramatically shortened by treatment with an auxin such as 2,4,5-trichlorophenoxyacetic acid (2,4,5-T) (Fig. 10.5). The response, however, has been attributed to ethylene gas produced in the tissue in response to auxin treatment, and can be induced by treatment with ethylene.

Shape of seeded fruits can be modified by growth regulator treatment. Application of gibberellins to one side of a young apple fruit results in greater enlargement on the treated side. 'Delicious' apples grown in Washington State are conic in shape with prominent calyx lobes at the blossom end, a response to cool nights during the early period of fruit development.

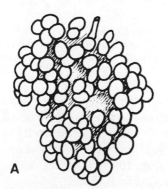
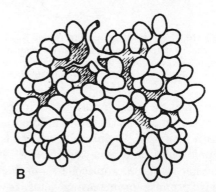

A   B

Fig. 10.4. Response of 'Thompson Seedless' grapes to gibberellin A_3 (40 mg L^{-1}) applied shortly after bloom. (A) control, untreated; (B) treated. Redrawn from Weaver (1972); used by permission.

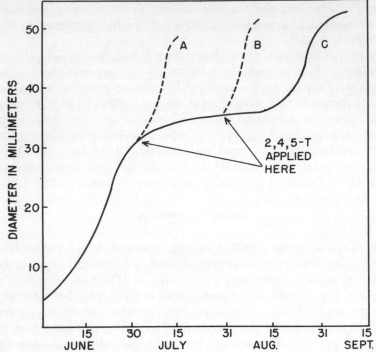

Fig. 10.5. Effect of applying 2,4,5-T on growth of fig fruits. From Crane and Blondeau (1949), by permission.

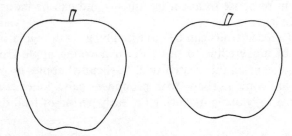

Fig. 10.6. Effects of gibberellins A₄ and A₇ plus 6-benzyladenine (100 mg L⁻¹ each of gibberellin mixture and cytokinin) on fruit shape of 'Red Delicious' apple. Fruit on left from blossom sprayed during bloom with growth regulators, that on right untreated. Redrawn from Stembridge and Morrell (1973); used by permission.

Fruits of the same variety grown in the Carolinas, where night temperatures are warm, are nearly round in longitudinal section. Treatment of blossoms with a mixture of gibberellins A_4 and A_7, both of which occur naturally in apple seeds, and benzyladenine, a synthetic cytokinin, results in a shape similar to that of apples grown in Washington (Fig. 10.6). These chemicals have promise for increasing yields, as well as altering fruit shape.

Growth retardants such as daminozide affect apples by limiting size and changing shape. Such chemicals are sometimes used to reduce the size of fruits on overly vigorous young trees, and are used extensively to prevent preharvest fruit abscission and to increase resistance to bruising.

MECHANISMS CONTROLLING FRUIT GROWTH

Because of the similar effects of seeds and growth regulators upon fruit development, early investigators speculated that seeds contained chemicals which controlled the growth of the ovary tissue. In 1939 F. G. Gustafson prepared extracts of placental tissue (containing the fertilized ovules), inner ovary wall, and outer ovary wall of several species, including tomato, and tested each for auxin content. He found the greatest amount of auxin in the seeds and the lowest in the outer wall of the ovary, suggesting that the auxin diffused from the seeds and thereby controlled the growth of the ovary. Additional evidence came from the work of J. P. Nitsch (1950) on strawberry, for he could replace the growth promoting effects of the achenes (see above) by substituting for them a lanolin paste containing the synthetic auxin naphthoxyacetic acid (Fig. 10.7). This result strongly suggested that the achene produced an auxin that stimulated fruit growth.

In general, however, subsequent workers have been unsuccessful in correlating growth regulator content of the ovary with its growth rate. Although immature seeds of many species are rich in auxins, gibberellins, and (presumably) cytokinins, the levels do not rise until after ovary growth has commenced in most species. Even then, growth regulator content often reaches a maximum when growth of the ovary is minimal, as during Stage II of peach fruit development (Fig. 10.8). The fact that the seeds can be removed from fruits when they are half-grown without markedly affecting

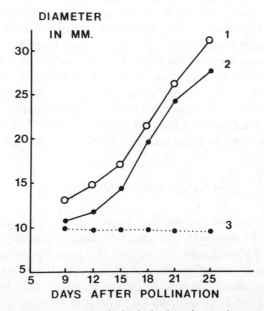

Fig. 10.7. Effect of achene removal, and of substitution of an auxin paste for the achenes, on the growth of strawberry fruits. (1) control, no treatment; (2) achenes removed, replaced with paste containing 2-naphthoxyacetic acid; (3) achenes removed, no further treatment. Redrawn from Nitsch (1950) by Nitsch (1965), used by permission.

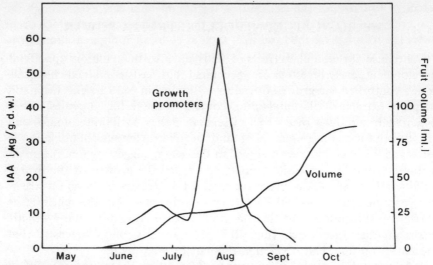

Fig. 10.8. Growth promoter content in microgram equivalents of the auxin, indole-3-acetic acid (IAA), of developing peach seeds vs. growth of entire fruit from shortly after bloom until harvest. Redrawn from Powell and Pratt (1966), used by permission.

subsequent growth indicates that their growth regulator content has little effect upon development at this time.

The poor correlations obtained between growth regulator content of seeds and fruit development do not in any way negate the importance of seeds; rather, they suggest that either the wrong compounds are being measured or the data are being misinterpreted. As noted in the section on fruit set, both fruits and growth regulators mobilize nutrients. Seeds are centers of metabolic activity and exert effects over and above those of the fruits which contain them, as for example in cucumber. When both seeded and seedless fruits exist on the same plant, the former have less tendency to abscise and grow more rapidly than the latter. Furthermore, certain "female" (gynoecious) cucumber cultivars which form seedless fruits produce many more fruits per vine than do seeded cultivars. These observations indicate that seeds play an important, if still obscure, role in fruit growth.

RIPENING AND SENESCENCE

Some fruits are of commercial value before they reach maturity. Cucumbers may be harvested at a range of sizes to produce pickles, or at full size for slicing cucumbers. If left on the vines too long, they ripen, becoming soft and unmarketable. Green beans and sweet corn are also picked before ripening occurs, while navy beans and field corn are harvested only after fruit ripening has occurred and the plant itself has senesced. Thus, horticulturists and agronomists distinguish between maturation, ripening, and senescence. Maturation may refer to the attainment of full size, or the size desired for commercial use, even though the fruit may not be ripe. Ripening includes the qualitative changes which occur after the fruit has reached full size, such as changes in pigmentation, firmness, and chemical constituents. Senescence is closely related to ripening, but consists of the catabolic phase of fruit development, during which the tissues deteriorate and the organ dies. Fruits such as tomatoes ripen *and* senesce. The fruits of grain plants, on the other hand, which consist mainly of seed tissues, only ripen; senescence of such fruits would lead to extinction of the species. The failure of fruits to continue growing indefinitely has been a source of considerable interest, but little is known as to the reasons for the stoppage of growth. The concentrations of growth substances in fruit tissues are low at maturity, and this may limit further growth. Gibberellin application to lemon (*Citrus limon* L.) and lime [*Citrus aurantiifolia* (Christm.) Swingle] fruits which have begun to yellow on the tree due to loss of chlorophyll causes the fruits to "regreen" and continue growing, a fact which allows commercial growers to store fruits "on the tree" when market conditions are poor. Similar, but less dramatic, effects have been reported with other fruits. In the great majority of cases, however, growth regulator treatment does not appear to be the key to continued growth. In fact, application of auxins stimulates the ripening of apple and pear fruits (see below).

Physiological Changes

Several processes characterize ripening. In some fruits, a marked rise in the rate of respiration, termed the climacteric, occurs during ripening, and such fruits are called climacteric fruits, as opposed to nonclimacteric fruits, in which this rise does not occur (Fig. 10.9). Some examples of climacteric fruits are apple, fig, pear, banana, and tomato; of nonclimacteric fruits cherry (*Prunus avium* L.), cucumber, grape, citrus fruits, pineapple (*Ananas comosus* Merr.), and strawberry. The latter normally ripen on the parent plant, and contain little or no starch, while immature fruits of the former contain considerable starch or other reserve storage materials and all are capable of ripening after removal from the plant. Climacteric fruits also produce appreciable quantities of ethylene gas while nonclimacteric fruits do not.

Color changes often accompany ripening, either because of chlorophyll degradation, production of new pigments, or both. Tomato fruits change from green to red as a result of accumulation of the pigment lycopene and concomitant disappearance of chlorophyll; apple fruits develop anthocyanins, chiefly idaein in the skin; yellow carotenoid pigments appear in oranges as the chlorophylls disappear. These changes are confined to the surface of the fruit in apple, but the fruit flesh itself is also pigmented in tomato, pepper, and other fruits. In pears, almost all of the apparent increase in yellow pigmentation merely reflects the loss of chlorophyll.

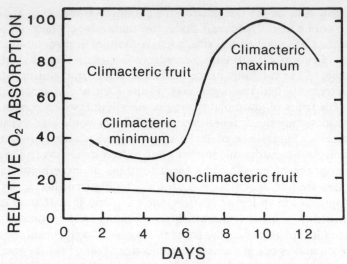

Fig. 10.9. Terminal respiratory patterns for climacteric (Class A) and nonclimacteric (Class B) fruits (schematic). From Biale (1960), used by permission.

Chemicals giving fruits their characteristic flavors accumulate as fruits ripen. Such chemicals are usually volatile components present in minute quantities. Aside from ethylene, which is odorless, volatile production by apple fruits stored at 3°C has been estimated to be only 10 mg per day per tonne of tissue. Because of man's sensitivity to such volatiles, however, a little goes a long way in terms of improving odor and taste. In citrus fruits, essential oils in the peel are responsible for most of the flavors. These are released on peeling or crushing the fruit and hence affect the flavor of the edible portion. The "foxy" flavor of 'Concord' grape (*Vitis labrusca* L.) juice is largely due to methyl anthranilate, a volatile component.

Fruits soften as they ripen as a result of enzymatic breakdown of the pectins which cement the cell walls together. Some degree of softening is desirable, as anyone who has eaten a green pear can attest. However, if ripening is allowed to proceed too long, the fruits become mealy and unpalatable. Some fruits, such as apple and banana, contain considerable starch in the immature stage. This starch is converted to sugar as ripening proceeds. No starch occurs in mature citrus fruits, only a little in avocado fruits, but the lipid content of avocados averages 21% of the fresh weight. Little change occurs in these components during ripening.

Chemical Treatment

Certain chemical compounds are capable of either hastening or delaying fruit ripening. The effects of gibberellin in delaying color changes in lemon and lime fruits have already been noted. Gibberellin delays color development in orange as well, an undesirable response in economic terms. However, it can be beneficial in delaying rind softening—a process which makes the fruits more susceptible to physiological disorders and attack by fungi.

Cytokinins such as N^6-benzyladenine (BA) retard senescence in vegetable crops, such as broccoli (*Brassica oleracea* var. *italica* Plenck) (Dedolph et al., 1962), but commercial application has been limited. Growth retardants, including daminozide and chlormequat (2-chlorethyl trimethylammonium chloride), delay senescence in many of these plants and in some florist crops, including carnation (*Dianthus caryophyllus* L.). Their effects upon fruits vary depending upon species, for daminozide hastens ripening in stone fruits (Unrath et al., 1969) while delaying it in pome fruits (Williams et al., 1964). This chemical is used commercially on both cherry and apple. Auxins such as naphthaleneacetic acid are used commercially by orchardists to prevent abscission of apple fruits prior to harvest (Gardner et al., 1939; Batjer, 1943). Although harvest can therefore be temporarily postponed, the fruits actually ripen more rapidly following treatment, and immediate sale rather than storage is recommended. These chemicals probably stimulate the production of ethylene in the fruit tissue, which in turn stimulates ripening. Ethephon, a chemical which releases ethylene gas within the treated tissue, stimulates both ripening (Russo et al., 1968) and abscission, and has considerable potential as an aid in the mechanical harvesting of fruits such as apple (Edgerton and Blanpied, 1968), cherry (Bukovac, et al., 1969), blueberry (*Vaccinium corymbosum* L.) (Eck, 1970; Howell et al., 1976) and grape (Hale et al., 1970), and in concentrating the harvest of tomatoes (Robinson et al., 1968; Sims, 1969). Again, storage life is shortened by the treatment. Ethylene gas itself is used commercially to hasten ripening of "green-ripe" (i.e., mature) tomatoes and bananas; shipments of ripe fruits would lead to overmaturity and excessive bruising. Ethylene is applied in closed rooms at concentrations of approximately 1000 mg L^{-1}. Oranges are "degreened" by similar treatment although ripening per se is not markedly affected.

Control of Ripening after Harvest

Very young fruits generally fail to ripen when removed from the plant. As maturity approaches, fruits ripen more rapidly and uniformly after picking. An extreme case is the avocado, which fails to ripen as long as it is attached to the plant (Burg, 1962). Once harvested, ripening can be either promoted or delayed by various treatments. Because of the role of ethylene in fruit ripening (see below), treatments which reduce its levels in the fruit or in the surrounding air markedly affect storage life. High CO_2 levels inhibit not only respiration (see below), but ethylene production as well. Certain materials, such as charcoal or paper saturated with mineral oil, absorb ethylene and other volatile gases, and can be used in filters on air circulation equipment, or as individual fruit wraps. A recent development which may revolutionize storage procedures is low pressure storage—holding of fruits under less than atmospheric pressure in order to favor the diffusion of ethylene gas from them (Dilley, 1972). This procedure greatly extends storage life. The primary roadblock to commercial use is the cost of building structures capable of withstanding the difference in pressure between outside and inside—a difference which may be as great as 826 g cm^{-2} (12 lb in^{-2}).

Temperature also affects the rate of ripening. Cold storage has been used effectively for generations to prolong the life of fruits, first in the form of root cellars, later in mechanically refrigerated storages. Its effectiveness depends upon the fact that the rate of respiration is proportional to temperature (Fig. 10.10), a concept expressed mathematically in the Q_{10} for respiration. The Q_{10} is obtained by dividing the rate of respiration at a given temperature, T, into the rate at the temperature T + 10°C. If the rate at 10°C is 15 mL CO_2 released per kg fresh weight per hour, and the rate at 20°C is 30 mL, then $Q_{10} = (30 \div 15) = 2$, indicating that a 10°C rise in temperature has doubled the rate. Thus the effect of temperature is exponential. The more rapid the rate of respiration, the sooner stored carbohydrates and lipids are exhausted, and the shorter the storage life of the fruit. Storage life of grains, as well as fleshy fruits, is prolonged by cold temperatures. However, longevity is improved by keeping relative humidity low, as in the National Seed Stocks Storage laboratory in Colorado where many seeds of grasses and legumes are stored at low temperature and humidity. In contrast, high humidity is required to prevent water loss and shriveling of fleshy fruits.

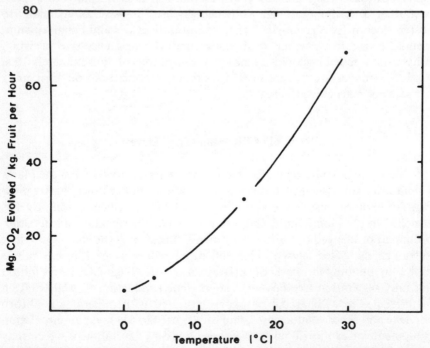

Fig. 10.10. Effect of temperature on respiration rate of 'Grimes Golden' apple fruits. Based on data of Magness et al. (1926); used by permission.

A number of limitations exist to the use of cold storage to prolong fruit life. Some tropical and subtropical fruits are injured at temperatures of 12 to 13°C, and must be stored at higher temperatures. Chilling injury takes the form of discoloration and loss of flavor and aroma. Certain fruits, such as avocados, will not ripen normally after storage for prolonged periods at temperatures approaching 5°C. On the other hand, several pear cultivars, such as the French 'Passe Crassane' must be stored for 11 to 15 weeks at temperatures near 0°C before ripening will occur at 18°C. This ripening is associated with ability to produce ethylene gas, and ethylene treatment can be used to replace chilling.

The gas mixture in which fruits are kept has a marked effect upon ripening rate, as might be expected from the overall equation for respiration of glucose:

$$C_6H_{12}O_6 + 6\,O_2 = 6\,CO_2 + 6\,H_2O + \text{heat}$$

From the fact that increasing the concentration of either a reactant or a product of the reaction will alter the rate of a reaction, one might predict that changing the concentration of oxygen or carbon dioxide would affect the rate of respiration. Both oxygen and carbon dioxide content of the air markedly affect storage life of fruits. Thus either lowering the oxygen level or raising the level of carbon dioxide reduces the rate of respiration and thereby prolongs storage life in a number of fruits. Commercial advantage is taken of this fact in so-called controlled atmosphere storage of apples—oxygen level is maintained at approximately 2 to 3% and carbon dioxide level at 5 to 10% in specially constructed gas-tight rooms. Levels vary depending upon cultivar, and deviations can result in injury to the fruits.

MECHANISMS CONTROLLING RIPENING

Burg and Burg (1965) pioneered in research on the relationship between ethylene gas and fruit ripening. Many scientists now accept their proposal that the gas is a "ripening hormone." The effect of ethylene in stimulating fruit ripening of climacteric fruits has been known for many years, but it was not until the gas chromotograph, an instrument used to measure very small amounts of gases, became readily available in the late 1950s and early 1960s that the minute levels of ethylene present in plant tissues could be detected and quantified. Since then, the Burgs and others have demonstrated that the ethylene content of fruits, such as apple and avocado, rises prior to the onset of the climacteric to the threshold level required to induce ripening—as little as 0.1 μL L^{-1} is sufficient in many fruits (Fig. 10.11). This is equivalent to a pinch of salt (10 mg) in 100 L (25 gallons) of water.

In climacteric fruits, ability to produce ethylene appears to be a prerequisite for ripening. Fruits must have reached a certain stage of maturity

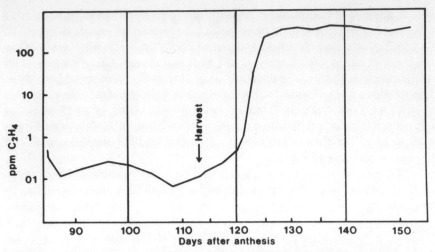

Fig. 10.11. Internal levels of ethylene in ripening apple fruits while attached to the tree and after removal (arrow). Redrawn from Sfakiotakis and Dilley (1973); used by permission.

before ripening will occur following harvest; fruits which are unable to ripen also lack the capacity to produce ethylene. In pear fruits which require a period of chilling to ripen, the chilling treatment apparently removes a block to the synthesis of ethylene. Removal of ethylene by keeping fruits under low atmospheric pressure prevents or considerably delays ripening, while addition of low levels of ethylene to the system restores the ripening response. Controlled atmosphere storage, with elevated levels of CO_2 and reduced levels of O_2, inhibits ethylene production by the fruits while prolonging storage life. All these facts strongly suggest that ethylene is the pivotal factor in the control of fruit ripening. However, its mechanism of action remains a puzzle.

Several compounds other than ethylene have been implicated in ripening and senescence. Because of the effects of cytokinin application in delaying senescence, there is speculation that endogenous cytokinins may play a similar role. However, experimental evidence is lacking on this subject. Immature fruits do not ripen when removed from the plant, even after exposure to ethylene, suggesting either that ripening inhibitors occur in the tissues, or that the cells do not possess the necessary mechanisms for ripening. This may be only a semantic difference, as the inhibitors could conceivably block the expression of genes which in turn control the synthesis of the required enzymes. While the evidence for such inhibitors is strong, little is known concerning their chemical natures or mechanism of action.

SUMMARY

Fruits serve one purpose for the plant—reproduction—but several functions for man, including food, ornamentation, and drugs. They may consist entirely of ovary tissues or of both ovary and accessory tissues, depending upon species, but all fruits are associated with ovules—those structures which give rise to seeds following fertilization. Some fruits develop without pollination or fertilization, but most are dependent upon one or both, and seed development is generally required for normal fruit growth. Fruit set and growth are affected by numerous environmental factors, light, water, and nutrient supply being the most limiting under normal circumstances. Growth promoting chemicals such as auxins and gibberellins are useful in some species for increasing fruit set or stimulating growth, and several types of naturally occurring promoters may play a role in normal fruit development.

At some point in their development, fruits lose the capacity for further growth and the process of ripening begins. This in turn leads to senescence and death of the tissues. Ripening involves changes in coloration, metabolic processes, and chemical constituents characteristic of species and cultivar, and can be affected by numerous treatments, particularly temperature and gas mixture. The role of ethylene gas as a "ripening hormone" is well documented, and treatments which affect ripening probably do so through their effects on the production of ethylene by the affected tissues.

REFERENCES

Batjer, L. P. 1943. Harvest sprays for the control of fruit drop. USDA Circular 685.

Biale, J. B. 1960. Respiration of fruits. Encycl. Plant Physiol. New Ser. 12(2):536–592.

Bukovac, M. J,, F. Zucconi, R. P. Larsen, and C. D. Kesner. 1969. Chemical promotion of fruit abscission in cherries and plums with special reference to 2-chloroethyl-phosphonic acid. J. Am. Soc. Hort. Sci. 94:226–230.

Burg, S. P. 1962. Postharvest ripening of avocados. Nature 194:398–399.

----, and E. A. Burg. 1965. Ethylene action and the ripening of fruits. Science (Washington, D.C.) 148:1190–1196.

Crane, J. C., and R. Blondeau. 1949. Controlled growth of fig fruits by synthetic hormone application. Proc. Am. Soc. Hort. Sci. 54:102–108.

Dedolph, R. R., S. H. Wittwer, V. Tuli, and D. Gilbert. 1962. Effect of N⁶-benzylaminopurine on respiration and storage behavior of broccoli (*Brassica oleracea* var. italica). Plant Physiol. 37:509–512.

Dilley, D. R. 1972. Hypobaric storage—a new concept for preservation of perishables. Annu. Rep. Michigan State Hort. Soc. 102:82–89.

Eck, P. 1970. Influence of ethrel upon highbush blueberry fruit ripening. HortScience 5:23–25.

Edgerton, L. J., and G. D. Blanpied. 1968. Regulation of growth and fruit maturation with 2-chloroethane phosphonic acid. Nature 219:1064–1065.

Erdman, M. H., and M. W. Adams. 1978. Row width, plant spacing and planting depth. *In* L. S. Robertson and R. D. Frazier (ed.) Dry bean production—Principles and practices. Michigan State Univ. Ext. Bull. E-1251.

Fitting, H. 1909. Die Beeinflussung der Orchideenbluten durch die Bestaubung und durch andere Umstande. Z. Botan. 1:1–86.

Gardner, F. E., C. P. Marth, and L. P. Batjer. 1939. Spraying with plant-growth substances for control of pre-harvest drop of apples. Proc. Am. Soc. Hort. Sci. 37:415–428.

Gustafson, F. G. 1936. Inducement of fruit development by growth-promoting chemicals. Proc. Natl. Acad. Sci. USA 22:628–636.

――――. 1939. Auxin distribution in fruits and its significance in fruit development. Am. J. Bot. 26:189–194.

Hale, C. R., B. G. Coombe, and J. S. Hawker. 1970. Effects of ethylene and 2-chloroethyl-phosphonic acid on the ripening of grapes. Plant Physiol. 45:620–623.

Howell, G. S., Jr., B. G. Stergios, S. S. Stackhouse, and H. C. Bittenbender. 1976. Ethephon as a mechanical harvesting aid for highbush blueberries (*Vaccinium australe* Small). J. Am. Soc. Hort. Sci. 101:111–115.

Kraus, E. J., and H. R. Kraybill. 1918. Vegetation and reproduction with special reference to the tomato. Oregon Agric. Exp. Stn. Bull. 149.

Magness, J. R., H. C. Diehl, M. H. Haller, and W. S. Graham. 1926. The ripening, storage and handling of apples. USDA Bull. 1406.

Nitsch, J. P. 1950. Growth and morphogenesis of the strawberry as related to auxin. Am. J. Bot. 37:211–215.

――――. 1963. Fruit development. p. 361–394. *In* P. Maheshwari (ed.) Recent advances in the embryology of angiosperms. Int. Soc. of Plant Morphologists, Univ. of Delhi, Delhi, India.

――――, C. Pratt, C. Nitsch, and N. J. Shaulis. 1960. Natural growth substances in Concord and Concord seedless grapes in relation to berry development. Am. J. Bot. 47:566–576.

Powell, L. E., and C. Pratt. 1966. Growth promoting substances in the developing fruit of peach (*Prunus persica* L.). J. Hort. Sci. 41:331–348.

Robinson, R. W., H. Wilczynski, and F. G. Dennis, Jr. 1968. Chemical promotion of tomato fruit ripening. Proc. Am. Soc. Hort. Sci. 93:823–830.

Russo, L., Jr., H. C. Dostal, and A. C. Leopold. 1968. Chemical regulation of fruit ripening. BioScience 18:109.

Sfakiotakis, E. M., and D. R. Dilley. 1973. Internal ethylene concentrations in apple fruits attached to or detached from the tree. J. Am. Soc. Hort. Sci. 98:501–503.

Sims, W. L. 1969. Effects of ethrel on fruit ripening of tomatoes. . .greenhouse, field and postharvest trials. Calif. Agric. 23:12–14.

Stembridge, G. E., and G. Morrell. 1972. Effect of gibberellins and 6-benzyladenine on the shape and fruit set of 'Delicious' apples. J. Am. Soc. Hort. Sci. 97:464–467.

Unrath, C. R., A. L. Kenworthy, and C. L. Bedford. 1969. The effect of Alar, succinic acid 2,2-dimethylhydrazide, on fruit maturation, quality and vegetative growth of sour cherries, *Prunus cerasus* L., cv. 'Montmorency'. J. Am. Soc. Hort. Sci. 94:387–391.

Weaver, R. J. 1972. Plant growth substances in agriculture. W. H. Freeman and Co., San Francisco.

Williams, M. W., L. P. Batjer, and G. C. Martin. 1964. Effects of N-dimethyl amino succinamic acid (B-Nine) on apple quality. Proc. Am. Soc. Hort. Sci. 85:17–19.

Wittwer, S. H. 1943. Growth hormone production during sexual reproduction of higher plants with special reference to synapsis and syngamy. Missouri Agric. Exp. Stn. Res. Bull. 371.

――――, and R. Robb. 1964. Carbon dioxide enrichment of greenhouse atmospheres for food crop production. Econ. Bot. 18(1):34–56.

Yasuda, S. 1934. Parthenocarpy caused by the stimulus of pollination in some plants of Solanaceae. [In Japanese with English summary.] Agric. Hort. 9:647–656.

SUGGESTED READING

Bollard, E. G. 1970. The physiology and nutrition of developing fruits. p. 387–425. *In* A. C. Hulme (ed.) The biochemistry of fruits and their products. Vol. 1. Academic Press, New York.

Crane, J. C. 1969. The role of hormones in fruit set and development. HortScience 4:108–111.

Dilley, D. R. 1969. Hormonal control of fruit ripening. HortScience 4:111–114.

Lieberman, M. 1975. Biosynthesis and regulatory control of ethylene in fruit ripening. A review. Physiol. Veg. 13:489–499.

Mapson, L. W. 1970. Biosynthesis of ethylene and the ripening of fruit. Endeavour 29:29–33.

Moore, T. C., and P. R. Ecklund. 1975. Role of gibberellins in the development of fruits and seeds. p. 145–182. *In* H. N. Krishnamoorthy (ed.) Gibberellins and plant growth. John Wiley & Sons, Inc., New York.

Luckwill, L. C. (ed.) 1981. Growth regulators in fruit production. Acta Hort. 120.

Nitsch, J. P. 1965. Physiology of flower and fruit development. Encyclopedia of Plant Physiol. 15(1):1537–1647.

----. 1970. Hormonal factors in growth and development. p. 427–472. *In* A. C. Hulme (ed.) The biochemistry of fruits and their products. Vol. 1. Academic Press, New York.

----. 1971. Perennation through seeds and other structures: Fruit development. p. 413–501. *In* F. C. Steward (ed.) Plant physiology, A treatise. Vol. VIA. Academic Press, New York.

Wellensiek, S. J. (ed.) 1973. Growth regulators in fruit production. Acta Hort. 34(1).

11 Genetics and Use of Physiological Variability in Crop Breeding

DONALD C. RASMUSSON AND
BURLE G. GENGENBACH
Department of Agronomy and Plant Genetics
University of Minnesota
St. Paul, Minnesota

As civilization developed, humans began to improve the crops that provided their food. This plant improvement started unknowingly when they grew their food plants from seeds. As with many ancient practices that have progressed with civilization, the methods have changed substantially. The rapid changes in plant breeding and crop improvement during the past 50 years closely parallel increased knowledge about genetics and plant development. The fact that yields have not always increased proportionately with increased knowledge about plants underscores the difficulty of translating this knowledge into crop improvement. However, to paraphrase S. W. Johnson's proposal in his 1868 book *How Crops Grow,* one must know how crops grow to accomplish effective improvement.

As discussed in the preceding chapters, plant growth is a continuous series of individual, interacting biochemical and physiological processes that are influenced by environmental conditions. The primary value of plant growth is the yield of grain, forage, fruits, vegetables, chemicals, or fiber. Historically, plant breeders have primarily selected varieties on the basis of yielding ability in performance trials. Crop improvement also includes the incorporation of resistance to various stresses, such as diseases, insects, and drought, into varieties with high yield potential. This chapter

Published in *Physiological Basis of Crop Growth and Development,* © American Society of Agronomy—Crop Science Society of America, 677 South Segoe Road, Madison, WI 53711, USA.

emphasizes physiological traits of plant growth and development that can be measured or evaluated and that directly or indirectly influence yield. Our definition of physiological traits is broad and includes more plant characteristics than are generally considered to be physiological by plant breeders or plant physiologists.

GENES AND GENE FUNCTIONS

Each plant characteristic and the potential for growth and high yield are determined by the genotype (total genetic information) of the plant. Plant improvement involves both physiological gene action and gene inheritance. The physiological action of genes determines every characteristic of a plant, such as its morphology, response to environmental conditions, and yielding ability (Fig. 11.1). Physiological gene action reflects gene differences that provide the basis for selecting desirable genotypes in plant breeding. Efficient recovery and maintenance of desirable genes transmitted from selected plants to their progeny requires knowledge about gene inheritance. Therefore, it is desirable to briefly examine basic aspects of gene function and inheritance and their importance in plant breeding.

Physiological Gene Action

Ability to identify and select plants having genes for desirable traits is important in plant breeding. As shown in Fig. 11.1, physiological processes are under genetic control and are influenced by the environment during the life cycle of a plant. Physiological gene functions determine the manner and extent of the genotypic contribution of the phenotype (outward appearance and other measurable characteristics) of the plant.

To understand genetic control of physiological processes we must examine how genes confer specific capabilities to a plant. In general, the deoxyribonucleic acid (DNA) sequence of a gene provides the information for synthesis of an enzyme enabling the plant to carry out a particular metabolic process. Enzymes are proteins that act as biological catalysts to initiate specific changes in one or more cellular metabolites (substrates). An estimated 1000 to 10 000 enzymes are contained in a typical plant cell with a corresponding number of potential enzyme reactions (Bonner, 1976). Some enzymatic reactions are required periodically while others are required continuously during plant growth and development. All plant processes such as photosynthesis, nutrient metabolism, protein synthesis, and cell division result from enzyme actions.

The sequential action of many individual enzymes in biochemical pathways converts raw products such as CO_2 from the air into end products like carbohydrates and amino acids and provides energy for plant growth.

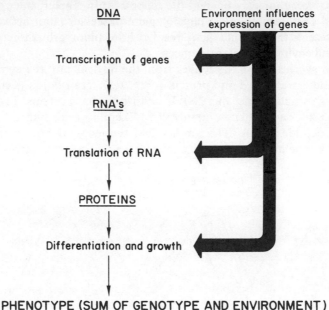

Fig. 11.1. Physiological gene action.

This cellular machinery is composed of many small pathways each synthesizing one or a few specific products as the plant or cell requires them for growth. These products may be used directly to form complex compounds, such as endosperm proteins or starch, or they may become starting materials for synthesis of new products by another biochemical pathway. In many cases, the interrelationship and interdependency of the pathways in the total plant metabolic system make it difficult to assess the impact on growth caused by changing the activity or the characteristics of one enzyme.

The function of an enzyme is specified by the DNA of the gene that codes for the enzyme. The first step in enzyme (protein) synthesis is the transcription of the DNA of the gene into ribonucleic acid (RNA) molecules (Fig. 11.2). These RNA's are macromolecules containing four ribonucleotide bases: guanine (G), cytosine (C), adenine (A), and uracil (U). In RNA synthesis, the DNA bases G, C, A, and thymine (T) code for the specific complementary RNA bases C, G, U, and A, respectively. Thus the DNA base sequence ACTGAC would code for an RNA segment with the sequence UGACUG.

Several kinds of RNA's are essential for protein synthesis (Fig. 11.3), and all RNA's contain bases copied from the appropriate gene DNA base sequences. Messenger RNA's (mRNA's) contain the specific information necessary for synthesizing proteins; the other two kinds of RNA, ribosomal (rRNA) and transfer (tRNA), assist in this process. Genetic, environmental, physiological and developmental factors influence RNA synthesis and the subsequent steps leading to protein synthesis. Descriptions of these factors and their interactions are beyond the scope of this chapter, but you should be alerted that regulation of physiological processes is an active area of plant research that should help resolve how plant growth responds to genetic and environmental differences.

After synthesis, mRNA moves from the nucleus into the cytoplasm of the cell and is translated into protein (Fig. 11.3). Translation occurs on the ribosomes, which consist of rRNA's and ribosomal proteins. Proteins are large molecules and often contain 75 to 150 amino acids linked linearly by covalent peptide bonds. The amino acid sequence of an enzyme is determined by the mRNA base sequence much as the mRNA base sequence is

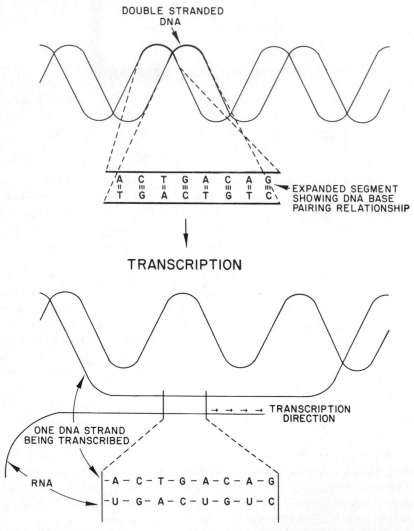

Fig. 11.2. The DNA of a gene and the relationship to the RNA synthesized during transcription.

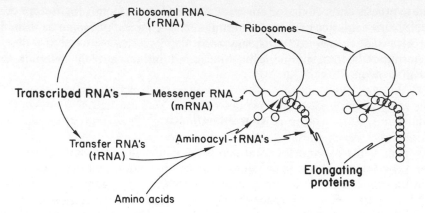

Fig. 11.3. Translation of mRNA and synthesis of protein.

determined by the DNA base sequence. Specific sets of three bases (codons) in the mRNA are required to specify each amino acid of the enzyme. Each amino acid is specified by a unique codon or set of codons. For instance, the codon UGG specifies the amino acid tryptophan, while both GAA and GAG specify glutamic acid.

In cooperation with the mRNA and ribosomes, tRNA's insert the correct amino acid into the protein chain in the order specified by the mRNA codons. There are over 20 different tRNA species, each transferring only one of the 20 different amino acids found in the enzymes. Each tRNA has a specific sequence of three bases (anticodon) that complement one of the codons found in mRNA. The sequential incorporation of amino acids into proteins is determined by mRNA codons pairing with complementary tRNA anticodons. For example, the tryptophan codon is UGG. Thus, the complementary anticodon is ACC, and it is found only in the tryptophan tRNA species. The movement of the ribosome along the mRNA coordinates the base pairing of codons and anticodons so that the proper amino acid is attached by a peptide bond to the growing end of the protein. The process is repeated along the entire mRNA molecule ensuring that each amino acid is inserted in the proper order.

After release of the protein from the ribosomes, interactions between the amino acids determine the final three-dimensional shape. This shape is determined primarily by the amino acid sequence and specific properties of the different amino acids. The most important part of an enzyme is its active site(s), which reacts with and causes a chemical change in one or more substrates. The shape and charge of the active site(s) determine which substrate is acted upon by an enzyme and how the substrate is changed. Through the reactions they catalyze, enzymes collectively determine total metabolic activity and thereby determine much of the genetic variability in plant traits needed by plant breeders.

Protein or enzyme synthesis is much more complicated than the above simplified discussion indicates. Its purpose is to point out how genes control the synthesis and activity of enzymes. In summary, the information in the DNA base sequence of a gene determines the sequence of amino acids in a protein and, consequently, its enzymatic activity. The genotype of a plant determines the total enzymatic machinery and thus has a role in determining plant growth.

Gene Inheritance

Gene inheritance is the transmission of genetic information to succeeding generations. This transmission involves replication and distribution of chromosomes both to daughter cells in mitosis and to gametes in meiosis and thus to succeeding progeny. During mitosis, the chromosomes are duplicated and then separate, ensuring that daughter somatic cells receive exact copies of each chromosome and have the same genotype as the progenitor cell.

Inheritance of plant traits is based on Mendel's classic laws of segregation and independent assortment of genes. Meiosis reduces the somatic chromosome number to the gametic number (haploid) during production of male and female gametes. Fertilization of an egg by a sperm nucleus restores the original somatic chromosome number in the zygote. If the genotypes of the gametic parents are different, new gene combinations will occur in the zygote. Knowledge of the behavior of chromosomes during meiosis and of the linear arrangement of genes on chromosomes enables geneticists to observe the segregation of alleles (different forms of the same gene) and to predict the assortment of different genes located on the same or different chromosomes.

The enormous number of plant enzymes is controlled by an equally large number of genes. Since there are very many genes and only a few chromosomes in a plant cell, it follows that many genes must be located on the same chromosome. If two genes are located close together on the same chromosome, they are said to be linked. Linkage reduces the chances for recombination of the genes during production of gametes. Recombination of linked genes results from exchange of DNA between homologous chromosomes.

The effect of linkage on the probability of obtaining a new gene combination is described in Fig. 11.4. If two parents, each with a desirable recessive trait, are crossed to incorporate the two traits into one genotype, the frequency of the desired type in the F_2 progeny is a function of linkage intensity. As shown in Fig. 11.4, 6.25% of the F_2 progeny are of the desired double recessive type in the absence of linkage. With linkage and 20% recombination, the frequency drops to 1%. The effect of linkage depends on the crossing methods used to recover the desirable types and on the level of recombination between the two genes. For instance, if the recombination frequency were 2% instead of 20%, the frequency of the *aabb* genotype

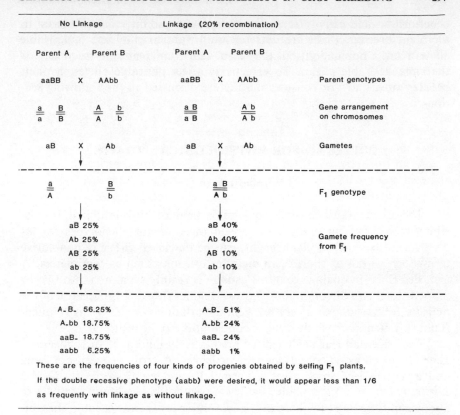

No Linkage		Linkage (20% recombination)		
Parent A	Parent B	Parent A	Parent B	
aaBB X	AAbb	aaBB X	AAbb	Parent genotypes
$\dfrac{a}{a}\dfrac{B}{B}$	$\dfrac{A}{A}\dfrac{b}{b}$	$\dfrac{a\ B}{a\ B}$	$\dfrac{A\ b}{A\ b}$	Gene arrangement on chromosomes
aB X	Ab	aB X	Ab	Gametes

$\dfrac{a}{A}$	$\dfrac{B}{b}$	$\dfrac{a\ B}{A\ b}$	F_1 genotype

aB 25%	aB 40%	
Ab 25%	Ab 40%	Gamete frequency
AB 25%	AB 10%	from F_1
ab 25%	ab 10%	

A_B_ 56.25%	A_B_ 51%	
A_bb 18.75%	A_bb 24%	
aaB_ 18.75%	aaB_ 24%	
aabb 6.25%	aabb 1%	

These are the frequencies of four kinds of progenies obtained by selfing F_1 plants.
If the double recessive phenotype (aabb) were desired, it would appear less than 1/6
as frequently with linkage as without linkage.

Fig. 11.4. Effect of linkage on gamete frequency and recovery of desired recombinant progeny.

would be 1 out of 10 000 (0.01%) instead of 1%. Thus the plant breeder must grow many more progeny to recover desirable types when the genes are closely linked. The problems linkage can create in plant breeding programs will be mentioned again.

An allele is one of two or more alternate forms of a gene occupying the same locus on a particular chromosome. A diploid plant has only two alleles for the same gene, one allele on each chromosome of a pair of chromosomes. In contrast, a population of plants collectively can have many alleles for the same gene. Alleles represent mutations (changes) in the DNA that have been preserved in populations by transmission to succeeding progeny. Many alleles are possible at each locus because a mutation of one base in a gene potentially produces a new allele. Mutations within a gene can cause the substitution of a new amino acid for the original in the protein. The location and type of amino acid substitution determines the effect of the mutation on the protein, such as an altered enzyme activity.

The function of an enzyme in plant growth determines how readily different alleles for that enzyme can be detected. Different alleles can have beneficial or deleterious effects on plant growth and on traits of interest to the plant breeder. Plant breeders not only want to eliminate undesirable alleles from a population, but they also want to increase the frequency of the most desirable alleles. To effectively select desirable alleles, a plant breeder must know or consider the factors discussed in the following sections.

BREEDING FOR PHYSIOLOGICAL TRAITS

Inheritance

The breeder must consider how the particular trait is inherited, that is, whether one, few, or many genes are involved. Plant phenotypes for individual traits, such as plant height, can be described either in qualitative terms such as tall or short or in quantitative units such as centimeters. A trait described in qualitative terms usually is readily separated into two or more distinct and discontinuous classes. Distinct classes arise because of the major effects of one or a very few genes. Variation caused by environment is not sufficient to mask the genetic contribution to the trait.

Leaf size and leaf angle can be qualitatively inherited. In soybean,[1] a gene identified as *na* conditions narrow leaves. A cross between a normal and a narrow-leaf stock gives a normal F_1 and an F_2 segregation of 3 normal:1 narrow. A gene called liguleless, symbol *li*, conditions presence or absence of a ligule at the base of the leaf in barley. Another interesting effect of *li* is that it causes leaves to be more erect than in normal barley. By selecting plants with or without ligules in an F_2 population, plants can be obtained with normal or erect leaves, respectively.

The *B* gene locus determines whether coumarin is formed by sweet-clover tissue (Schaeffer et al., 1960). The trait cannot be evaluated by visual observation of the plant, but distinct phenotypes can be identified by chemical procedures. Enzyme assays show that the *B* allele codes for an enzyme that catalyzes the final step in the production of coumarin, whereas the *b* allele codes for the protein that has no detectable enzymatic activity for this reaction. Production of a precursor of coumarin early in the pathway also is controlled by one gene (Goplen et al., 1957). The *Cu* allele at this locus causes high precursor synthesis, while the *cu* allele results in much less precursor synthesis. Combining both recessive alleles into one homozygous genotype, *cucubb,* reduces coumarin synthesis at each of the two steps. The resulting low-coumarin type is desired because coumarin reduces the feeding value of sweet clover forage. This example illustrates the important principle that careful study of a character may reveal distinctive phenotypes conditioned by single genes with major effects. Without careful study or appropriate techniques, the character may appear to be quantitative and controlled by many genes.

---

[1] Scientific names of important crop plants are given in Table 1.1, Chapter 1.

The majority of physiological characters are quantitatively inherited. The varied expressions of these characters are continuous and cannot be fitted into discrete classes. Yield and primary physiological processes such as photosynthesis, respiration, translocation, and transpiration are quantitative characters. Such characters are often conditioned by many genes with small individual effects, and there is often a sizable environmental effect. Because variation is continuous, identifying different genotypes on the basis of their phenotypes often is difficult. For example, most variation in plant height can be attributed to genetic control when the plants are grown under carefully controlled conditions in a phytotron. However, when the same genotypes are grown under variable conditions in the field, a major portion of plant-to-plant variability may be attributable to environmental causes.

Heritability, Genetic Variability, and Genetic Gain

Distinguishing between genetic variability, which is usable in a breeding program, and variability caused by environment, which is not usable, is important in plant breeding programs. The total variability observed in a trait is called phenotypic variability. Phenotypic variability is caused by genetic and environmental components and the interaction between them. It is customary to express this relationship as a linear function showing that the phenotype of a plant or a family of plants is due to its genotype and environment and their interaction: Phenotype = Genotype + Environment + (Genotype × Environment).

The genetic proportion of the phenotypic variability commonly is expressed as a ratio called heritability (H), written in simplest terms as:

$$H = \frac{\text{Genetic Variance}}{\text{Phenotypic Variance}}$$

This relationship is often used to describe the inheritance of quantitative traits. For instance, an H value of 0.20 indicates that, for the population tested and environment utilized, only 20% of the total variation is genetically determined. In comparison, a 0.60 value indicates a much higher proportion of genetic variation. The higher the heritability, the greater the likelihood for genetic improvement.

Heritability can be increased by increasing the genetic component or by reducing the environmental component of the phenotypic variation. In some cases, both components might be altered to increase heritability, but the plant breeder is especially interested in genetic variability for the trait under selection. Improvement is possible only when there is genetic variability in the breeding material. Therefore, the breeder is concerned with collecting genetically variable germ plasm for the trait in question.

Collecting variable germ plasm may be easy or difficult, depending on which trait or traits are wanted. For many important crops large germ plasm collections, including more than 10 000 genetically distinct seed lots, have been made and are available on request. The National Seed Storage Laboratory at Fort Collins, Colorado maintains diverse genetic stocks of many important crops. Several specialized state and federal plant introduction centers located throughout the USA handle a small number of similar crops. In some cases they have been evaluated, and it is possible to request germ plasm of a desired type. For example, spring barley with long awns and six rows of kernels or awnless winter barley with two rows of kernels can be obtained from the world barley collections.

Germ plasm useful in breeding for physiological traits frequently can be obtained from the germ plasm collections or from other research workers. Most researchers freely share their germ plasm, and a literature search often uncovers reports giving detailed information on a character of interest. In other instances, a careful screening of several hundred or even thousands of different lines may be necessary to find the desired germ plasm.

Knowledge of heritability and the amount of genetic variability is necessary to predict the amount of progress or genetic gain that can be achieved from selection for a trait. The gain from selection (Gs) is expressed as: $Gs = (H)(SD)$, where SD, the selection differential, is the difference between the average value of selected plants and the average of all plants in the population. The Gs formula indicates that significant gains are likely when heritability is high and there is sufficient genetic variation in the population to result in a large selection differential. Heritability must be high enough so that genetic gain is likely; otherwise a breeding program should not be started.

Choice of Physiological Traits

In years past, plant breeders emphasized breeding for yield per se and breeding to reduce the hazards of production. Breeding for yield has focused on yield trial performance, and the most productive entries have become varieties. Disease and insect resistance, tolerance to weather conditions, and resistance to shattering and lodging reduce the hazards of production and often have been very beneficial in obtaining and maintaining high yields.

The decision to undertake a breeding program to increase yield by breeding for a physiological trait is a very important one for the plant breeder. A breeding effort on a new trait generally requires a commitment of resources for several years; a single breeding cycle seldom leads to a new variety. The current varieties in many crops are the culmination of 30–50 years of breeding effort in which numerous gene combinations were tried and utilized if beneficial or discarded if not. Furthermore, genes for the desired trait are often found in exotic stocks that are poorly adapted and low-yielding in the geographical area served by the breeding program. This is a problem because the breeder may introduce genes for many traits that reduce overall performance of the progeny.

Will selection for a physiological trait such as leaf angle or photosynthesis increase yield? This is the major question a breeder wants answered before proceeding with a breeding program on a physiological trait. An absolute affirmative answer seldom is available, but information on which to base a decision can be obtained from several sources. Such a decision often is best made through the collaboration of plant physiologists and plant breeders. Together they can identify traits of metabolic and developmental importance, methods for their evaluation, and the most appropriate breeding approach. Available evidence must be examined carefully. A published report may have indicated, for example, that an erect leaf trait increased yield. However, varying the conditions, such as the genetic material or geographic location, may affect the value of erect leaves. Leaf angle may be of limited importance in regions where sunlight is intense or in varieties with an especially small leaf area. Often, the necessary information has to be obtained for the specific breeding situation.

The contribution of a character to yield can be assessed with specially developed stocks called isolines. Isolines are alike genetically for nearly all genes except those controlling the character being investigated. If one line of an isoline pair yields more than the other, the character by which they differ will usually be important in determining yield. California researchers (Qualset et al., 1965) determined the effect of awn length on barley yield by using four backcross-derived isolines, each having a different awn length (Table 11.1). Fully awned and half-awned plants yielded more than quarter-awned and awnless plants. The data showed that longer awns contributed to greater kernel weight and thus to higher yields. Therefore, awns are desirable under California growing conditions, and barley breeders there select plants with long awns.

When isolines are not available, a useful alternative is to make comparisons among a large number of lines that have contrasting phenotypes for the trait. If sufficient lines that differ for the trait are not available, they can be derived by crossing parents chosen for their differences and then selecting lines with contrasting phenotypes. The researcher hopes that the

Table 11.1. Mean performance of awn-length isogenic lines of Atlas barley.[†]

Genotype	Yield	Kernel wt	Kernel no./spike	Spike no./plant
	g/plant	mg		
Fully awned	41.1	39.9	57.8	25.8
Half-awned	43.8	36.5	61.8	26.2
Quarter-awned	38.6	33.8	60.8	26.3
Awnless	35.6	32.7	57.8	26.9

† From Qualset et al. (1965).

lines selected for comparison have approximately equal merit in all other genes, except those controlling the character under investigation. A pitfall in this approach is that a set of lines or progenies from a cross likely have hundreds of genes that distinguish one line from another, and these can affect performance. For example, disease reaction, standing ability, maturity, or photoperiod response can be more important in conditioning differences in yield than the trait being investigated. It is especially risky to draw conclusions about the role of a trait in determining yield if the lines in the experiment are not adapted to the area where they are being tested.

Another way to determine whether a trait influences yield is to observe what happens to yield when environmental factors such as light, temperature, moisture, or CO_2 are varied. For example, if yields are increased by increasing the incident light within a crop canopy, modification of leaf angle might be worthwhile. Another approach is to physically change the plant trait from its normal status and to observe the effect on yield. Altering leaf position, removing leaves at different stages of growth, and removing parts of the inflorescence are examples of this type of study. Such studies frequently are done in the laboratory, and the treatments often impose conditions on the plants that alter normal development. Hence the results must be interpreted with care.

SELECTION AND BREEDING METHODS

When initiating a breeding program for a physiological trait, the objective is to combine the physiological trait with other desired traits that have been accumulated in elite selections or commercial varieties. Consequently, the breeder commonly makes controlled crosses involving two or more parental varieties. For a comprehensive review of breeding methods used to identify and isolate the desired combination of characters, students are referred to the plant breeding texts cited at the end of this chapter. Three selection methods commonly used to manipulate plant populations in breeding programs are reviewed briefly in the following sections.

Mass Selection

Mass selection refers to a procedure in which individual plants are selected in heterogeneous populations (Fig. 11.5). The original populations may be obtained through artificial hybridization involving two or more parental varieties. Alternatively, they may be obtained by mixing seed of different sources and allowing interplant pollination to occur naturally. Seed from selected plants in each generation or cycle of selection is composited to form a new heterogeneous population and is used to grow the next generation. Selection may be for easily observed characters, such as height or leaf angle, or for characters that require harvest and seed analysis,

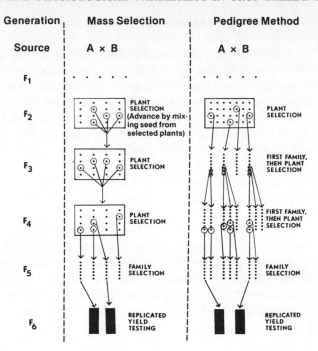

Fig. 11.5. Comparison of mass and pedigree selection methods used in self-pollinated crops.

such as protein or oil content. The population is expected to be improved with each generation of selection. The improved population may be utilized directly as a new variety, or plants possessing desired characteristics may be selected and their seed increased to become a new variety. A third alternative is to utilize the improved population as a source of inbred lines in a hybrid breeding program.

The likelihood of success using mass selection depends on many factors, including whether the crop is cross- or self-fertilized. In self-fertilized crops, little intercrossing occurs so the plants become homozygous at a rapid rate and new gene combinations are unlikely after a few generations. In contrast, crossing naturally occurs between plants of cross-fertilized crops; thus heterozygosity is maintained, and new gene combinations are formed each generation. However, effectiveness of selection is reduced by one-half in cross-fertilized as compared to self-fertilized species because seeds borne on cross-fertilized plants and used to advance to the next generation receive one-half of their genetic complement from a random (unselected) male parent.

Another very important factor in determining the success of mass selection is the heritability of the character under selection. Traits such as yield have low heritability and are not well suited to mass selection of individual plants. However, mass selection can be effective for highly heritable traits. For example, it should be possible to effectively change leaf area, leaf angle, or shape and size of inflorescence by mass selection.

Pedigree Selection

In pedigree selection both families and plants within families are evaluated. Seed for the next generation comes from individual plants, and individual plant progenies are kept separate throughout the several generations of selection (Fig. 11.5). The name pedigree is applied since it is usually possible to trace the lineage of a plant or line in an advanced generation back to the F_2 or F_1 plant from which it was derived. Pedigree selection is used extensively in breeding self-fertilizing crops and in developing inbred lines for use as parents of hybrid varieties.

Plant material (populations) for pedigree selection often is obtained by crossing parental varieties that possess between them the traits and genes desired in a new variety. When each parent is homozygous, the F_1 progeny plants are genetically similar, and selection is not effective in this F_1 generation. In the F_2 generation each plant is genetically different, and desirable plants are selected. Effectiveness of selection in the F_2 generation depends on many factors, including the amount of genetic variation for the character being selected and the influence of environment on the character. As a general rule selection in the F_2 generation is not done for traits of low heritability or for traits whose measurement is time-consuming or expensive.

The pattern of selection followed is similar in the F_3, F_4, and F_5 generations. In these generations, the best families are chosen first, and then individual plants within the elite families are selected. Seed from the selected plants is used to grow the families of the next generation. Selection on the family basis, which is frequently very effective, is an important advantage of pedigree selection compared to some other selection methods. Each generation affords a repeated opportunity for family selection. Pedigree selection is more expensive than mass selection but the gain from selection may more than offset the increased costs. Also, plant breeders have found ways to mechanize planting and harvesting so that costs can be kept low.

Following the F_5 generation, it is customary to begin replicated testing for yield and other lowly heritable traits such as milling and baking quality in wheat. Yield evaluation normally requires 3 to 4 years beginning with preliminary trials at one or two locations.

Backcross Method

The backcross method of breeding differs in objective and procedure from the mass and pedigree selection methods. The objective in most backcross programs is to improve an existing variety or inbred line by adding a desirable character not possessed by the variety. The procedure entails transferring the useful character from a donor line to the desirable variety or inbred line. The transfer is accomplished by making a cross between the desirable variety, called the recurrent parent, and the donor line. Repeated crosses (backcrosses) to the recurrent parent are made with progeny selected for the desired character from the donor. The proportion of the genes coming from the donor variety is reduced by one-half with each backcross to the recurrent parent; after n crosses and backcrosses the proportion is $(1/2)^n$. Thus, with six backcrosses after the original cross, the proportion of germ plasm from the donor parent theoretically is $(1/2)^7 = 1/128$. Except for the gene transferred, donor genes are rapidly replaced by genes of the recurrent parent.

The most extensive use of backcrossing has been to transfer single genes for disease resistance into standard varieties. These programs have shown that it is generally possible to transfer a character from one line to another when one gene controls the character being transferred. Backcrossing is less suitable for characters controlled by more than one or two genes. Another important consideration in backcross breeding is the extent to which all of the genes from the recurrent parent are recovered after repeated backcrosses. Failure to recover all of the genes of the recurrent parent is serious when undesirable genes are closely linked to the donor gene. The formula given above does not apply for genes linked to the gene being transferred.

Backcrossing may be of special interest in breeding for physiological traits controlled by one or two genes found in otherwise inferior genetic backgrounds. When this is the case, the backcrossing procedure would be helpful in incorporating the desired trait into a good genetic background.

ACCOMPLISHMENTS IN BREEDING FOR PHYSIOLOGICAL TRAITS

Plant breeders have devoted considerable attention to breeding for yield. These programs usually have emphasized selection for higher yields based on yield testing results. It is also true, however, that plant breeders have for decades sought to improve yield by manipulating what are often termed physiological components of yield. In this section we give two examples of improvements in crop plants resulting from selection of desirable plant types. These examples deserve special attention because they involve wheat and rice, which rank first and second in the world in providing food for the human population. In these crops, breeders have selected new gene combinations and new plant types that are highly productive under a wide range of growing conditions, especially when fertility is high.

Rice Breeding

The development of short-stature rice varieties at the International Rice Research Institute (IRRI) in the Philippines began in the 1960s following observations that short-strawed varieties yielded more than the taller, leafier indica varieties when given nitrogen fertilizer (Jennings, 1964). A number of characters were associated with generally higher yields and the nitrogen response. The responsive selections were short and had many tillers and good resistance to lodging. Their leaves were short, thick, relatively narrow, erect, dark green, and remained functional until shortly before harvest. Subsequent genetic studies revealed that a single gene controlled many of the leaf characters as well as height and tillering. Two additional characters, early maturity and high floret fertility, though not controlled by the above mentioned gene, were incorporated into the genetic package that contributed to high yield under nitrogen fertilization.

In breeding for new plant types at IRRI, a well-adapted tropical indica varieties were crossed with introduced varieties of either japonica or indica types, which served as sources of the desirable new traits. Observation of populations resulting from these crosses indicated that the desirable, short, less leafy plants were not competitive and hence were being eliminated from the breeding populations. The importance of this association is demonstrated by the yield data in Table 11.2. The short, less leafy plants compete very poorly in the mixture with tall plants, but in pure stands the short plants yield more than the tall. Consequently, a mass selection procedure is used to increase the proportion of desireable short plants in the breeding populations. The populations are planted sparsely so that the tall, leafy plants can be eliminated from the population. Seed harvested from the remaining plants is used for the next generation.

Selection against the tall leafy plants is continued for several generations until all the remaining plants are short. Then the seed from individual plants is increased and conventional yield testing is initiated. Many of the short plants with the combination of useful characteristics described above have been highly productive. One plant, IR-8, gained worldwide recognition because of its very high yields, especially in response to nitrogen fertilization.

Table 11.2. Changes in percent composition in a mixture of five rice varieties and relative yields of the five varieties in pure stands.†

Variety	Description	Survival				Grain yield
		1963	1964	1965	1966	
		%				t/ha
TN1	Short-statured; small, erect leaves	20	3.7	0.2	0.04	6.46
Ch. 242	Similar to TN1	20	2.1	0.06	0.0	5.53
M6	Tall, leafy, few tillers	20	1.7	0.02	0.0	4.99
MTU	Tall, leafy, many tillers	20	25.7	9.3	5.7	3.53
BJ	Similar to MTU	20	66.9	90.5	94.3	3.11

† From Jennings and Jose de Jesus (1968).

Semidwarf Wheat

In 1935 a wheat selection of hybrid origin, 'Norin 10,' was released for use in Japan. Norin 10 was 1/2 to 2/3 as tall as common wheat varieties and had more heads. In the USA, Norin 10 was not useful as a variety, but as a parent in crosses it provided genes that started a revolution in wheat breeding. One breeder, O. A. Vogel, a USDA plant breeder at Pullman, Washington, crosses Norin 10 with ordinary wheats such as 'Brevor.' From these crosses came the first highly productive semidwarf wheats in the USA. The first new semidwarf, 'Gaines,' and a closely related semidwarf, 'Nugaines,' soon achieved great popularity, and other wheat breeders began to concentrate on semidwarf wheat. N. E. Borlaug, winner of a Nobel Peace Prize in part for his research on wheat, made extensive use of the Washington wheats and the Norin 10 genes. By 1968 semidwarf wheats were being grown successfully in the USA and in more than 20 foreign countries.

As many as a dozen characters potentially contribute directly to high yield in short-stature rice. However, less is known about the physiological basis for high productivity in the semidwarf wheats. Semidwarf wheats are more resistant to lodging than their taller counterparts, but lodging resistance is not the entire reason for the yield difference because semidwarf varieties are often superior to tall varieties when lodging does not occur. Many of the semidwarf wheats have more heads per unit area than taller wheats and a higher ratio of grain to straw (higher economic to biological yield). The leaves of many productive semidwarf wheats are generally similar to the droopy, wide, and long leaves of taller wheats. Consequently, the productive semidwarf wheats and short-stature rices differ greatly in leaf characteristics. Whatever the basis for the higher yields in the new rice and wheat varieties, their superiority suggests that crops can be made more productive by modifying their general form.

OPPORTUNITIES IN BREEDING FOR PHYSIOLOGICAL TRAITS

Accumulation of fundamental information about how plants grow, improved techniques and instrumentation, recognition of the opportunities in interdisciplinary research, and continuing interest in higher yields all enhance the potential for crop improvement through breeding for physiological components of yield. In this section we describe physiology-genetics-breeding research that has been done on several physiological components of yield. Emphasis is on traits that appear to be important in determining yield, on their genetic variability and inheritance, and on evidence indicating their relationship to yield.

Photosynthesis

Photosynthetic Rate

Because of the fundamental relationship between photosynthesis and yield, there is considerable interest in enhancing photosynthesis capacity through breeding. Research has concentrated both on direct measurements of photosynthetic rates and on characters such as leaf area index, leaf angle, leaf orientation, and stomatal frequency. As discussed in earlier chapters, these characters affect light utilization and CO_2 entry into the plant and photosynthesis.

Varietal differences in photosynthetic rates have been reported in many crops, including corn, soybean, wheat, barley, ryegrass, sugarcane, and red kidney bean (Wallace et al., 1972). An example of the differences observed is given in Fig. 11.6, which shows net CO_2 exchange of two bean varieties, 'Michelite-62' and 'Red Kidney,' at five light intensities. The consistent difference shows that selection in a breeding program should be possible over a wide range of intensities, at least in genetic stocks like Michelite-62 and Red Kidney. Izahr and Wallace (1967) concluded that the genetic mechanism controlling the varietal difference in net CO_2 exchange rate is quantitative, although relatively few genes may be involved.

Heritability estimates for photosynthesis are available in only a few crops. In one study, Wilson and Cooper (1969) crossed six strains of ryegrass and evaluated their progeny. They obtained relatively high estimates of heritability, encouraging results to those interested in breeding for higher photosynthetic rates. Even so, the opportunities for direct improvement of

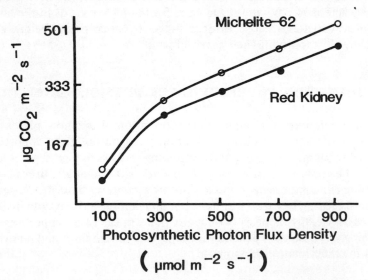

Fig. 11.6. Net CO_2 exchange at five light intensities for 'Red Kidney' and 'Michelite-62' cultivars of field bean. From Izhar and Wallace (1967).

photosynthesis seem limited at present. Measurements of photosynthesis on a single leaf at one stage of growth may not be indicative of photosynthetic rates throughout the life of a plant or of photosynthesis in a field canopy. To date, replicated photosynthesis measurements on crop canopies in the field are impractical for routine breeding purposes. However, a limited number of genotypes might be evaluated in the laboratory or field to identify parents to use in crosses. When such parents are used, genes favorable to high rates of photosynthesis might be fixed by chance in their progenies without selection for photosynthesis.

Leaf Angle

In many plant species, individual leaves become light-saturated at relatively low light intensities. Examples of such species are barley, wheat, oat, soybean, tobacco, and sugar beet. However, in field canopies, leaves are shaded so they often are not light-saturated. In these situations, changes in the canopy architecture could reduce shading and increase canopy photosynthesis. In other crops, such as corn, sorghum, and most tropical grasses, neither individual leaves nor field canopies are light-saturated at normal light intensities. These circumstances appear to afford a good opportunity to increase photosynthesis by breeding for increased light penetration. Modifications of leaf angle and leaf area are potential ways of increasing canopy light penetration.

Genetic variation in leaf angle is common in most crop species. In some barley and corn lines, upright leaf angle is conditioned by a single gene. In rice, the gene that causes short stature also conditions upright leaf angle. On the other hand, upright leaf angle can be quantitatively inherited and controlled by several genes. Breeding is more difficult when several genes are involved. Some studies indicate, however, that heritability of upright leaves is adequate for breeding progress. In liguleless barley, heritability of leaf angle was as high as 40% on the F_2 plant basis and 60% on the F_3 family basis (Barker, 1970).

Since breeding for upright leaf angle is feasible, the crucial question becomes whether it is worthwhile. Theoretical considerations (see chapter 6) suggest that it should be. Other positive evidence is the success of the short-stature, erect-leaf varieties of rice discussed above and the results of Pendleton et al. (1968) who studied the relationship of leaf angle and grain yield of corn in the field. In one experiment these workers used genetic isolines. One member of the isoline pair was homozygous for the recessive lg_2 allele and had upright leaf angle; the other line was homozygous for the Lg_2 allele and had normal leaves. In a companion experiment, leaves of a high-yielding hybrid were mechanically positioned to obtain three leaf-angle orientations. Table 11.3 shows that upright leaf-angle plants had a yield advantage in these two experiments. In the comparison of isogenic lines, erect-leaf plants had a yield advantage of 2567 kg ha^{-1}, an increase of 41% over the normal

Table 11.3. Grain yields of corn plants with normal and upright leaves.†

Comparisons	Yield
	kg/ha
Genetic isolines of hybrid C103 × Hy	
Normal leaf	6 202
Upright leaf	8 769
Mechanical manipulation of leaf angle of Pioneer 3306	
Normal (untreated)	10 683
All leaves positioned upright	11 386
Leaves above ear positioned upright	12 202

† From Pendleton et al. (1968).

line. In the mechanical manipulation experiment, yields were increased significantly either when all leaves or when only leaves above the ear were positioned upright. Other investigators working on corn, however, have reported no or little yield advantage due to upright leaves.

Leaf orientation for some plants is determined by the direction and intensity of the light source. Individual leaflets of soybean and field bean bend at their point of attachment to the petiole, thus changing their orientation in response to direction and intensity of light. On bright days the upper leaves of some soybean varieties are oriented to reduce the light intercepted to about 50% of the potential, whereas the leaflets assume a more horizontal position on overcast days. It has not been determined if plants with this capacity have a yield advantage over those that lack it, but breeding to obtain optimum leaflet orientation might be a worthwhile objective.

Leaf Area

Genetic variation in leaf area and in its primary components, leaf number and leaf size, has been reported for many crops. In grasses, including cereals, genetic variation for leaf size is larger than for leaf number. In legumes such as alfalfa, soybean, and field bean, variation is large for both leaf size and number. Genetic studies indicate that inheritance of leaf size may be either qualitative, as discussed previously for soybean, or quantitative. In studies of several populations of barley, Fowler and Rasmusson (1969) found that heritability of leaf area on an individual plant basis ranged from 18 to 73% suggesting that selection would be effective in barley populations.

The effect of leaf area on soybean yield was studied by Hicks et al. (1969) using isolines having the normal (Na) or narrow leaflet (na) character. More light penetrated into the narrow leaflet canopies than into canopies of the normal type, but the two leaflet types did not differ in yield. The relationship between yield and leaf area is complex, as discussed in earlier chapters, and it is not now possible to predict how changes in leaf area will affect yield.

Stomatal Frequency

Genetic variation exists for number, size, and time of opening and closing of stomates. These factors influence entry into the plant of the CO_2 essential to photosynthesis. The results of Miskin and Rasmusson (1970) are typical of data reported for stomatal frequency and size. Stomatal frequency on the lower surface of the flag leaves of 649 cultivars from the world collection of barley ranged from 36 to 98 stomates mm^{-2}, with a mean of 64 mm^{-2} (Fig. 11.7). Among cultivars examined for stomatal size, average guard cell length varied from 41 to 56 μm. Miskin et al. (1972) crossed cultivars differing in stomatal frequency and obtained inheritance information in the F_2 and F_3 generations. In all populations, estimates of heritability exceeded 25%, and one estimate was 74%.

Water Utilization

Many plant breeders are concerned with increasing yields in environments where water is a significant limiting factor. As you would expect, much attention has been given to breeding for drought resistance. Much less breeding effort has been devoted to increasing water use efficiency in crops, that is, attaining higher yields from a given amount of water.

Water use efficiency can be improved by using less water to produce a constant yield and by increasing the yield from a constant amount of water. Breeding emphasis could be placed on efficiency, total water requirement, or total dry matter production. In high-rainfall climates and under irrigation, total yield would be more important than efficiency or water requirement, while the relative importance would be reversed under dry conditions.

Two approaches can be used in attempting to modify water use efficiency, direct selection for efficiency or selection for traits that affect efficiency. The direct approach seems possible because research shows that plant species differ greatly in efficiency. Results from one experiment given in Table 11.4 show that temperate and tropical grasses produced about the same amount of dry matter. However, the tropical grasses required much less water per unit of leaf area and hence were almost twice as efficient as the temperate grasses. The superior efficiency of the tropical grasses (all are C_4 species, see Chapter 5) suggests that the first step in a breeding program should be to select among species, assuming that various species meet requirements for the area. Differences in water use efficiency within a species are smaller than differences between species but seem to be large enough to justify selection. Water use efficiency of six clones of blue panicgrass ranged from 371 to 575 g of water required to produce one g of forage (Dobrenz et al., 1969).

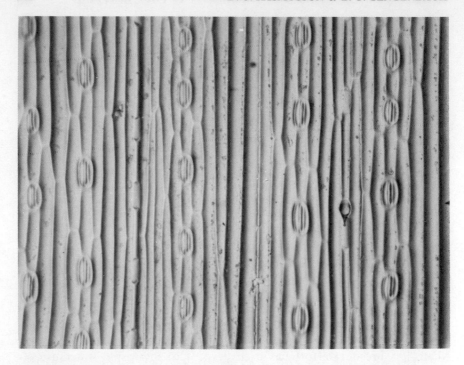

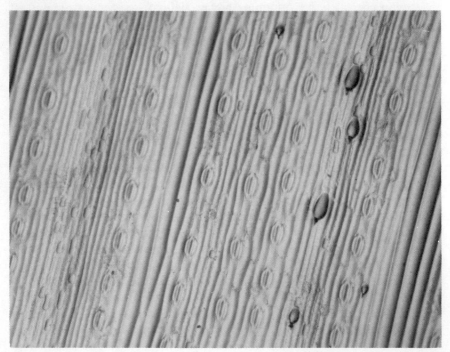

Fig. 11.7. Photomicrographs ($\times$ 150) of the lower surface of leaves of barley cultivars with low (left) and high (right) stomatal frequencies. The low line had 39 stomates per mm^2 and the high line 96. From Miskin and Rasmusson (1970).

Table 11.4. Dry matter produced, water used, and water efficiency of temperate and tropical grasses (at 25°C).†

Species	Dry Matter (D.M.)	Water Used (W.U.)	D.M./W.U. × 10⁴
	—— gm dm⁻² week⁻¹ ——		
Temperate grasses			
Oat	0.62	380	16
Annual ryegrass	0.67	485	14
Wheat	1.04	507	21
Average:	0.78	457	17
Tropical grasses			
Proso millet	0.63	171	37
Pennisetum clandestinum	0.87	298	29
Sudangrass	0.60	219	27
Average:	0.70	229	31

† From Downes (1969).

The alternative to direct selection for water use efficiency is to breed for traits that affect efficient use. Rooting traits such as depth, degree of branching, and ability to grow in dry soil might affect efficiency and qualify for a breeding effort. Other plant traits possibly affecting water use efficiency are stomate frequency and size, duration of various growth stages, presence or absence of awns, and ratio of economic to biological yield.

In considering ways to alter a plant's ability to control water loss, attention is drawn to the critical control that stomates exert over water movement through plants. The amount of water loss through the cuticle is unknown for many plants, but it is clear that stomates are more important in controlling water loss. A complication in breeding for optimum stomate frequency, size, or time and degree of opening and closing is that stomates are the major site of CO_2 entry into the plant. If selection is for stomates that are small, infrequent, or that only open partially for short periods, not only might water loss be reduced, but water use efficiency also could be reduced if CO_2 entry into the plant were slowed and photosynthetic rates were reduced. The available evidence, however, indicates that stomate frequency exerts greater control over water loss than over CO_2 entry.

Regardless of the effect on other plant traits, genetic modification of stomatal frequency is possible. Research on several crops has shown that a wide range of variability exists for number of stomates. Heritability of stomate frequency is sufficiently high to encourage early-generation selection among individual plants. In backcross programs, for example, it has been possible to select plants with desired high or low frequencies of stomates after three generations of backcrossing. Evidence obtained in barley (Miskin et al., 1972) indicates that transpiration can be reduced by lowering stomatal frequency (Fig. 11.8). The eight barley lines used in the experiment were obtained by crossing two varieties differing in stomate frequency and then selecting the extreme high and low stomate lines. Additional research is needed to learn how modifying stomatal frequency through breeding will affect efficiency of water use.

Photoperiod Response

Many crops are affected by the light-dark cycle, or photoperiod regime, in which they are grown. As discussed in previous chapters, the duration of darkness is the determining factor in photoperiodic responses, although the effect is usually described in terms of day length. An important photoperiod response is the difference in time required for flower initiation under various day lengths. Short-day crops such as soybean and sorghum begin flowering much earlier under short than under long days. Other important crops such as wheat and barley, however, initiate flowering earlier when days are long rather than short and are called long-day plants.

Photoperiod responses are important in determining the portion of a plant's life cycle spent in vegetative versus reproductive growth. This factor, in turn, influences yields obtained in various latitudes. In the USA, daylength increases during the summer growing season as you go north. Thus, a highly productive soybean variety in the South may not be nearly as productive in the North because reproductive growth is not initiated until the short daylengths of the fall season. When the photoperiod requirement was studied in soybean of different maturity groups, some of the early-maturing northern soybean varieties were found to be photoperiod insensitive (Polson, 1972). That is, date of flowering was little affected by daylenths ranging from 15 to 22 hours.

In wheat the inheritance of photoperiod-insensitivity was found to differ among the varieties studied (Klaimi and Qualset, 1973). For example, two genes with major effects explained the differences in heading date between two insensitive and two sensitive varieties grown under short daylengths. Insensitivity was conditioned by a dominant gene in one variety and by a homozygous recessive gene in the other. Other genes with minor effects also influenced the response. Under short-days, insensitive varieties headed

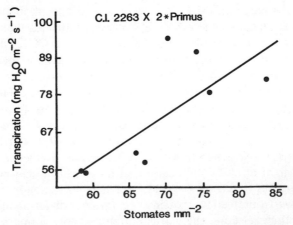

Fig. 11.8. Relationship between stomatal frequency and transpiration in eight barley backcross lines. From Miskin et al. (1972).

within a 7-day interval, whereas sensitive varieties headed over a 40-day interval. This indicates that environmental effects on different genotypes can differ in magnitude—a situation with which breeders often must work.

Four genes that affect maturity and response to long daylengths have been described in sorghum by Quinby (1972). In this work, the recessive allele of each gene conditioned a degree of insensitivity to longer daylengths (Table 11.5). Under short days (10 hours), flower primordia were initiated at the same time for the four genotypes, but genotypes homozygous for one or more of the recessive alleles initiated flower primordia much earlier than the dominant genotype under long days (14 hours).

Genetic variability for photoperiod response provides one basis for changing the environmental adaptive range of many crops. Varieties can be selected for the photoperiod response to match a particular environment, and genes for photoperiod insensitivity can be used to increase a variety's range of adaptation.

Mineral Nutrition

The availability of minerals can greatly affect plant growth. This effect can range from increased yield resulting from increasing the supply of a limiting essential element to reduced growth resulting from a toxic level. Examples of genetic differences for mineral toxicity are found in several crops. For instance, the wheat variety 'Atlas 66' was more tolerant to an acidic, aluminum (Al) toxic soil than 'Thatcher' (Foy et al., 1965; Foy et al., 1976). Inheritance of tolerance can be controlled by a single gene. For example, a single dominant gene controls Al tolerance in certain barley populations and results in higher grain yields in acid, Al-toxic soil (Reid, 1970). Also, genetic differences in ability to grow under conditions of mineral deficiency are known for many crops. Efficient utilization of iron (Fe) in soybean is conditioned by a single gene that affects Fe uptake by the roots (Weiss, 1943). These examples illustrate the possibility of breeding crop plants to overcome a specific soil-regulated mineral toxicity or deficiency rather than having to change the soil environment.

The accumulation of minerals by crops is of interest because of the possible relationship between yield and accumulation. Heritability estimates for variation in phosphorous (P), potassium (K), calcium (Ca), and mag-

Table 11.5. Sorghum genotype responses to photoperiod.[†]

Genotype	Days to floral initiation	
	10-hour days	14-hour days
$ma_1\,Ma_2\,ma_3\,Ma_4$	19	35
$Ma_1\,ma_2\,ma_3\,Ma_4$	19	38
$Ma_1\,ma_2\,Ma_3\,Ma_4$	19	44
$Ma_1\,Ma_2\,Ma_3\,Ma_4$	19	70

[†] From Quinby (1972).

nesium (Mg) accumulation in F_3 wheat families were sufficiently high to expect progress from selection (Rasmusson et al., 1971). No relationship was found, however, between ability to accumulate these minerals and yield.

The metabolism of nitrogen (N), an essential element, is quite important because N is incorporated into the structure of many cellular constituents, primarily proteins and nucleic acids. In plants that do not fix N symbiotically, N is taken into the plant through the roots as NO_3^-, which is then reduced to NH_4^+ before incorporation into carbon compounds (see Chapter 7). The initial step in this conversion process is carried out by the enzyme nitrate reductase (NR), which reduces NO_3^- to NO_2^-. The rate of NO_3^- reduction to NO_2^- determines how fast subsequent steps are carried out, that is, nitrate reductase is rate limiting for the initial metabolism of N in plants. Because nitrate reductase occupies a rate-limiting position in the metabolism of the very important element N, it could be important in determining growth and yield potential.

Extensive work on nitrate reductase by Hageman et al. (1967) has centered on determining the genetic control of nitrate reductase activity (NRA) and its relation to grain and protein yields of several crops. Their work with corn indicates a wide range of NRA among different inbred lines. Crosses between inbreds generally resulted in F_1's with NRA values intermediate between the parental values. However, a significantly higher NRA was obtained in the F_1 of a cross between two inbreds with low NRA. The higher NRA in the F_1 was attributed to two genes, one affecting the rate of NR synthesis and the other affecting the rate of NR inactivation. Initial studies also showed that NRA was related positively to water-soluble protein content of leaves. When NRA was measured throughout the growing season and expressed as total seasonal NRA or as input of reduced N, positive relationships with grain protein and yield were obtained. However, to be useful in selection work, a single sampling date is desired. When varietal rankings obtained from seasonal NRA were compared with rankings obtained from one sampling at the postanthesis stage, the same positive correlation was obtained. This indicates that NRA assays from a single sampling date might be useful in selecting genotypes capable of utilizing the higher NRA for increased grain and grain protein yields.

The work with nitrate reductase provides an opportunity to point out two requirements when attempting to use enzyme activity as a selection criterion in breeding work. First, a simple assay must be available, and a sampling system must be determined that accurately reflects genetic differences and not just environmental or seasonal effects. Second, the characteristics of the enzyme must be known. For instance, NO_3^- within the plant tissue is necessary before nitrate reductase is synthesized by the plant. Therefore, an adequate and continuous supply of NO_3^- is needed for genetic differences in NRA to be expressed as differences in yield or protein.

Optimum Plant Type

The success of semidwarf wheat and short-stature rice, which are new plant types, underscores the opportunities in breeding for optimum or model plant types. Theoretical reasons why modifications in plant type should increase grain yields have been advanced by many researchers including Donald (1968), who urged that plant breeding be extended to include breeding of model plants. These model plants can be specified by a few important characters or by considerable detail involving numerous characters. Table 11.6 contains a listing of several characters which have received attention in model breeding in barley, oat, rice, and wheat. In this section we will consider how some of these plant type characters might be utilized to increase yield.

Kernels Per Head

Because grain yield is the product of kernels per head, kernel weight, and number of heads per unit area, the number of kernels in a head can vary greatly even among high-yielding cereal grain varieties. Breeding for these individual components would not be advantageous if the benefit derived from genetically increasing one component were offset by a simultaneous reduction in one or both of the others. Fortunately, this is not always the case. Though much of the evidence is indirect, it appears that a large sink or storage capacity provided by many kernels per head is advantageous in obtaining high yields. Kernels per head is an attractive breeding objective because it is relatively easy and inexpensive to score, and genetic variation is available or easily obtained. Kernels per head is a quantitative character, but heritability is high enough that good progress in increasing kernels per head can be expected in a breeding program.

Table 11.6. Characters with potential for model breeding in cereal grains.

Leaf characters	Culm characters	Type of canopy
Leaf size	Head number	Height of plants
Leaf angle	Survival of culms	Harvest index
Leaf vein frequency	Diameter of culms	Angle of ear
Number of leaves	Number of ears	
Duration of leaves	Vascular bundles	
Thickness of leaves		
Specific leaf weight		
Inflorescence characters	Root characters	Other
Awn length	Volume	Photoperiod response
Awn number	Depth	Length of growth stages
Kernels per head		
Kernel weight		

Head Number

Head number potentially can be used to increase yield because a change in head number modifies the leaf area or photosynthetic source, and a change in head number modifies the sink capacity of the plant. Some interesting evidence indicates that a crop community of single-culm plants should maximize production due to the absence of internal competition between developing heads and young tillers. This point is illustrated by new high-yielding oat varieties in Great Britain that have fewer panicles than older varieties. In contrast, new high-yielding barley varieties in Great Britain have more heads than the older varieties. Survival of tillers is important in determining whether more or fewer heads per plant will maximize yield. Simmons et al. (1982) described a barley genotype that has high tiller number and low tiller mortality. Breeding for low mortality as well as higher or lower head number may be beneficial.

Cereal grain varieties possess tremendous capacity to adjust number of heads to fit the environment. Single, isolated plants often tiller sufficiently to produce 20 heads, whereas, in a commercial planting, the average number of heads per plant may be as low as two or three. Heritability for head number is often low because of the large environmental influence. Nevertheless, genetic variation for head number is substantial, and good progress from selection for tiller and head number can be made in experiments designed to reduce the environmental effect.

Harvest Index

Harvest index is the ratio of economic yield (grain weight in cereals) to biological yield (total plant dry weight). Part of the genetic improvement in yield of several crops is derived from a higher percentage of the biological yield being partitioned into plant parts comprising economic yield. Economic yields can be increased by increasing biological yields without changing the harvest index and by partitioning more of the dry matter production into economic yield. The latter can be very important as discussed by Wallace et al. (1972). They noted that the harvest index had increased from 32% for wheat varieties grown in the early 1900s to 49% for current high-yielding semidwarf varieties.

Wallace et al. recommend that a breeding program regularly consider harvest index as well as economic yield and biological yield. One challenge is to find the balance between biological yield and harvest index that maximizes economic yield. Biological yield should be evaluated in a breeding program because it is closely related to light interception, relative growth rate, net carbon exchange, and other factors that should be kept high. Data on both biological yield and economic yield can be obtained from routine yield trials.

Leaf Vein Frequency

In 1968 it was reported that accumulation of assimilate in an illuminated leaf may reduce the net photosynthesis of that leaf. We know that assimilate accumulation and photosynthesis frequently are negatively correlated, but the physiological explanation still is not clear. One way to minimize assimilate buildup in the leaf is to increase the number and size of the vascular bundles (veins) that translocate photosynthate from the leaf. Twofold differences in leaf vein frequency (2.4 to 4.5 veins mm^{-1}) have been observed among varieties of barley (Hanson and Rasmusson, 1975). Heritability for leaf vein frequency was 20% in the F_2 generation and 45% in the F_3 generation. The essential requirements for genetic alteration—genetic variation and a reasonably high heritability—therefore exist in barley.

SUMMARY

In this chapter, we described the genetic, physiological, and breeding basis for using physiological plant traits to improve crops. We considered a physiological trait to be any measurable characteristic involved in plant growth and development. Several points should be emphasized in summarizing the utilization of such traits in crop breeding.

1) Past selection of improved varieties has been accomplished mainly on the basis of direct selection for yielding ability. Through this approach, it is certain that physiological traits with direct effects on yield have been improved concurrently with yield. The emphasis is reversed when selecting for altered physiological traits. For this approach to be worthwhile, however, the change in a physiological trait must result in a yield increase, and it must be less costly than direct breeding for yield.

2) The extent to which differences in physiological traits are expressed, and are measurable, depends partly on the genetic potential of the plant. This potential is manifested through the interrelationships among genes, enzymes, and plant growth. A gene contributes the information for biosynthesis of an enzyme that functions in a particular metabolic reaction. The combined effect of many genes, through their control of enzymes, results in physiological traits contributing to plant growth, development, and yield.

3) Variation in a physiological trait has both genetic and environmental origins. A plant breeder works with the genetic component to change the genetic potential of the crop for a particular trait; environmental variation hinders this effort. Both adequate genetic variability and heritability are required to obtain genetic changes by breeding.

4) Through breeding the expression of nearly all physiological traits can be changed. But because time and resources are limited in breeding programs, there should be experimental evidence or at least a theoretical basis

for expecting a positive effect on yield or performance before a program is started. The expression for a trait that maximizes its contribution to yield rarely has been determined, and no plant trait likely has a single optimum expression encompassing all possible combinations of environment and genotype. The optimum expression of many traits will differ from crop to crop, variety to variety within a crop, and from one geographical area to another.

5) Plant breeders, plant physiologists, and others should work together to obtain a better understanding of physiological traits and their relationship to yield so that productivity of crop plants can continue to increase.

REFERENCES

Barker, R. E. 1970. Leaf angle inheritance and relationships in barley. M.S. Thesis. Univ. of Minnesota, St. Paul.

Bonner, J. 1976. Cell and subcell. *In* J. Bonner and J. E. Varner (eds.) Plant biochemistry. p. 3–14. Academic Press, New York.

Briggs, F. N., and P. F. Knowles. 1967. Introduction to plant breeding. Reinhold Publishing Corp., New York.

Dobrenz, A. K., L. N. Wright, A. B. Humphrey, M. A. Massengale, and W. R. Kneebone. 1969. Stomate density and its relationship to water-use efficiency of blue panicgrass (*Panicum antidotale* Retz.). Crop Sci. 9:354–357.

Donald, C. M. 1968. The breeding of crop ideotypes. Euphytica 17:385–403.

Downes, R. W. 1969. Differences in transpiration rates between tropical and temperate grasses under controlled conditions. Planta 88:261–273.

Fowler, C. W., and D. C. Rasmusson. 1969. Leaf area relationships and inheritance in barley. Crop Sci. 9:729–731.

Foy, C. D., W. H. Armiger, L. W. Briggle, and D. A. Reid. 1965. Differential aluminum tolerance of wheat and barley varieties in acid soils. Agron. J. 57:413–417.

––––, A. L. Fleming, G. R. Burns, and W. H. Armiger. 1967. Characterization of differential aluminum tolerance among varieties of wheat and barley. Agron. J. 31:513–521.

Goplen, B. P., J. E. R. Greenshields, and H. Baenziger. 1957. The inheritance of coumarin in sweetclover. Can. J. Bot. 35:583–593.

Hageman, R. H., E. R. Leng, and J. W. Dudley. 1967. A biochemical approach to corn breeding. Adv. Agron. 19:45–84.

Hanson, J. C., and D. C. Rasmusson. 1975. Leaf vein frequency in barley. Crop Sci. 15:248–251.

Hicks, D. R., J. W. Pendleton, R. L. Bernard, and T. J. Johnston. 1969. Response of soybean plant types to planting patterns. Agron. J. 61:290–293.

Izhar, S., and D. H. Wallace. 1967. Studies of the physiological basis for yield differences. III. Genetic variation in photosynthetic efficiency of *Phaseolus vulgaris* L. Crop Sci. 7:457–460.

Jennings, P. R. 1964. Plant type as a rice breeding objective. Crop Sci. 4:13–15.

––––, and Jose de Jesus, Jr. 1968. Studies on competition in rice. I. Competition in mixtures of varieties. Evolution 22:119–124.

Johnson, S. W. 1868. How crops grow. Orange Judd & Co., New York, N.Y.

Klaimi, Y. Y., and C. O. Qualset. 1973. Genetics of heading time in wheat (*Triticum aestivum* L.). I. The inheritance of photoperiodic response. Genetics 74:139-156.

Miskin, K. E., and D. C. Rasmusson. 1970. Frequency and distribution of stomata in barley. Crop Sci. 10:575-578.

----, ----, and D. N. Moss. 1972. Inheritance and physiological effects of stomatal frequency in barley. Crop Sci. 12:780-783.

Pendleton, J. W., G. E. Smith, S. R. Winter, and T. J. Johnston. 1968. Field investigations of the relationships of leaf angle in corn (*Zea mays* L.) to grain yield and apparent photosynthesis. Agron. J. 60:422-424.

Poehlman, J. M. 1959. Breeding field crops. Henry Holt and Co., Inc., New York.

Polson, D. E. 1972. Day-neutrality in soybeans. Crop Sci. 12:7/3-776.

Qualset, C. O., C. W. Schaller, and J. C. Williams. 1965. Performance of isogenic lines of barley as influenced by awn length, linkage blocks and environment. Crop Sci. 5:489-494.

Quinby, J. R. 1972. Influence of maturity genes in plant growth in sorghum. Crop Sci. 12:490-492.

Rasmusson, D. C., A. J. Hester, G. N. Fick, and I. Byrne. 1971. Breeding for mineral content in wheat and barley. Crop Sci. 11:623-626.

Reid, D. A. 1970. Genetic control of reaction to aluminum in winter barley. Barley Genetics II. Proc. 2nd Inter. Barley Genetics Symp., Washinghton State Univ. Press, Pullman.

Schaeffer, G. W., F. A. Haskins, and H. J. Gorz. 1960. Genetic control of coumarin biosynthesis and β-glucosidase activity in *Melilotus alba*. Biochem. Biophys. Res. Commun. 3:268-271.

Simmons, S. R., D. C. Rasmusson, and J. V. Wiersma. 1982. Tillering in barley: genotype, row spacing, and seeding rate effects. Crop Sci. 22:801-805.

Wallace, D. H., J. L. Ozbun, and H. M. Munger. 1972. Physiological genetics of crop yield. Adv. Agron. 24:97-146.

Weiss, M. G. 1943. Inheritance and physiology of efficiency in iron utilization in soybeans. Genetics 28:253-268.

Wilson, D., and J. P. Cooper. 1969. Diallel analysis of photosynthetic rate and related leaf characters among genotypes of *Lolium perenne*. Heredity 24:633-649.

Subject Index

323

Fruit—continued
  seed and, 265, 279–280
  seedless, 265, 273, 280
  senescence, 280–286
  set, 268–277
    C-N ratio, 269
    chemical treatment, 271
    climate and, 270
    cytokinin effect on, 272
    definition, 268–269
    drought and, 270
    environmental control, 269–270
    hormones, 271–272
    humidity, 270
    light and flowering effects on, 270–271
    mechanisms controlling, 271–272
    moisture stress, 270
    N effects on, 269
    naphthoxyacetic acid (NOA), 271
    nutrition, 269
    shade effects on, 270–271
    temperature effects on, 270
    volatiles in, 282
    water, 270
  size as related to yield, 273
  starch in, 282
  storage, 283–286
    humidity and, 284
    O_2, 284–285
Fruiting, 165
Fuchsia, 243, 250
Fusarium, 86

GADA, 62
Gamete, 296, 297
Gametophytes, 53
Garner, W. W., 238, 240
Gene, 292–298
  assortment, 296
  function, 292–298
  Hily (high lysine), of barley, 38
  inheritance, 296–298
  opaque-2 (dull). *See* Opaque-2 (dull) gene
  physiological, action, 292–296
  segregation, 296
  transfer, 305
Genetic gain, 299–300
Genetic regulation, 31
  amylopectin, 31
  amylose, 31
  dent corn, 31
  opaque endosperm, 31
  shrunken endosperm, 31
  sugary endosperm, 31
  waxy corn, 31
Genetic variability, 299–300
Genetics, 291
Genotype, 293
Germination, 53, 59, 63–73, 76, 82, 253

anabolism, 73
catabolism, 73
denaturation, 68
environment, 53
enzyme activation, 73–75
failure, 68
hydration, 73–75
imbibition, 64, 68, 70
  water, 73
inhibition, 72
light, 71–73
moisture for, 65, 71
osmotic pressure, 82
oxygen, 64
physiology, 73–78
phytochrome, 71
  role in, 253
  spectrum for, 254
promoters, 70
promotion, 72
rate, 70
regulation
  dormancy, 63
  light, 71–73
  oxygen, 70–71
  temperature, 66–70
  water, 64–66
seed moisture, 65, 73
soil water, 82
uniformity, 70
water imbibition, 73
Germplasm, 300
Gibberellin
  effect
    flowering, 252–253, 261–262
    fruit growth, 277, 279, 281
    fruit ripening, 281–282
    fruit set, 271–272
    sex expression, 253
  in seeds, 51
    bound, 51
    changes during germination, 51
    de nova synthesis, 51
    stimulate enzyme synthesis, 51
Glutamate dehydrogenase, 183, 184
Glutamate synthase, 183
Glutamic acid decarboxylase, 62
Glutamine synthetase, 183
Glycolipids, 43
Glycolic acid, 149
Glyoxysomes, 76
Grafting, 239, 260–261
Grain
  filling, 162
  production
    efficiency, 202, 203, 226
    time, 203, 224, 226
  storage, 284
  yield, 162
  yield components
    seed number, 209, 223, 224, 226